BIOLOGICAL AND MEDICAL PHYSICS, BIOMEDICAL ENGINEERING

For further volumes:
http://www.springer.com/series/3740

BIOLOGICAL AND MEDICAL PHYSICS, BIOMEDICAL ENGINEERING

The fields of biological and medical physics and biomedical engineering are broad, multidisciplinary and dynamic. They lie at the crossroads of frontier research in physics, biology, chemistry, and medicine. The Biological and Medical Physics, Biomedical Engineering Series is intended to be comprehensive, covering a broad range of topics important to the study of the physical, chemical and biological sciences. Its goal is to provide scientists and engineers with textbooks, monographs, and reference works to address the growing need for information.

Books in the series emphasize established and emergent areas of science including molecular, membrane, and mathematical biophysics; photosynthetic energy harvesting and conversion; information processing; physical principles of genetics; sensory communications; automata networks, neural networks, and cellular automata. Equally important will be coverage of applied aspects of biological and medical physics and biomedical engineering such as molecular electronic components and devices, biosensors, medicine, imaging, physical principles of renewable energy production, advanced prostheses, and environmental control and engineering.

Eugenijus Kaniusas

Biomedical Signals and Sensors I

Linking Physiological Phenomena and Biosignals

With 125 Figures

Springer

Ao. Univ.-Prof. Dipl.-Ing. habil. Dr. Eugenijus Kaniusas
Head of research group 'Biomedical Sensors'
Vienna University of Technology
Institute of Electrodynamics, Microwave and Circuit Engineering
Gusshausstr. 27–29,
1040 Wien, Austria
E-mail: kaniusas@tuwien.ac.at

Biological and Medical Physics, Biomedical Engineering ISSN 1618-7210
ISBN 978-3-642-24842-9 ISBN 978-3-642-24843-6 (eBook)
DOI 10.1007/978-3-642-24843-6
Springer Heidelberg Dordrecht London New York

Library of Congress Control Number: 2012930477

Printed on acid-free paper

Springer is part of Springer Science+Business Media (www.springer.com)

Preface

The present two volume set focuses on the interface between physiologic mechanisms and diagnostic human engineering. A multitude of biomedical sensors are commonplace in clinical practice today. The registered biomedical signals, which will be referred to as biosignals, reflect vital physiologic phenomena and are relevant not only for the pre-screening and diagnosis of maladies but also for therapy and follow-up treatment. For instance, the diagnosis of sleep apnea, i.e., abnormal cessation of respiration during sleep, requires the monitoring of a complete set of sleep and respiratory variables with at least eight different sensors distributed over the entire body.

In order to adequately apply biomedical sensors and reasonably interpret the corresponding biosignals, a proper understanding of the physiologic phenomena involved, their influence on the registered biosignals, and the technology behind the sensors is critical. Moreover, a nearly unlimited diversity of biosignals emphasizes the need for a strategic approach in the genesis of biosignals, including a profound understanding of fundamentally different mechanisms in a biosignal's generation.

From a strategic point of view, biosignal generation involves the biosignal formation path from the biosignal source at the physiological level, to biosignal propagation in the body, to biosignal transmission in the sensor up to its conversion to a, usually electric, signal. To give an example, heart sounds, an acoustic biosignal, are created by the closure of heart valves, which constitutes the biosignal source. Sound attenuation in the thoracic tissue represents the propagation mechanism. Amplification and filtering of the heart sounds in the chestpiece (of the stethoscope) reflect biosignal transmission effects in the sensor, with biosignal conversion being performed by a microphone at the output of the chestpiece.

The first volume is focused on the interface between physiologic mechanisms and the resultant biosignals, whereas the second volume is devoted to the interface between biosignals and biomedical sensors. Unlike other contributions, this book deals differently on the subject of either specific physiologic mechanisms or specific engineering aspects pertaining to particular biomedical sensors, since it emphasizes the *interface* between them. Both volumes systematically describe basic mechanisms of biosignal formation while electric, acoustic, optic, and mechanic biosignals are considered in depth.

In the given volume, the physiologic mechanisms determining biosignals are described from the basic cellular level—as the place of origin of each and every biosignal—up to their advanced mutual coordination level, e.g., during sleep. It allows a physiologically accurate interpretation and comprehensive analysis of the biosignals. The resultant biosignals are discussed within the scope of vital and common physiologic phenomena to foster their understanding and comprehensive analysis.

This book is directed primarily at graduate and postgraduate students in biomedical engineering and biophysics. It should also appeal to those who are studying or are interested in physical, engineering, and life sciences, since expected background knowledge is minimal and many basic phenomena are explained in depth within the numerous footnotes. Furthermore, the book should serve engineers and practitioners who have an interest in aspects of biomedical engineering. This book attempts to provide a blinding glimpse of the obvious, in spite of the issues that appear rather complex at first glance.

It is important to note that this book was mainly inspired by my lectures entitled “Biomedical Sensors and Signals,” “Biomedical Instrumentation,” and “Biophysics” which constitute a significant part of a master’s degree program “Biomedical Engineering” at the Vienna University of Technology.

Vienna, Austria *Eugenijus Kaniusas*

Acknowledgments

A number of personalities have shaped me and my educational background within the scope of the present work. First of all, I wish to express my appreciation to Univ.-Prof. Helmut Pfützner from the Institute of Electrodynamics, Microwave and Circuit Engineering (EMCE), Vienna University of Technology, who has guided me in the field of biomedical engineering and magnetism. I gratefully acknowledge the support from Univ.-Prof. Adalbert Prechtl from the EMCE for unlimited scientific advice in the field of electrical engineering.

I would like to express my sincerest thanks to Univ.-Prof. Giedrius Varoneckas from the Institute of Psychophysiology and Rehabilitation (IPR), Medicine University of Kaunas, Lithuania, for an almost infinite number of fruitful discussions, inspiration, and close cooperation in clinical investigations. Furthermore, I am grateful for the cooperation of Univ.-Prof. Bernd Saletu from the Department of Psychiatry, University of Vienna, who has supported me in an unrestricted way concerning numerous clinical investigations.

Indeed my understanding and appreciation of biomedical issues have been boosted by collaborative research and interdisciplinary studies. Here I would like to place on record my gratitude to my colleagues from the EMCE as Dr. Lars Mehnen, Dr. Jürgen Kosel, Dr. Karl Futschik, Dr. Stefan Traxler, and Dr. Peter Schönhuber. Special thanks go to Linas Zakarevicius and Audrius Alonderis from the IPR. In particular, I thank my students Stefan Kampusch, Florian Thürk, and Jillian Haac for careful proofreading. The book has significantly benefited from countless small and large projects in which numerous diligent students of mine have been involved. I give sincere thanks to all of them.

I would like to thank my family which has been the source of inspiration and recharging all the time. While supporting me with their love and affection, they have had to bear the loss of my time and effort at home. I express my deep gratitude to my parents and other relatives for providing me with an inner strength and solid background to meet challenges and achieve goals.

Acknowledgements

Contents

Symbols and Abbreviations

Note: Variables used within limited contexts are not listed, for they are described within the relevant section. The different types of biosignals are separately listed below.

A	Surface area, signal amplitude
A_M	Maximum cross section of the artery
ATP	Adenosine triphosphate
c	(Molar) concentration, constants
C	Capacitance
C'	Length related capacitance
C''	Area related capacitance
CBT	Core body temperature
CHA	Central sleep hypopnea
CSA	Central sleep apnea
d	Membrane thickness
D	Axon diameter, electric flux density
D_F	Diffusion coefficient
e	Elementary charge
E	Electric field, Young's modulus
EDR	Electrocardiogram derived respiration
E_P	Pressure-strain modulus
f	Frequency
f_A	Activation rate
f_C	Heart rate
f_R	Respiratory rate
F	Force
g	Arbitrary function
G'	Length-related electrical conductance
G''	Area-related electrical conductance
h	Wall thickness
HF	High frequency
HRV	Heart rate variability

i	Current
i_C	Capacitive current
i_E	(Electric) ionic current
I	Current amplitude, intensity of sensation
I_A	Augmentation index
I_T	Threshold current amplitude
j_D	Chemical diffusion rate
j_E	Electric diffusion rate
J	Current density
J_C	Capacitive current density
J_D	(Chemical) diffusion current density
J_E	(Ionic) electric current density
k	Index
l	Vessel/tube length, propagation distance
LF	Low frequency
m	Ionic mobility
MHA	Mixed sleep hypopnea
MSA	Mixed sleep apnea
N_A	Avogadro constant
NREM	Nonrapid eye movement
OHA	Obstructive sleep hypopnea
OSA	Obstructive sleep apnea
p	Power spectral density, membrane permeability, probability, (intraarterial) blood pressure
P	(Complex) blood pressure amplitude
p_D	Diastolic blood pressure
p_E	External pressure outside the blood vessel
p_{O_2}	Partial pressure of oxygen in blood
p_{CO_2}	Partial pressure of carbon dioxide in blood
p_I	Incident pressure wave
p_{IF}	Inflection point in pressure wave
p_R	Reflected pressure wave
p_S	Systolic blood pressure
$p_{S,D}$	Systolic–diastolic deflection of the blood pressure
p_T	Transmural pressure
pH	pH value
PNS	Parasympathetic nervous system
PSG	Polysomnography
q	Blood flow, air flow, cardiac output, charge
Q	(Complex) blood flow amplitude, electric charge
q_I	Incident blood flow wave
q_R	Reflected blood flow wave
r	Blood vessel radius, (ion) radius
r_D	Diastolic artery radius
r_M	Maximum radius of the artery
r_S	Systolic artery radius
r_T	Artery radius at zero transmural pressure
R	Fluid/vascular longitudinal resistance, gas constant
R_T	Total peripheral resistance

R'	Length related electrical resistance
RR	Interbeat interval from electrocardiogram
REM	Rapid eye movement
s	Biosignal, see below
$s_{S,D}$	Systolic–diastolic deflection of the cardiac component
S	Hemoglobin oxygen saturation, stimulus strength
SNS	Sympathetic nervous system
t	Time
T	Absolute temperature, duration, period
u	Voltage, blood flow velocity
U	(Membrane) voltage amplitude
U_R	Resting (membrane) voltage amplitude
v	Pulse wave velocity, nerve conduction (propagation) velocity, drift velocity
V	Volume
V_S	Left ventricular stroke volume
VLF	Very low frequency
W	Energy
x	Coordinate, distance
y	Coordinate
z	Valence
Z	(Complex) electrical impedance, (complex) longitudinal vascular impedance, vascular impedance
Z_0	(Complex) characteristic vascular impedance
Z_I	(Complex) input vascular impedance
α	Attenuation coefficient
Γ	Reflection factor
ε	Dielectric permittivity
ϑ	Temperature
κ	Module of volume elasticity
λ	Wavelength, (membrane) length constant
μ	Dynamic viscosity of the liquid
ρ	Specific resistance, density
σ	Standard deviation, mechanical stress
τ	Pulse running (arrival, transit) time, (membrane) time constant
τ_A	Membrane time constant for axial currents
τ_D	Diastolic transit time
τ_S	Systolic transit time
τ_{PW}	Width of the pulse wave
τ_R	Membrane time constant for radial currents
γ	Phase angle, electrical conductivity
γ_I	Electrical conductivity of the intracellular medium
γ_M	Electrical conductivity of the membrane
φ	Phase angle, electric potential
ψ	Poisson ratio
ω	Angular frequency

Symbols of Biosignals

The types of biosignals discussed and their short descriptions.

Symbol	Name		Biosignal class	Phenomena reflected
s_{BCG}	Barocardiogram signal	permanent	Mechanic	Arterial blood pressure
s_{ECG}	Electrocardiogram signal	permanent	Electric	Electrical excitation of heart muscles
s_{MRG}	Mechanorespirogram signal	permanent	Mechanic	Circumference changes of the abdomen or chest during breathing
s_{MSG}	Mechanospirogram signal	permanent	Mechanic	Air flow through the mouth
s_{PCG}	Phonocardiogram signal	permanent	Acoustic	Sounds emitted by consecutive closures of heart valves
s_{TG}	Thermogram signal	permanent	Thermal	Skin temperature from proximal and distal body regions
s_{TRG}	Thermorespirogram signal	permanent	Thermal	Air temperature in front of the nostrils during breathing
s_{BG}	Barogram signal	induced	Mechanic	Pressure in the cuff on the upper arm
s_{OPG}	Optoplethysmogram signal	induced	Optic	Pulsatile blood absorption of artificial light
s_{PG}	Phonogram signal	induced	Acoustic	Sounds emitted by local turbulence in the blood flow of the brachial artery (Korotkoff sounds)

Chapter 1
Fundamentals of Biosignals

If you sharpen your electrical sense to generators in the body,
If you listen to body sounds emerging from depths of the body,
If you look through a fragment of the body,
If you feel the skin pulsation of the body,
You are to gain a valuable knowledge of the body's well-being...

Sensing technologies in physiology gain a lot of importance for the assessment of the human functional state. The registered biomedical signals—referred to as *biosignals* here—are important not only for timeless classical applications concerning medical diagnosis and subsequent therapy, but also for future applications such as daily driver monitoring.

Thus, this chapter starts by giving a definition of biosignals and its very *general model*, considering biosignal generation, propagation, and its conversion for application-specific analysis. This model offers a solid basis for each type of biosignal, which will accompany us throughout the book. Then the very beginning steps of biosignal registration, the *history of biosignal* assessment, are discussed. The problems encountered (at that time) are described, as well as applied methods to solve them, with some of these methods having outlasted many centuries and are in use even today. Possible *classifications* of commonly used biosignals (state of the art biosignals) are introduced in this chapter to perceive a nearly unlimited diversity of biosignals. Lastly, a few *ubiquitous applications* of biosignal assessment are given, followed by *future trends* in biosignal monitoring.

1.1 Definition and Model of Biosignals

Within the scope of biomedical signals and sensors, a *biosignal* can be defined *as a description of a physiological phenomenon*, irrespective of the nature of this description. Since there is a nearly unlimited number of physiological mechanisms

E. Kaniusas, *Biomedical Signals and Sensors I*, Biological and Medical Physics, Biomedical Engineering, DOI 10.1007/978-3-642-24843-6_1,

of interest, the number of possible biosignals is very large. In the broadest sense, the variety of biosignals extends from a visual inspection of the patient (Sect. 1.2) up to signals recorded from the human body using sensors, e.g., electrocardiography, compare Fig. 1.1. The huge diversity of biosignals can be best demonstrated by the fact that there are numerous kinds of biosignal classification, as discussed later in Sect. 1.3.

To give an *example of a biosignal* from its generation up to its registration, Fig. 1.2 depicts the formation of *acoustic biosignals* which are used, for instance, for the assessment of cardiorespiratory pathologies. The corresponding biosignal *source* in the heart is given by the periodic closure of heart valves, which yields

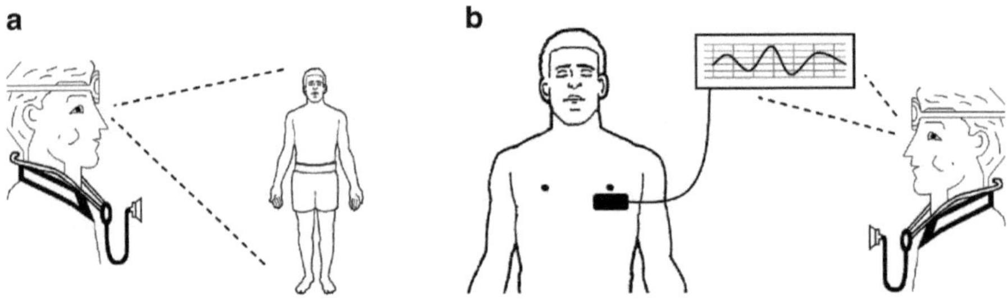

Fig. 1.1 Basic procedures for biosignal assessment from (**a**) visual appraisal of patient by a physician to (**b**) application of a biomedical sensor on the patient

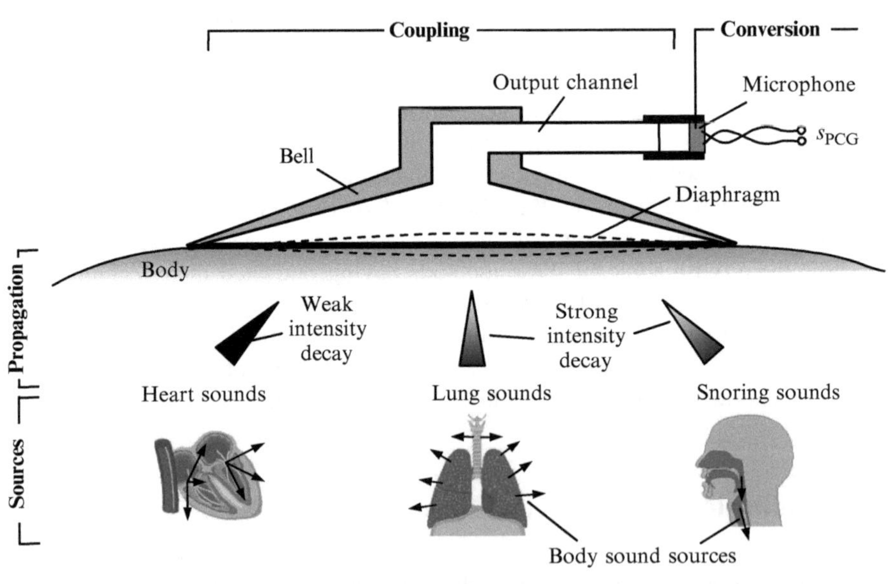

Fig. 1.2 The biomedical sensor on the chest for the registration of body sounds. The generation phenomena of the acoustic biosignals are depicted, along biosignal's propagation, coupling, and registration

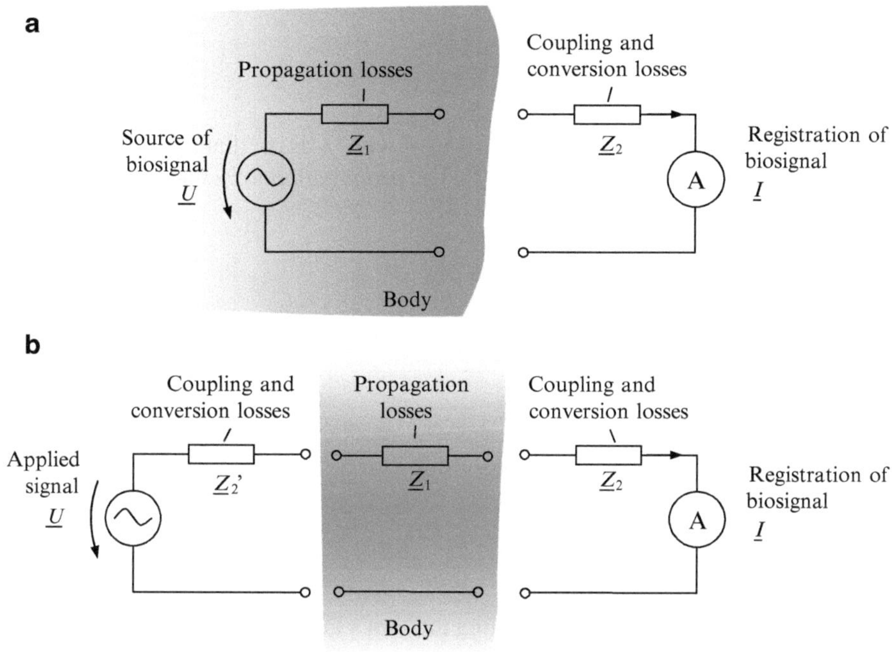

Fig. 1.3 Model of biosignal generation, propagation, coupling, and registration. (**a**) Permanent biosignal. (**b**) Induced biosignal

heart sounds. In addition, the lung sounds are generated by air turbulences in the branching airways of the lung, whereas the snoring sounds arise in the upper airways due to elastic oscillation of the pharyngeal walls. The sounds *propagate* throughout the tissue and undergo attenuation due to increasing distance from the source and damping by the medium itself. As indicated in Fig. 1.2 by intensity decay, the attenuation is different for different sounds, since their spectral components differ. In particular, the attenuation is less for the heart sounds than for the lung and snoring sounds, since the latter sounds exhibit more high-frequency components facing a stronger damping. The *coupling* (and amplification) of sounds is performed by a stethoscope chestpiece with an oscillating diaphragm and a resonating volume. Lastly, the *conversion* of the acoustical pressure vibrations into an electric signal is carried out by an electroacoustic transducer, a microphone.

Thus, the principle behavior in the formation of an arbitrary biosignal can be *modeled as an equivalent circuit* according to Fig. 1.3a. That is the *source of the biosignal* is represented by a sinusoidal[1] voltage source $u(t) = U \cdot \cos(\omega t + \varphi_U)$

---

[1]Usually the source of the biosignal exhibits nonsinusoidal behavior. However, the *nonsinusoidal* waveform can be represented as a sum of *sinusoidal* functions (according to Footnote 150), thus the equivalent circuit from Fig. 1.3a is also applicable here.

with complex amplitude

$$\underline{U} = U \cdot e^{j\varphi_U}, \tag{1.1}$$

magnitude U, angular frequency ω ($= 2\pi \cdot f$ with f as oscillating frequency), and phase φ_U, satisfying $u(t) = \mathrm{Re}[\underline{U} \cdot e^{j\omega t}]$. The *propagation losses* are represented by a series impedance

$$\underline{Z_1} = Z_1 \cdot e^{j\varphi_1}, \tag{1.2}$$

the *coupling and conversion losses* by another series impedance

$$\underline{Z_2} = Z_2 \cdot e^{j\varphi_2}, \tag{1.3}$$

and the *registered biosignal* by the resulting current $i(t) = I \cdot \cos(\omega t + \varphi_I)$ with complex amplitude

$$\underline{I} = I \cdot e^{j\varphi_I}, \tag{1.4}$$

satisfying $i(t) = \mathrm{Re}[\underline{I} \cdot e^{j\omega t}]$. According to Ohm's law,[2]

$$\underline{I} = \frac{\underline{U}}{\underline{Z}_1 + \underline{Z}_2}. \tag{1.5}$$

In other words, the higher the losses, e.g., the magnitudes $Z_1(\neq 0)$ and $Z_2(\neq 0)$ of usually capacitive-ohmic losses, the weaker the registered biosignal will be, i.e., the magnitude I. In general, $\varphi_I \neq \varphi_U$ provided that $\varphi_1 \neq 0$ or $\varphi_2 \neq 0$; likewise, if all losses can be modeled by real resistances then $\varphi_I = \varphi_U$ and $I = U/(Z_1 + Z_2)$. It should be noted that *physiological phenomena of interest are hidden* not only in $\underline{U}$ but also in $\underline{Z}_1$, for the propagation may influence the resulting $\underline{I}$ in a significant and even advantageous way (Sect. 5).

If the *acoustic biosignal* (Fig. 1.2) is considered in the light of the above *model* (Fig. 1.3a), the temporal behavior of an acoustical source can be described by $u(t)$ and its intensity by U. The strength of the propagation losses of the body sounds can be given as Z_1 (1.2) while the capacitive behavior of the propagating

[2]Georg Simon Ohm (1789–1854) was a German physicist after which *Ohm's law* was named. The law states that the strength of electric current I through a conductor is directly proportional to the voltage U across the conductor divided by the impedance Z of the conductor, if a constant Z is given, e.g., over conductor temperature or oscillation frequency of the current. For complex values, it can be written as

$$\underline{I} = \frac{\underline{U}}{\underline{Z}}.$$

For the continuum form of Ohm's law see Footnote 45.

medium can be described by the corresponding phase angle $\varphi_1(\neq 0)$. Alternatively, the strength of the coupling and conversion losses in the acoustical sensor can be defined as Z_2, whereas the corresponding $\varphi_2(\neq 0)$ can describe the time delay in the chestpiece and the conversion delay in the microphone (1.3). The output $s_{\mathrm{PCG}}(t)$ of the microphone—as schematically shown later in Fig. 1.15c—corresponds then to $i(t)$ [compare (1.5)].

While the model from Fig. 1.3a applies to permanent biosignals with their source already inside the body, Fig. 1.3b depicts a *model of an induced biosignal* (Sect. 1.3). Here, the biosignal is generated outside the body with an artificial signal source with its complex amplitude $\underline{U}$. After coupling and conversion losses $\underline{Z}'_2$ on the input side, the induced signal undergoes propagating losses $\underline{Z}_1$ in the body, which are *modulated by a physiologic phenomena* of interest. On the output side, the coupling and conversion losses $\underline{Z}_2$ co-determine the resulting induced biosignal $\underline{I}$ according to

$$\underline{I} = \frac{\underline{U}}{\underline{Z}_1 + \underline{Z}_2 + \underline{Z}'_2}. \tag{1.6}$$

To give an example, $\underline{U}$ could characterize an incident *artificial light* beam coupled into a finger, whereas $\underline{Z}_1$ varies by the changing light absorption due to pulsating blood volume (Sect. 6). Since blood pulsations carry cardiac and respiratory information, the transmitted light characterized by $\underline{I}$ reflects cardio-respiratory activity, as depicted later in Fig. 1.15c, which can be used advantageously in clinical applications.

In accordance with the origin of the biosignals, the biosignals are used in both diagnosis and therapy. While the *diagnosis*[3] is concerned with an assessment of health status based on biosignals (Fig. 1.3), the *therapy*[4] utilizes the biosignals as an objective feedback for selecting appropriate therapeutic measures, continuously monitoring their impact, and improving their efficiency, as depicted in Fig. 1.4. In the latter case, the biosignal registered by a diagnostic device and represented by $\underline{I}$ controls a therapeutic device by adjusting its stimulus given by $\underline{U}$.

From a practical point of view, the aforementioned acoustic biosignals (Fig. 1.2) could serve as an example for the *diagnostic application of biosignals*, as will be discussed in Sect. 5 in detail. The *therapeutic application of biosignals* could be demonstrated by functional muscle stimulation (e.g., on the leg) or functional nerve stimulation (e.g., on the ear auricle). While the stimulation (i.e., therapy) is performed by the use of electric impulses in both cases—compare Fig. 1.4—the respective feedback is given, for instance, by electromyography or force/torque measurement to assess the muscle response in the former case and by heart rate

[3]Generally, the *diagnostic area* of biomedical technologies can be *classified* into functional evaluation of the physiological state, clinical evaluation, and bioimaging (Turchetti et al. 2010); compare Footnote 4.

[4]The *therapeutic area* of biomedical technologies can be *classified* into noninvasive treatments, invasive treatments (minimally invasive and surgical), artificial organs and prosthesis, and rehabilitation (Turchetti et al. 2010); compare Footnote 3.

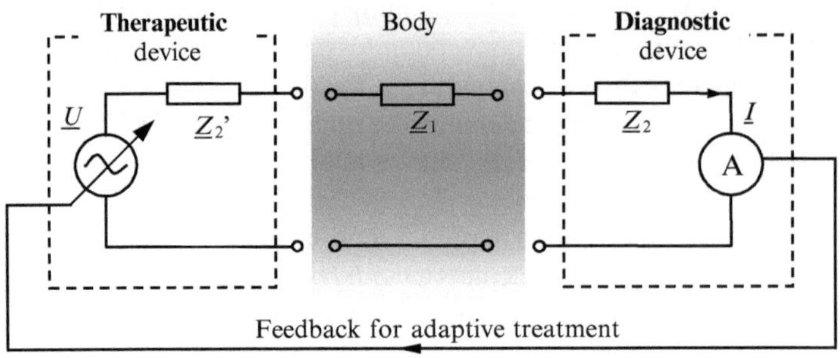

Fig. 1.4 Diagnostic application of biosignals (compare to Fig. 1.2b)

variability to assess the response of the autonomic nervous system (Sect. 3.1.1) in the latter case.

1.2 Historical Aspects

The registration of human biosignals underwent a long-lasting development over many centuries. It began with visual inspections without the use of any instruments, moved to the application of technical tools for signal registration, and is now in an implementation stage of pervasive, almost imperceptible, monitoring. Obviously this development has been driven by patient and physician needs as well as by problems that were encountered, interestingly not always relevant from a pure diagnostic point of view. As was recognized centuries ago concerning biosignal analysis in Mahomed (1872): "…surely it must be to our advantage to appreciate fully all it tells us, and to draw from it all that it is capable of imparting…."

1.2.1 The Very First Biosignals

The very first diagnoses were made on the patient's verbal account of his illness with the unaided senses. Forthcoming investigations yielded the first biosignals which were used for the diagnostic purposes only. The methods applied here encompassed mainly inspection, palpation, percussion, and auscultation (Fig. 1.5):

- *Inspection* (latin *inspectio* scrutiny) is the thorough visualization of the patient by the use of the naked eye. The physician may judge, for instance, body features, nutritional state, or skin color (Fig. 1.1a).
- *Palpation* (latin *palpare* feeling by touch) involves feeling the surface of the body with the hands to determine the size, shape, stiffness, or location of the organs beneath the skin (Fig. 1.5a). Often, applying a small amount of pressure to the surface of the skin or superficial artery to partially constrict it facilitates an easy observation of mechanical changes.

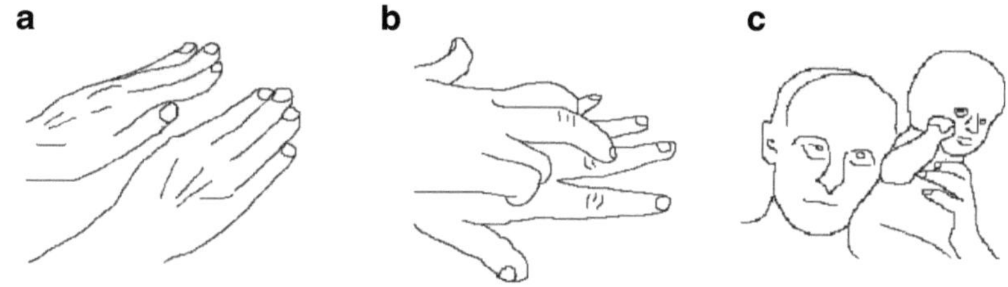

Fig. 1.5 Primary diagnosis methods besides inspection from Fig. 1.1a. (**a**) Palpation. (**b**) Percussion. (**c**) Auscultation

- *Percussion* (latin *percussio* striking) is a procedure that involves striking the body directly or indirectly with short, sharp taps of a finger or a hammer (Fig. 1.5b). The sounds produced display a resonant or dull character, indicating the presence of a solid mass or hollow, air-containing structures, respectively. The sounds are helpful in determining the size and position of various internal organs, in localizing fluid or air in the chest and abdomen, and in aiding in the diagnosis of certain lung disorders.
- *Auscultation* (latin *ausculto* hear attentively) describes a diagnostic procedure in which the physician listens to inner body sounds to detect pathologies or the state of health (Fig. 1.5c). The body sounds may be comprised of heart sounds due to closure of the heart valves or lung sounds due to air turbulences in the branching airways.

Hippocrates of Cos (around 460 BC–377 BC), ancient Greek physician regarded as the father of medicine, emphasized a simple *visual inspection*: "It is necessary to begin with the most important things and those most easily recognized. It is necessary to study all that one can see, feel, and hear, everything that one can recognize and use" (Castiglioni 1941). For instance, he noted that good humor, quiet sleep, clear mind, and mobility were descriptive of a favorable prognosis. By contrast, lying with the mouth and eyes open with legs spread apart, insomnia, and intense movements, indicated an unfavorable prognosis (Marinella 2008).

Palpation was also used by Hippocrates as a method for clinical examination, as demonstrated in Fig. 1.6. For instance, in his work "Diseases of Women" he writes "...And if you then palpate the uterus...." In particular, palpation of the *arterial pulse* has been recognized from antiquity as the most fundamental sign of life,[5] a periodic expansion of an artery (e.g., radial artery on the wrist) is felt in response to a periodic rise in blood pressure. Galen of Pergamum (around 129–200), Greek physician and philosopher, was one of the first great authorities on the pulse,

[5]Erasistratus (about 310 BC–250 BC), Greek physician, regarded by some as the "father of physiology," already used the pulse in clinical diagnosis. As a curiosity, the lover's pulse or love-sickness became a well-documented clinical entity and an integral part of pulse lore through the centuries. The love-sickness was described as pulse quickening in the presence of a beloved person (Hajar 1999).

Fig. 1.6 Hippocrates is pictured palpating a young patient (painting from Christian Medical College 2008)

admired by his patron, the emperor Marcus Aurelius. He described the pulsation as "The feeling of the artery striking against the fingers" and characterized it in many details as "the worm-like pulse, feeble and beating quickly; the ant-like pulse that has sunk to extreme limits of feebleness" (Hajar 1999).

Centuries later, Dr. Leopold Auenbrugger (1722–1809), Austrian physician, introduced the *percussion* technique as a diagnostic tool in medicine in 1761 in Vienna, Austria. Percussion was described as "a slow tapping with the fingers, brought close together and extended, on the fingers of the other hand laid on the chest" (Auenbrugger 1761). However, this technique was widely disseminated only decades later by Dr. Jean-Nicolas Corvisart (1755–1821), French physician and primary physician of Napoleon Bonaparte, who translated Auenbrugger's book into French (Auenbrugger and Corvisart 1808) in 1808, as illustrated in Fig. 1.7.

The *direct auscultation* of body sounds (Fig. 1.8) was also already employed more than twenty centuries ago, as suggested in Hippocrates work "de Morbis": "If you listen by applying the ear to the chest..." (Rappaport and Sprague 1941). However, only at the beginning of the nineteenth century did the body sounds gain adequate relevance and recognition among physicians.

A few decades later, after the wide acceptance of the percussion, which also involves an auscultation of artificially produced sounds, the *auscultation* technique was *fundamentally improved* by Dr. Rene Theophile Hyacinthe Laennec (1781–1826). The French internist and a student of Dr. Corvisart made in 1816 an epoch making observation with a wooden cylinder, which was primarily sought to avoid embarrassment. "I was consulted," says Laennec, "by a young women who presented some general symptoms of disease of heart, in whose case the application of the hand and percussion gave but slight indications, on account of her corpulency. On account of the age and sex of the patient, the common modes of exploration (i.e., immediate application of the ear) being inapplicable, I was led to recollect a well

NOUVELLE MÉTHODE

POUR RECONNAITRE

LES MALADIES INTERNES

DE LA POITRINE

PAR LA PERCUSSION DE CETTE CAVITÉ,

PAR AVENBRUGGER,

Médecin ordinaire de la Nation Espagnole dans l'Hôpital impérial, à Vienne en Autriche.

OUVRAGE TRADUIT DU LATIN ET COMMENTÉ

PAR J. N. CORVISART,

Premier Médecin de S. M. l'Empereur et Roi, Offic.r de la Lég. d'honneur, Commandeur de l'Ordre Royal de Hollande; Professeur hon.re de médecine clinique à l'École de Paris, Professeur hon.re de médecine au Collège de France; Médecin de l'Hôp. de la Charité; Associé hon.re de l'Académie Imp.le Joséphine de Vienne, de la Soc. Roy.le de Naples, de la Société Méd.le d'Emulation de Paris; et Membre de la plupart des Sociétés savantes de l'Empire.

32,541

Insonuêre cavæ....
VIRG., ÆNEID.

A PARIS,

DE L'IMPRIMERIE DE MIGNERET

1808.

Fig. 1.7 Title page of Corvisart translation about percussion as a diagnostic tool (Auenbrugger and Corvisart 1808)

known acoustic phenomenon... I took a quire of paper which I rolled together as closely as possible, and applied one end to the precordial region; by placing my ear at the other end, I was agreeably surprised at hearing the pulsation of the heart much more clearly and distinctly than I had ever been able to do by the immediate application of the ear" (Rappaport and Sprague 1941; Abdulla 2001).

A *precursor of the stethoscope* (greek *stetos* chest and *skopein* explore) was born—as shown in Fig. 1.9—viewed by many as the very symbol of medicine, for conduction of the sounds generated inside the body between the body surface and the ears, as depicted in Fig. 1.10. An oil painting is shown in Fig. 1.11 with Laennec among students holding his stethoscope in the hand, while applying his ear to the chest of a patient.

Later, in 1894, A. Bianchi introduced a rigid diaphragm over the part of the (wooden) cylinder, i.e., the chestpiece, that was applied to the chest (Hollins 1971; Rappaport and Sprague 1941), compare Fig. 1.2. The *modern stethoscope* consists of a bell-type chestpiece for sound amplification (Welsby et al. 2003; Abdulla et al. 1992), rubber tube for sound transmission, and earpieces for conducting the sound into ears (Ertel et al. 1971).

Fig. 1.8 Direct auscultation of body sounds

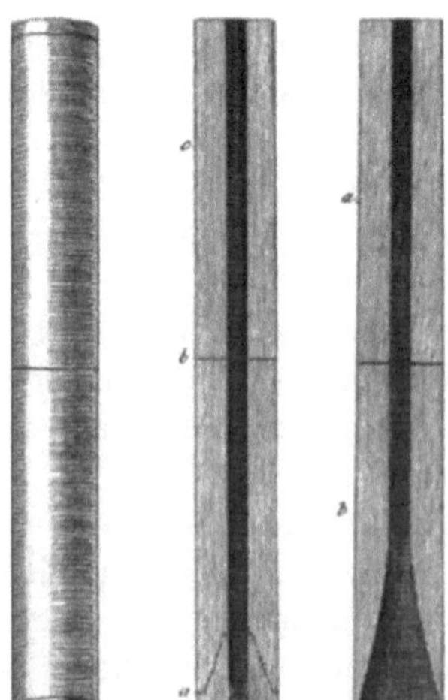

Fig. 1.9 Drawings of the original Laennec's stethoscpe (Laennec 1819)

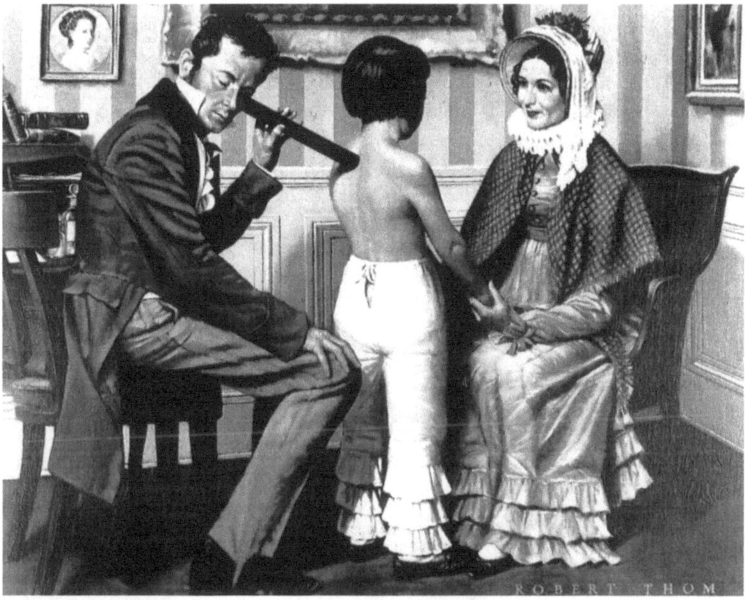

Fig. 1.10 Indirect auscultation of body sounds with Laennec's stethoscope (Thom 1954)

At the end of the nineteenth century, Laennec's stethoscope was still not used on a regular basis. The introduction of the stethoscope *forced physicians* to undergo a *cardinal reorientation*, for the stethoscope altered both the physician's perception of disease and his relation to the patient. Despite the clear superiority of the instrument in the sound auscultation, it was accepted with some *antagonism* even by prominent chest physicians. Among others, the amusing critics concluded that "The stethoscope is a largely decorative instrument insofar as its value in diagnosis... Nevertheless, it occupies an important place in the art of medicine. Apprehensive patients with functional complaints are often relieved as soon as they feel the chestpiece on their pectoral muscles..." or physicians complained that "they heard too much" (Loudon and Murphy 1984).

1.2.2 Problems and Solutions

The main *problems faced* by the original biosignal acquisition methods—inspection, palpation, percussion, and auscultation (Sect. 1.2.1)—were related to an *objective evaluation* of the diagnostic results. In particular,

- *Proof* of biosignals
- *Analysis* of biosignals
- *Comparison* of biosignals
- *Circulation* of biosignals

Fig. 1.11 Laennec, inventor of the stethoscope, applies his ear to the chest of a patient (Chartran 1849–1907)

were impossible due to the subjective nature of the diagnosis. In other words, *reproducibility* of the biosignal observation was not possible because of the observer's variability and no means for the biosignal's archival storage, as is self-evident for today's applications. *Analysis* of the biosignals was restricted to an instantaneous impression by the physician, with the impression being strongly affected by the physician's personal experience. The classification of biosignals was impeded by nomenclature difficulties. The *comparison* of two biosignals was hardly possible, as they were restricted to a single physician and recent impressions. *Circulating* the accumulated biosignal data was also impossible because of the lack of archives.

Obviously the above problems and limitations were recognized early, with an attempt to circumvent them in a contemporary manner. The most notable

Table 1.1 Approaches to objectify the biosignals from a historical point of view

Description	Subjective impact	Quality
Verbal	Strong	Qualitative
Musical notes	Weak	Qualitative and quantitative
Technical	No	Quantitative

approaches to objectify and characterize the attained biosignals, given roughly in chronological order, were

- *Verbal* descriptions
- *Musical* notes
- *Technical* tools

As summarized in Table 1.1, the *verbal descriptions* had the most subjective impact from the author of the description, since it is purely *qualitative.* A variety of qualifying adjectives were used as well as vague subjective terms. Avicenna (980–1037), Muslim polymath and Islam's "Prince of Physicians," ingeniously compares more than 50 identifiable pulses with natural objects and human actions: "irregular pulse as the flight of a gazelle; stone bullet shot out of a crossbow; scattered leaves" (Hajar 1999). In order to accommodate the difficulties in describing lung sounds, familiar sound descriptions (at that time) were chosen to clarify the distinguishing characteristics (Loudon and Murphy 1984). Descriptive and illustrative sounds were used such as "crepitation of salts in a heated dish," "noise emitted by healthy lung when compressed in the hand," "bass note of a musical instrument," "wet, dry, crackling sound," or even "cooing of wood pigeon." As another example, percussion sounds were described as being sonorous, morbid, or dull (Murray and Neilson 1975). The *difficulties in the verbal description* could be best viewed in terms of Laennec's observation that the sounds heard with this "cylinder" were *easier to distinguish than to describe* (Loudon and Murphy 1984), yielding a need for better methodologies.

The proposed use of *musical notes* obviously reduced the subjectivity and provided—for the first time—a quantitative means to objectify biosignals (Table 1.1). While the height of the note could be used for a *qualitative* coding of biosignals, the rhythm of the successive notes could be used for a *quantitative* coding. A very nice example is given by notable attempts to *objectively* describe the pulsatile behavior of the blood pressure with music rhythm. The flute teacher Francois Nicolas Marquet(1687–1759) made pulse to a natural metronome (Marquet 1769), as demonstrated in Fig. 1.12. Up to 30 different pulses were documented by music notes.

The last and clearly most successful approach implements the use of *technical tools* (Table 1.1). Historical progress in technical tools is shortly but conclusively summarized in Geddes and Roeder (2009). They eliminate subjective influence from the observer, since the approach is intrinsically based on quantitative data. With each new advance in these novel techniques, new vistas with previously *unforeseen opportunities* became exposed. Since there is an *enormous diversity* of technical

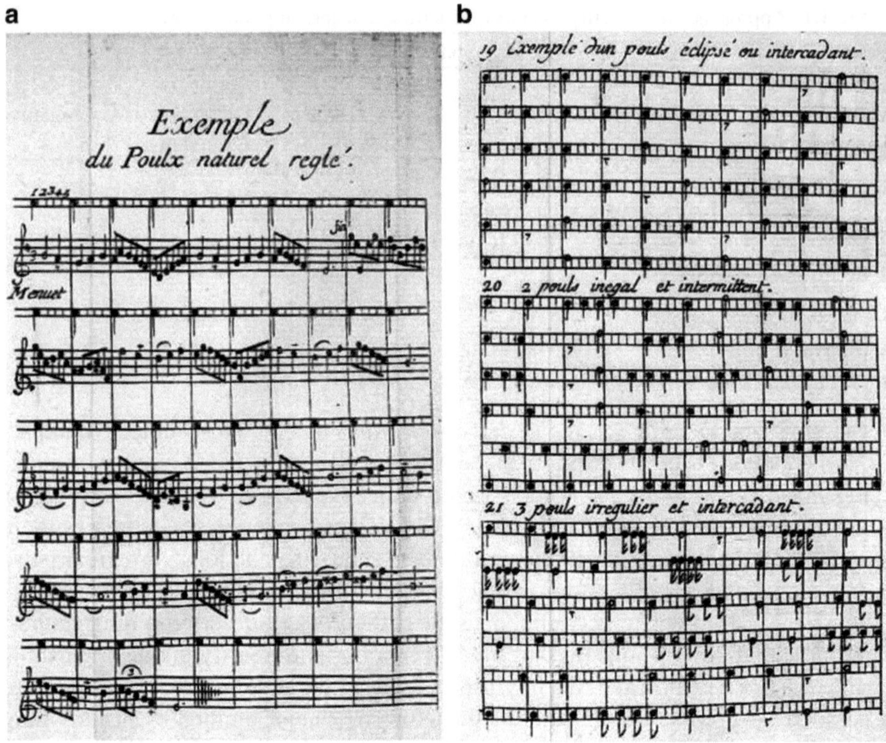

Fig. 1.12 Coding of heart pulses with musical notes (Marquet 1769). (**a**) Natural regulated pulse. (**b**) Three different abnormal pulses including, from *top* to *bottom*, discontinuous pulse, irregular intermittent pulse, and irregular pulse arising in between normal pulses

tools being introduced, only two historically relevant developments will be shortly mentioned.

Representative of tools applied in *clinical* praxis, Fig. 1.13 demonstrates an ancestor of a sphygmomanometer (greek *sphygmos* pulse, *manometer* pressure measuring device) used for recording *pulse* and blood pressure on, e.g., radial artery. The device stems from developments in the nineteenth century and is acknowledged as the *first diagnostic instrument* introduced for artificial palpation of the pulse if the thermometer and stethoscope are regarded as clinical aids only. It used cuff-based recording, the methodology that is still nearly unrivaled up to current times, see Sect. 3.1.3.1.

The *advent of portable* technical tools for diagnosis is demonstrated by a sphygmograph (greek *sphygmos* pulse, *grapho* write), as shown in Fig. 1.14. This instrument was devised by Dr. Robert Ellis Dudgeon (1820–1904) for graphically recording features of the radial pressure pulse, which was beautifully compact and found its way into medical practice around the world. It consists of a lever with an elastic spring placed on the radial artery. The other end of the lever carries a stylus for recording of the pulse on a moving smoked paper.

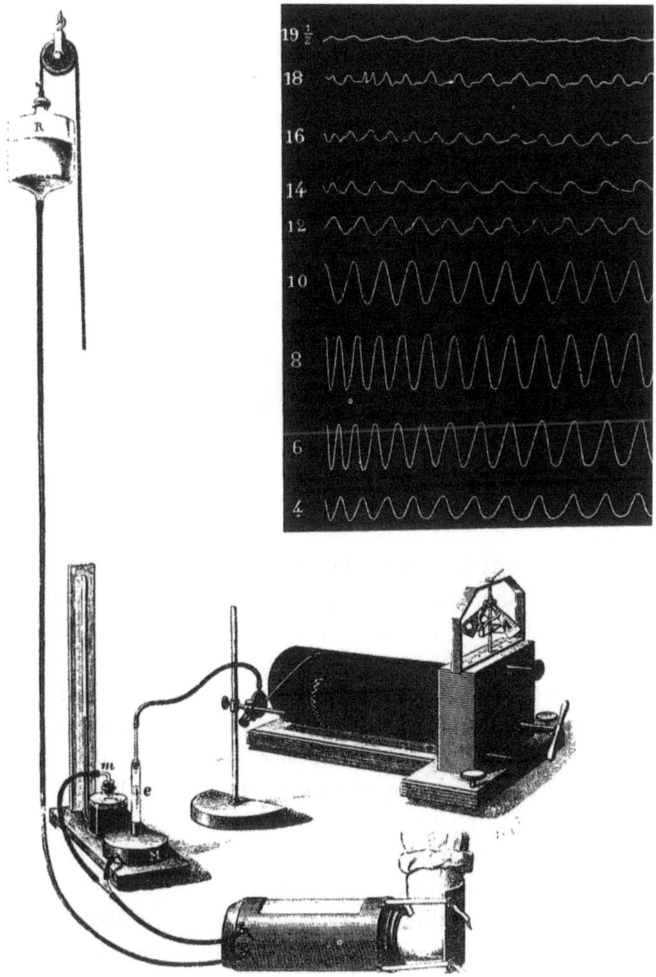

Fig. 1.13 The ancestor of the sphygmomanometer for clinical applications (Marey 1858)

1.3 Classification of Biosignals

The variety of biosignals is nearly unlimited, as shown in Sects. 1.1 and 1.2. This circumstance makes a unique classification of biosignals impossible. However, there are at least three ways of defining their (overlapping) strategic classification, as demonstrated in Fig. 1.15 and described below.

As a first classification method, the *existence of biosignals* could be taken as a basis for their classification. In particular,

- Permanent biosignals
- Induced biosignals

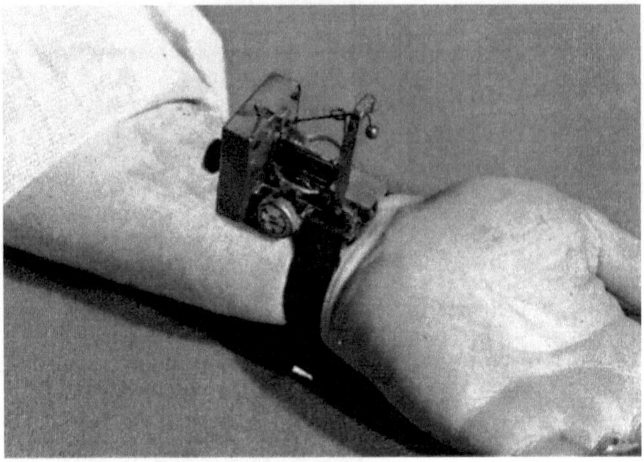

Fig. 1.14 A sphygmograph according to Dr. Dudgeon for portable application (Dudgeon 1882)

would comprise the corresponding classification groups. *Permanent biosignals* exist without any artificial impact, trigger, or excitation from outside the body and are available at any time (compare Fig. 1.3a). The source of the biosignal is already inside the body. To give some examples, an *electrocardiographic* signal (=electrocardiogram) induced by electrical heart muscle excitation (Sect. 4) with the typical peaks P–Q–R–S–T (Fig. 1.15a) and the aforementioned *acoustic* biosignal (= phonocardiogram) induced by the consecutive heart valve closures (Sect. 5) with the typical first and second heart sounds (Fig. 1.15c) belong to the group of permanent biosignals.

The group of *induced biosignals* considers biosignals that are artificially triggered, excited, or induced (compare Fig. 1.3b). In contrast to permanent biosignals, induced biosignals exist roughly for the duration of the excitation. That is, as soon as the artificial impact is over, the induced biosignal decays with a certain time constant determined by the body properties. The interaction of the tissue with the induced stimulus, irrespective of the stimulus nature, is then recorded as an induced biosignal. A corresponding example could be given by *electric plethysmography*, in which an artificial current is induced in the tissue and a voltage along the current path reflects tissue impedance changes (Sect. 4). The voltage is then registered as an induced biosignal (=electroplethysmogram) with discernible cardiac and respiratory components (Fig. 1.15a). Alternatively, *optical oximetry* uses artificially induced light while the transmitted light intensity is mainly governed by light absorption through local pulsatile blood volume (Sect. 6). The transmitted light is detected as an induced biosignal, showing a steep systolic increase and a slow diastolic decrease (Fig. 1.15c). In general, the origin of the induced stimulus, e.g., magnetic field from coils above the head for magnetic stimulation, may be different from that of the registered biosignal, e.g., generated electric potentials from electrodes on the head.

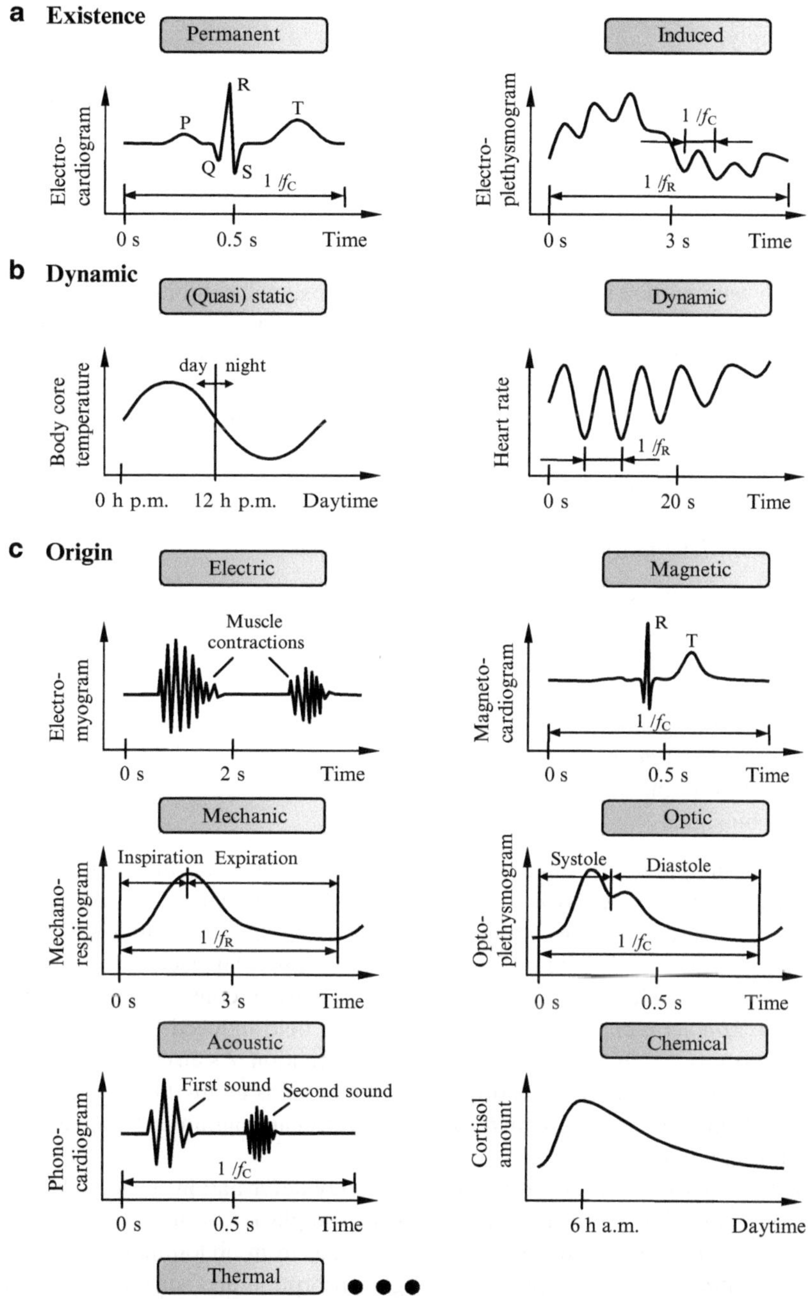

Fig. 1.15 The possible classifications of biosignals according to their (**a**) existence, (**b**) dynamic, and (**c**) origin, with indicated heart rate f_C, respiratory rate f_R, and additional information

The second classification method considers the *dynamic nature of biosignals*. Accordingly,

- (Quasi) Static biosignals
- Dynamic biosignals

can be differentiated. A *(quasi) static biosignal* carries information in its steady-state level which may exhibit relatively slow changes over time. By contrast, *dynamic biosignals* yield extensive changes in the time domain, with dynamic processes conveying the physiological information of interest. For instance, the *core body temperature* would be a (quasi) static biosignal, exhibiting relatively slow circadian changes over 24 h (Sect. 3.1.5). As shown in Fig. 1.15b, it increases during the morning hours and decreases before the onset of sleep (Sect. 3.2.4). On the other hand, the instantaneous *beat-to-beat changes of the heart rate* would constitute a highly dynamic biosignal (Sect. 3.1.1). The course of the heart rate (Fig. 1.15b) reveals respiratory related oscillation, i.e., an increase during inspiration and a corresponding decrease during expiration.

The third classification method uses the *origin of biosignals* as a basis for their classification. The most prominent origins encompass

- Electric biosignals
- Magnetic biosignals
- Mechanic biosignals
- Optic biosignals
- Acoustic biosignals
- Chemical biosignals
- Thermal biosignals
- Other biosignals

Correspondingly, *electric biosignals* comprise, for instance, the aforementioned *electrocardiogram* (Fig. 1.15a), *electroencephalogram*, which reflects electrical activity of neurons in the brain, or *electromyogram*, which reflects electrical activation of muscles. Figure 1.15c schematically depicts an electromyogram which shows bursts of electrical impulses yielding muscle contractions of different strengths. *Magnetic biosignals* reflect a magnetic field induced by usually nonstationary currents which convey physiological information. As an example, Fig. 1.15c shows a *magnetocardiogram* reading of magnetic fields emitted by currents during electrical heart excitation (compare peaks in electrocardiogram and magnetocardiogram in Fig. 1.15).

Mechanic biosignals reflect, for instance, body deformations or local body skin vibrations unveiling physiological data. An example is given in Fig. 1.15c by a *mechanorespirogram*, showing a respiratory cycle from abdominal circumference changes. *Optic biosignals* benefit from light absorption and scattering, which are related to propagation volume and medium, both changing in a physiologically relevant way. Here, an artificial light is used within the scope of induced biosignals, as already described. As demonstrated in Fig. 1.15c, cardiac pulsations with a clinically relevant time course can be clearly recognized in an *optoplethysmogram*.

Acoustic biosignals remain for the assessment of diverse body sounds, ranging from cardiac sounds to snoring sounds to swallowing sounds. A *phonocardiogram*, as shown in Fig. 1.15c and discussed earlier in Sect. 1.1, mirrors cardiac activity. It is comprised of two discernable heart sounds corresponding to two consecutive heart valve closures. The oscillation amplitude and frequency indicate the closure strength and the valve's stiffness, respectively. *Chemical biosignals* reflect chemical composition and its temporal changes in body solids, liquids, and gases. To demonstrate their relevance, Fig. 1.15c shows a typical course of *cortisol* (= stress hormone) over 24 h in humans, with a peak during the morning hours in order to prepare the body for awakening. Lastly, *thermal biosignals* usually assess highly heterogeneous mechanisms of heat loss and heat absorption in the body. For instance, the aforementioned body core temperature in Fig. 1.15b constitutes a thermal biosignal. For the sake of completeness, it should be mentioned that the above list of biosignals—classified according to their origin—is obviously not complete.

1.4 Trends in Biosignals Monitoring

Biosignals were first employed more than twenty centuries ago, as exemplified in Sect. 1.2.1, and became even more prominent in the twenty-first century. Though having been used since time immemorial, a further advancement of their acquisition, interpretation, and use in the diagnostic approaches was still never out of question. Their *developmental history is marked with revolutions* rather than continuous improvements, with revolutions usually followed by antagonism.[6] Even today, their proper assessment and analysis are the focal point of many research groups worldwide. The obvious reason for these *never ending improvements* in biosignal monitoring is that the biosignals reflect human health and wellbeing. Biosignals are essential for mankind and not just for increased comfort. In particular, biosignals detail vital physiological phenomena and are relevant not only for the pre-screening of the human functional state and diagnosis of illness but also for subsequent therapy, follow-up treatment, and appraisal of its efficiency.

Future trends in biosignal monitoring could be partly deduced from the *history* of biosignals and the current *state-of-the-art* technology, as aimed at in Fig. 1.16. From a technical point of view, a qualitative relation exists between the comfort of the sensor system, approximated as the number of applied sensors (horizontal axis),

[6]For instance, Kurt Karl Stephan Semm (1927–2003), German gynecologist, who performed the first appendicectomy in 1980 in a *laparoscopic way*, was heavily criticized by his colleagues and public. Later it was recognized that it not only helps patients recover faster and with less pain, but also prevents deaths in the operating room. Another example would be Ignaz Semmelweiss (1818–1865), Hungarian physician, who was largely ignored or ridiculed when in 1847 he suggested that childbed fever could be drastically reduced if doctors *sterilized their hands*.

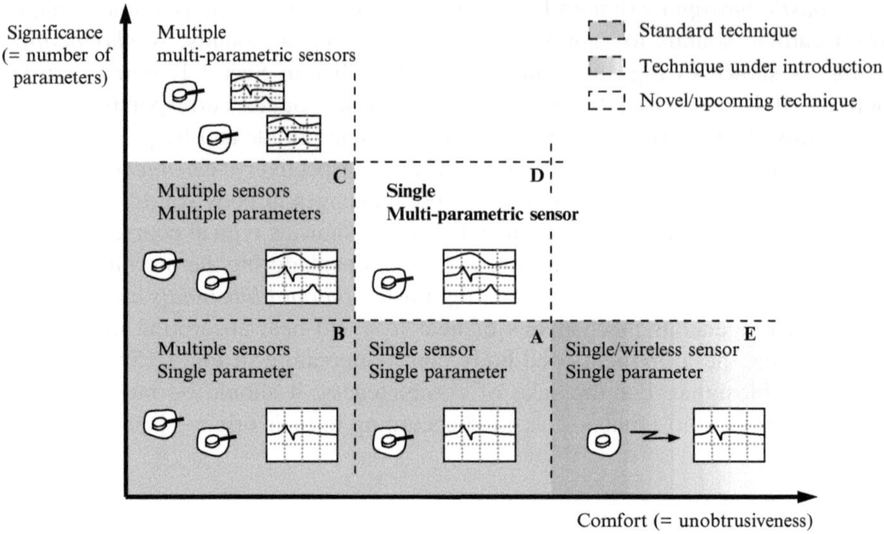

Fig. 1.16 Future vision of physiologic monitoring including standard and novel techniques. Qualitative relationship is given between the significance and comfort of the different monitoring systems, i.e., number of physiological parameters attained versus number of sensors needed, including novel multiparametric sensors. *Bold* letters refer to cases discussed in the text

and the significance of attained biosignals, the latter quantified as the total number of physiologic parameters available to derive (vertical axis).

Obviously the oldest and most commonly used systems follow the rule that a *single physiological parameter* is attained per *single sensor* (case A in Fig. 1.16). For instance, respiratory rate is usually assessed by a respiratory belt around the thorax, which monitors circumference changes related to breathing. In many cases, a single sensor may not be sufficient to determine a *single parameter*; thus, *two or more sensors* might be needed (= multisite recording), as depicted in case B. Here, the common arterial blood pressure recording could be an example, in which decreasing cuff pressure on the upper arm is recorded in parallel to sounds (= Korotkoff sounds) recorded by a microphone over the brachial artery; audible sounds arise due to blood flow turbulences at cuff pressure values corresponding to systolic and diastolic blood pressure. In comparison with case A, case B shows reduced comfort but the same significance because the assessed number of physiologic parameters is the same.

If two or *more single-parameter sensors* (from case A) are applied, then obviously *multiple parameters* are provided (case C). For instance, sleep monitoring in sleep labs includes the monitoring of a large number of brain, cardiac, and respiratory parameters with the use of the corresponding single-parameter sensors. Numerous parameters are needed here for a comprehensive sleep assessment, e.g., for sleep staging.

Consequently, the technique of *multiparametric monitoring* could be deduced from cases A and C if *multiple parameters* are derived through the use of a *single sensor*, namely, a multiparametric sensor (case D). The multiparametric sensor yields the comfort of a single sensor (case A) in combination with the significance of multiple parameters (case C), as demonstrated in Fig. 1.17. An acoustic body sound sensor on the chest offers this type of monitoring, yielding cardiac activity, respiratory activity, and breathing obstruction from a single spot. In order to achieve an adequate realization of the multiparametric monitoring, a *concerted effort* should be taken on the part of

- Novel sensor *concepts*, e.g., based on advances in technology as miniaturization
- Optimized sensor *location*, e.g., proximal instead of distal to increase physiological content of biosignal
- *Type* of recorded signals, e.g., optic instead of electric to get a higher spatial resolution
- Mutual *interrelations* and clinical correlations of physiologic parameters to derive, e.g., use of cardiorespiratory interrelations
- Advanced signal *processing* methods, e.g., decomposition of signals into its components based on their independence

Within this concept, a thorough understanding of the mechanisms of generation and transmission of biosignals, physiologic factors that affect them, a priori knowledge about biosignal characteristics and their appropriate decomposition are necessary.

Finally, as shown in case E in Fig. 1.16, the comfort of the subject is significantly increased by the use of *wireless data transfer* which bypasses the need of an electronic hook-up to the subject. In particular, portable devices for home monitoring profit from cable-free operation.

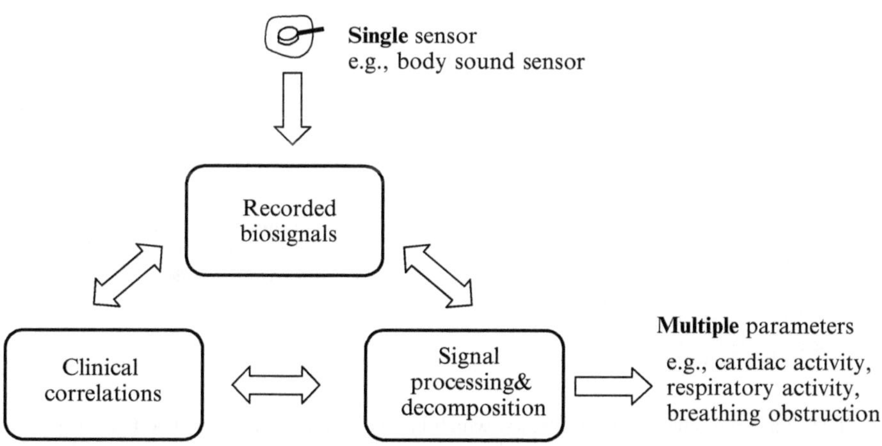

Fig. 1.17 Principle of multiparametric physiologic monitoring

In contrast to *technical* considerations in Fig. 1.16, paradigm changes from an *application* point of view are depicted in Fig. 1.18. As already discussed in Sect. 1.2, the registration of biosignals outlasted centuries,

- Beginning with a *basic inspection* (Fig. 1.18) without any (or with simple) instruments.
- Established *clinical applications* (Fig. 1.18) followed showing the highest reliability but requiring a large effort in all three: applied devices, attending physicians, and laboratory premises. Furthermore, the laboratory window of observation is limited in time, i.e., infrequent (usually vital) physiologic events are easy to miss.
- Then *portable applications* (Fig. 1.18) start to emerge which is not only sought in response to the above economic imperatives and need of improved access to diagnosis but also because it may provide a more realistic appraisal of 24-h pathology and more complete information about the physiologic state of the patient. In addition, unattended studies conducted in a home environment allow for improved comfort and familiarity. However, portable recording usually suffers from several problems, such as difficult hook-up of patients, poor assessment of signal quality and data loss, as well as insufficient experience required for proper interpretation of portable data records.

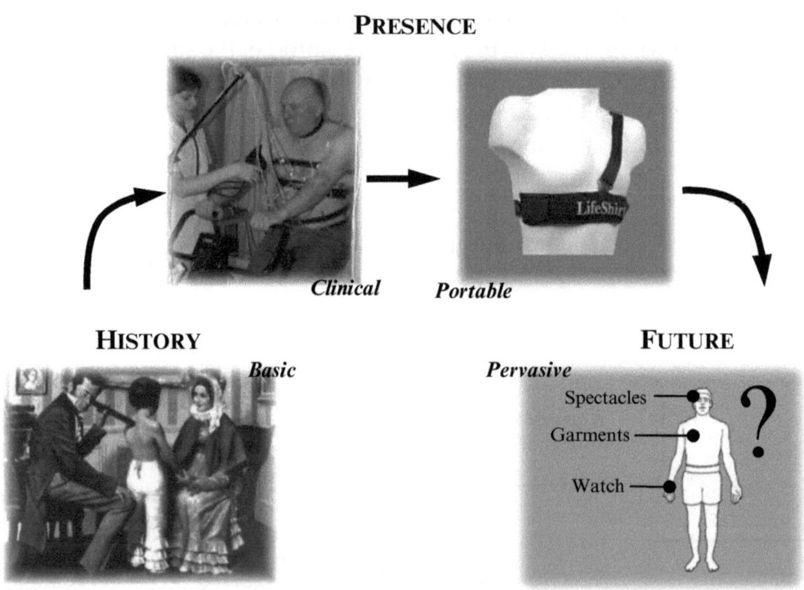

Fig. 1.18 Paradigm changes from history, which brought basic monitoring functions, to present times, which emphasize advanced functionality in both clinical settings and portable home applications, to the future, which may yield integrated biomedical monitoring not perceivable by patient but easily usable by the physician. The portable LifeShirt system shown is taken from RAE Systems (2011).

- Lastly, *pervasive applications* (Fig. 1.18) seem to govern the research trends in biomedical engineering. The goal of pervasive health care is to provide continuous personalized health monitoring of *patients and healthy individuals* at any time without constraints of space, time, and physician availability. Unobtrusive monitoring settings include not only daily activities but also demanding circumstances such as physical training or observation of medical treatment. During examination, the presence of the medical staff should be avoided, reducing involuntary stress of the individual and providing a realistic appraisal of pathology or process of recovery.

In order to realize a *pervasive monitoring system*, several

- *Hardware-related*
- *System-related*

requirements must be (ideally) met. The *hardware-related* requirements include *minimal obtrusiveness* and compactness, nonhazardous and *inexpensive* design of the system (Ahamed et al. 2006; Kollmann et al. 2006), resulting in a minimum number of spatially distributed sensors and avoiding tethering patients in a tangle of cables. An *inconspicuous* and *nonstigmatizing* design is needed to allow for long-term monitoring (Poh et al. 2010). In particular, *unnoticeable* monitoring is demanded, with capacitive, magnetic, and optical technologies being especially relevant because of their noncontact physical nature. It is imperative that recorded signals contain a large amount of *physiological information*. A compromise should be made between *long-term wearability* and *reliable sensor* application (Asada et al. 2003). The recorded biosignal should be *robust*, i.e., its resistance to prevalent environmental impacts such as body motions, temperature changes, or external interference (noise) while wirelessly communicating. In addition, a purposeful *preprocessing* of the biosignal, its storage and transmission under (very) *low-power* consumption comprise the most important design characteristics. *Differential architectures* gain attractiveness for attenuating external *interference*, with the architecture including one sensing unit for the biosignal and another one only for the environmental interference. Obviously *safety and security risks* should be accounted for and the *risks* should be acceptable in relation to an expected monitoring benefit and *health regulations* (Leitgeb 2010). For instance, economical energy efficient *encryption* techniques are a prerequisite for data transmission from/to the sensor (Kailas et al. 2010).

System-related requirements (design paradigms) include *real time*, robust, reliable, and sensitive data interpretation (besides fixed thresholds) to *minimize false alarms* which frighten the user and increase costs. In addition, *bidirectional data* transfer is necessary for sensing and (adaptive) therapy, e.g., diagnosis of cardiac state and urgent therapy by defibrillation if necessary. Interaction with the user has to be minimal and must result in a *meaningful representation* of the state of health. In particular, a *context-aware* health representation is needed (Kailas et al. 2010), whereas the collected data is presented in different ways to the physician (e.g., more details included) and the user (e.g., less details but personalized with a visual representation of health lifestyle tendencies).

A distinctive feature of *pervasive application* is, in contrast to all chronologically preceding applications (Fig. 1.18), that it should be readily accessible to *physicians*, patients, and even to healthy individuals. In particular, *high-risk patients* (e.g., with apneas given by a temporal cessation of breathing during sleep) and *chronic patients* (e.g., chronic heart failure) profit from pervasive monitoring, as well as *athletes* (interested in cardiorespiratory feedback during rest or training), the *elderly* (with restricted mobility), or even *specialized occupations* (e.g., professional drivers) forced to undergo preventive medical checkup to receive more timely treatment.[7] That is, the user-friendliness of pervasive systems would play an even more important role, for it is a more relevant issue for healthy individuals than for ill clinical patients. In addition, *demographic evolution* of the population will be a limitation on the physician's workload associated with diagnostic examinations and thus raises the need for smart tools in pervasive assistance.

Figure 1.18 indicates *possible realizations* of pervasive monitoring by hardware integration into spectacles, garment, watch, or mobile phone, i.e., by integration into indispensable objects of everyday use. For instance, an electrocardiographic system was newly developed which is integrated in a shirt and operates fully autonomously by thermal and optical energy harvesting from the ambient environment (Leonov et al. 2009).

Pervasive applications can be expected to reduce total medical *costs*,[8] increase *continuity*[9] and improve *availability*[10] of health care even in the leading European countries and facilitate the work of physicians. Interestingly, physicians, patients, and healthy subjects appear to accept information technology to assist in their decision making (Ahamed et al. 2006), to turn the physician's attention to the person if necessary (Kollmann et al. 2006), and to be an objective guide through the positive way of life (Connelly et al. 2006).

It appears that there is much room for radical *improvements* of the conventional physiological sensing and monitoring techniques because of the inflexible application of classical sensors as well as established signal processing. However, it cannot be expected that future revolutions or multiparametric sensors will replace established sensors anytime soon or even the stethoscope-bearing clinician, but

[7]There is data from the UK (Flemons et al. 2003) which suggest that the wait for investigation with the polysomnography (= comprehensive clinical monitoring in sleep lab, Sect. 3.2.4) versus portable monitoring was reduced from a median of 47 days to 18 days.

[8]As reported in Flemons et al. (2003), the portable monitoring was 30% up to 50% the cost of the polysomnography.

[9]For instance, patients measuring their own blood pressure by oneself may leave out unsatisfactory numbers (Asada et al. 2003).

[10]In Wisconsin (USA), 93% of women and 82% of men with moderate-to-severe sleep apnea, i.e., temporal cessation of effective respiration during sleep, did not receive diagnoses (Flemons et al. 2003).

they may significantly expand diagnostic capabilities in medicine and may have a *broader impact* on society, such as with improved access to sleep diagnosis.

References

R. Abdulla: The history of the stethoscope. Pediatric Cardiology 22(5), 371–372 (2001).

M. Abella, J. Formolo, D.G. Penney: Comparison of the acoustic properties of six popular stethoscopes. The Journal of the Acoustical Society of America 91(4 Pt 1), 2224–2228 (1992).

S.I. Ahamed, M. Sharmin, S. Ahmed, M. Haque, A.J. Khan: Design and implementation of a virtual assistant for healthcare professionals using pervasive computing technologies. Journal Springer e&i 123(4), 112–120 (2006).

H.H. Asada, P. Shaltis, A. Reisner, R. Sokwoo, R.C. Hutchinson: Mobile monitoring with wearable photoplethysmographic biosensors. IEEE Engineering in Medicine and Biology Magazine, 22(3), 28–40 (2003).

L. Auenbrugger, J.N. Corvisart: A new discovery that enables the physician from the percussion of the human thorax to detect the diseases hidden within the chest (in French: La nouvelle méthode pour reconnaître les maladies internes de poitrine par la percussion de cette cavité). Paris (1808).

L. Auenbrugger: Novel invention concerning striking human thorax in order to uncover concealed internal breast diseases (in Latin: Inventum novum ex percussione thoracis humani ut signo abstrusos interni pectoris morbos detegendi). Vienna (1761).

A. Castiglioni: The golden age of Greek medicine: Hippocratic medicine - biologic and synthetic concept. In E.B. Krumbhaar: A History of Medicine. New York, Alfred E. Knopf Publisher, 148–178 (1941).

T. Chartran, Oil painting from National Library of Medicine, Bethesda, Maryland, USA (1849–1907).

CMC: Painting from Christian Medical College, Vellore, India (2008).

K.H. Connelly, A.M. Faber, Y. Rogers, K.A. Siek, T. Toscos: Mobile applications that empower people to monitor their personal health. Journal Springer e&i 123(4), 124–128 (2006).

R.E. Dudgeon, Museum of History of Medicine, Paris, France (1882).

P.Y. Ertel, M. Lawrence, W. Song: Stethoscope acoustics and the engineer: Concepts and problems. Journal of the Audio Engineering Society 19(3), 182–186 (1971).

W.W. Flemons, M.R. Littner, J.A. Rowley, P. Gay, W.M. Anderson, D.W. Hudgel, R.D. McEvoy, D.I. Loube: Home diagnosis of sleep apnea: A systematic review of the literature: An evidence review cosponsored by the American Academy of Sleep Medicine, the American College of Chest Physicians, and the American Thoracic Society. Chest 124(4), 1543–1579 (2003).

L.A. Geddes, R.A. Roeder: Where do ideas come from? IEEE Engineering in Medicine and Biology Magazine 28(5), 60–61 (2009).

R. Hajar: The Greco-Islamic pulse. Heart Views 1(4), 136–140 (1999).

P.J. Hollins: The stethoscope. Some facts and fallacies. British Journal of Hospital Medicine 5, 509–516 (1971).

A. Kailas, C.C. Chong, F. Watanabe: From mobile phones to personal wellness dashboards. IEEE Pulse 1(1), 57–63 (2010).

A. Kollmann, D. Hayn, P. Kastner, G. Schreier: Mobile phones as user interface in the management of chronic diseases. Journal Springer e&i 123(4), 121–123 (2006).

R.T.H. Laennec: On mediate auscultation or treatise on the diagnosis of the diseases of the lungs and heart (in French: De l'auscultation médiate ou trait du diagnostic des maladies des poumon et du coeur). Paris (1819).

N. Leitgeb: Safety of electromedical devices. Springer Publisher (2010).

V. Leonov, T. Torfs, C.V. Hoof, R.J.M. Vullers: Smart wireless sensors integrated in clothing: an electrocardiography system in a shirt powered using human body heat. Sensors & Transducers Journal 107(8), 165–176 (2009).

R. Loudon, R.L.H. Murphy: Lung sounds. The American Review of Respiratory Disease 130(4), 663–673 (1984).

F.A. Mahomed: The physiology and clinical use of the sphygmograph. Medical Times and Gazette 1, 62–64 (1872).

E.J. Marey, Museum of History of Medicine, Paris, France (1858).

M.A. Marinella: On the Hippocratic facies. Journal of Clinical Oncology 26(21), 3638–3640 (2008).

F.N. Marquet: New, simple, and interesting method to discover pulse by music notes (in French: Nouvelle méthode facile et curieuse pour connaître le pouls par les notes de la musique). Paris (1769).

A. Murray, J.M.M. Neilson: Diagnostic percussion sounds: 1. A qualitative analysis. Medical and Biological Engineering and Computing 13(1), 19–28 (1975).

M.Z. Poh, N.C. Swenson, R.W. Picard: A wearable sensor for unobtrusive, long-term assessment of electrodermal activity. IEEE Transactions on Biomedical Engineering 57(5), 1243–1252 (2010).

Rae Systems: LifeShirt, Personal Life Sign Monitor from RAE Systems (2011).

M.B. Rappaport, H.B. Sprague: Physiologic and physical laws that govern auscultation, and their clinical application. The acoustic stethoscope and the electrical amplifying stethoscope and stethograph. The American Heart Journal 21(3), 257–318 (1941).

R. Thom, Oil painting (1954).

G. Turchetti, E. Spadoni, E. Geisler: Health technology assessment. IEEE Engineering in Medicine and Biology Magazine 29(3), 70–76 (2010).

P.D. Welsby, G. Parry, D. Smith: The stethoscope: Some preliminary investigations. Postgraduate Medical Journal 79, 695–698 (2003).

Chapter 2
Physiological and Functional Basis

An *introduction* will be given into physiological structures and fundamental mechanisms behind these structures that are involved in the *genesis of biosignals* in humans. We start with the cells, the smallest units of life, go over the larger structures of vital organs, and end up with the circulatory system which involves all the preceding structures.

Profound knowledge of the physiological situation is crucial for a proper understanding of a biosignal's generation phenomena and a correct *interpretation* of biosignals—assessed by technical means—from a physiological point of view. It should be noted that physiological and functional structure will be considered only from a *biosignal perspective*, since extensive literature is available on general physiological and functional structure in humans.

2.1 Cell

The *cell* is the smallest autonomous unit of life and represents the functional and structural *basic unit* in multicellular organisms, such as humans with about 10^{14} cells (Silverthorn 2009). In actuality, the cell represents *the only origin* of any *biosignal* in the widest sense. Thus an overview will be given about the cell's basic structure and the cell's most relevant bioelectric phenomena.

2.1.1 Functional Structures

The morphological structure of the cell is tightly related to its function. There is a huge variety of cells, as illustrated in Fig. 2.1. However, each cell shows a certain *basic structure*, as depicted in Fig. 2.2, that is given by:

E. Kaniusas, *Biomedical Signals and Sensors I*, Biological and Medical Physics, Biomedical Engineering, DOI 10.1007/978-3-642-24843-6_2,

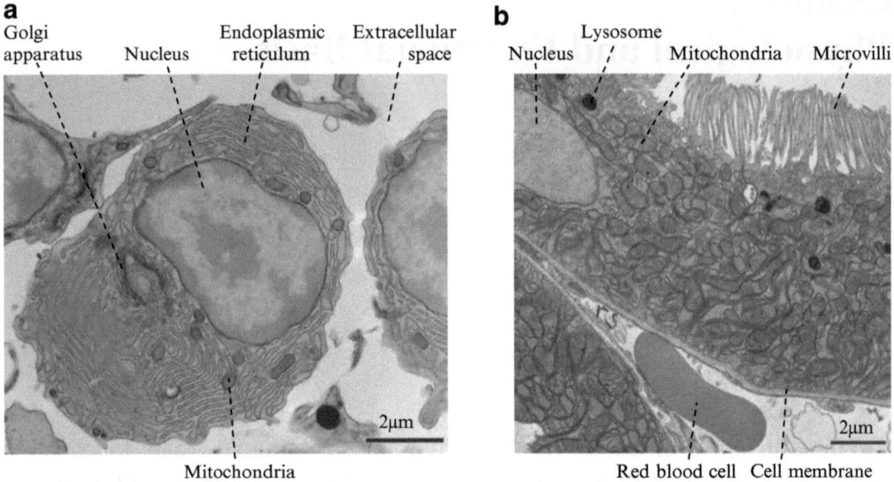

Fig. 2.1 Microscope images of different types of mammalian cells. (**a**) Monocyte in mouse spleen. (**b**) Epithelial cells from the proximal tubule of a mouse kidney. The photographs were taken by transmission electron microscopy (Wang and Sougrat 2011). The basic structure of the cell is indicated; compare with Fig. 2.2

- Outer cell *membrane* enclosing
- Cell content, i.e., *cytoplasm*, with different
- Specialized subunits within, i.e., *organelles*

The *size* of the human cell varies between 5 and 150 μm with a typical size of about 10 μm while the *shape* varies considerably from cell to cell. However, some cells, such as nerve cells (known as neurons), attain impressive dimensions of about 1 m if the cell's appendages (dendrites and axons) are considered. Figure 2.1 demonstrates two different types of mammalian cells, whereas Fig. 2.20 depicts a nerve cell embedded into a network of surrounding nerve cells via connecting dendrites and axons.

The *membrane* of the cell separates the interior of the cell (*intracellular* space) from the outside environment (*extracellular* space) and has a typical *thickness* of about 7–8 nm.[11] The membrane serves not only as a *barrier* between intracellular and extracellular space *anchoring* the cytoskeleton, but also controls *mass transfer* in and out of the cytoplasm and provides a means of *communication* between the cell and neighboring cells or its environment. A specific *electrical behavior* of the membrane and its components (e.g., channel proteins) plays a crucial role here (Sect. 2.1.2), especially for cell communication and induction of *biosignals*.

[11] It is interesting to note that the membrane thickness is less than the size of the cell by factor of about 1,000 so that the cell, in relation to its membrane and membrane fluidity, can be imagined as a *thin-walled balloon* filled with water.

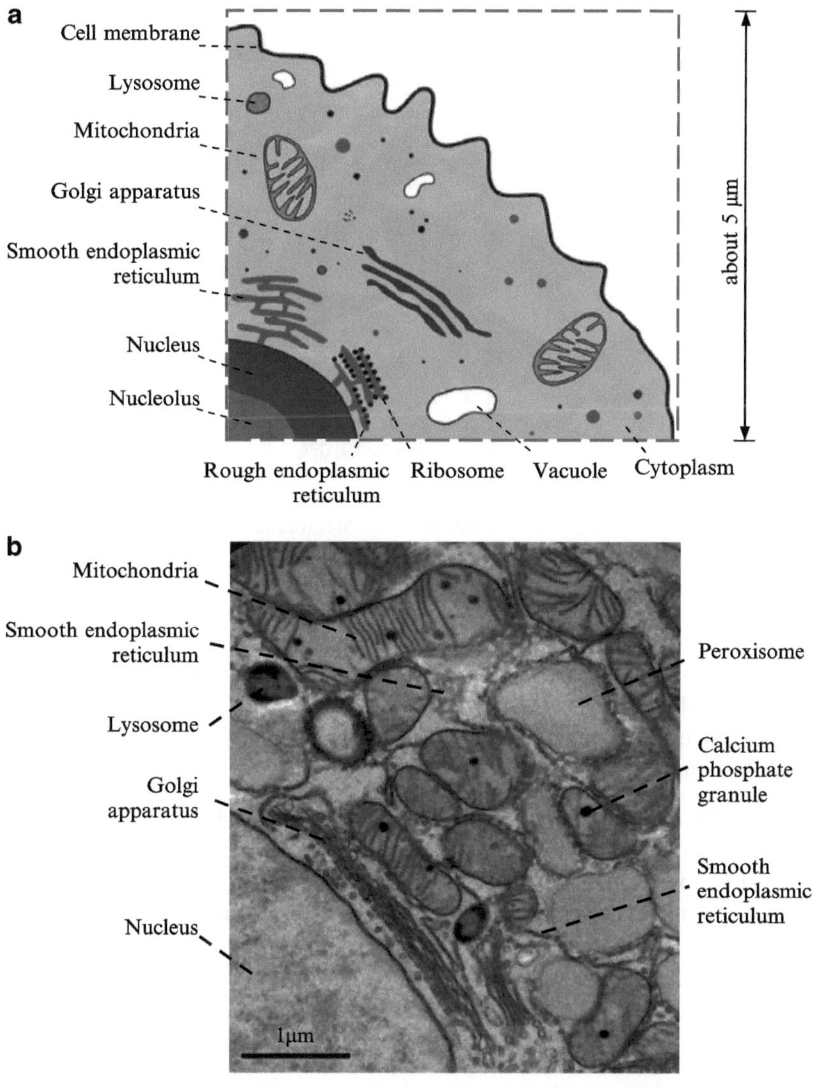

Fig. 2.2 (**a**) Simplified illustration of a section of a cell with different organelles floating in the cytoplasm and enfolded by the cell membrane. (**b**) Microscope image of a comparable cell section, namely, epithelial cell from the proximal tubule of a mouse kidney. The micrograph was taken by transmission electron microscopy (Wang and Sougrat 2011)

As schematically depicted in Fig. 2.3, *lipid molecules* make up the bulk of the membrane, arranged in two layers forming a *bilayer*. In fact, this bilayer structure *spontaneously* results from electrostatic interactions between

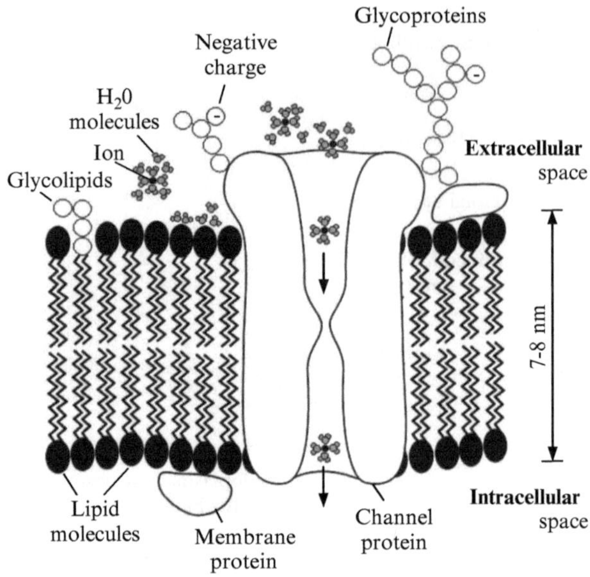

Fig. 2.3 Simplified illustration of cell membrane with an embedded channel protein (transport protein), adhered glycoproteins, and hydrated ions passing the channel

- Aqueous milieu full of *ional* and *polar structures*,[12] which enclose the membrane from both sides and
- Unique electrostatic behavior of the lipid molecule which includes both *polar* and *nonpolar structures*

More precisely, the lipid molecule has a polar, *hydrophilic head* (attracted to water) and two nonpolar, *hydrophobic tails* (repelled by water); see Fig. 2.4. When this molecule is exposed to an aqueous solution, its head is attracted to polar water molecules while its tails prefer the vicinity of nonpolar structures[13] excluding water

[12]*Ional structure* is given if the *total charge* of the structure is nonzero, whereas *polar structure* shows asymmetric *charge distribution* with dispersed balance points of positive and negative charges. In the ional structure, surplus electrons or protons induce a net negative or positive charge, respectively; e.g., a sodium ion Na^+ has a positive net charge. In contrast, the asymmetry in the polar structure induces positive and negative ends though the total net charge is zero; e.g., a *water molecule* H_2O has a negative pole on the oxygen side and two positive poles on the hydrogen side (compare Fig. 2.4).

[13]Simple energetical considerations offer a plausible basis for the *interaction* of *polar* and *nonpolar* structures (Pfützner 2003). Basically, an electrostatic attraction occurs between ional and polar structures of inverse sign, whereas the resulting *force magnitude* is proportional to the *energy* W of the *electric stray field* in between the attracting structures (Fig. 2.4) with the *electric field* magnitude E and *electric flux density* magnitude D. A stable and the most probable arrangement of polar and nonpolar structures shows *minimal* W given by

$$W = \int_V \frac{E \cdot D}{2}\,\mathrm{d}V = \int_V \frac{D^2}{2 \cdot \varepsilon}\,\mathrm{d}V.$$

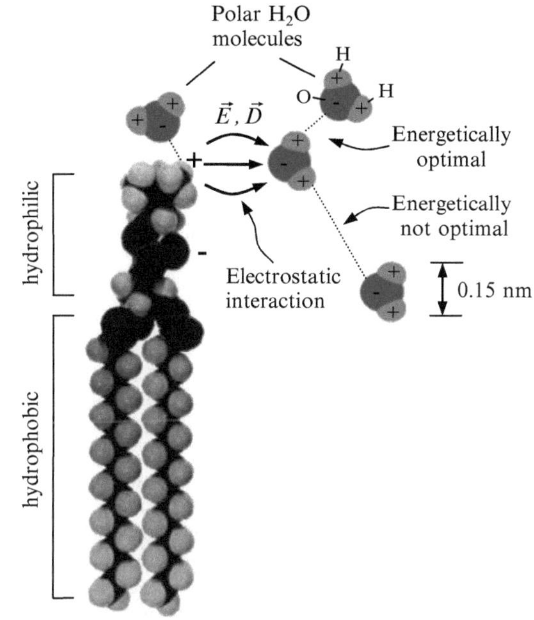

Fig. 2.4 Simplified illustration of a lipid molecule as a main building block of the membrane. Electrostatical interactions are indicated between the polar head of the lipid molecule and polar water molecules with $\vec{E}$ as the electric field and $\vec{D}$ as the electric flux density. *Dashed lines* indicate hydrogen bonds

and ions. In the case of numerous lipid molecules, the lipids spontaneously bury their tails together and leave the heads exposed to water molecules. In consequence, the lipid molecules *spontaneously* form a bilayer structure in an *aqueous medium outside and inside* the membrane, with the *heads on the outer side* of the membrane and the *tails on the inner side* (Fig. 2.3).

Here V is the spatial volume encompassing the electric field with nonzero D and ε is the constant dielectric permittivity of the medium. According to the equation, the *level of W* can be minimized either by decreasing D and V or increasing ε. The *level of D* is actually proportional to excess charge and inversely proportional to squared distance from the charge to the considered volume element $\mathrm{d}V$ (compare Footnote 21).
The level of V can be minimized through mutual saturation of electric stray fields, i.e., by minimizing the spatial extension of the electric stray field. Thus a *close opposition* of inversely charged *polar* structures and reduced distance in between contribute to a reduction of V and consequently to a reduction of W. Figure 2.4 demonstrates energetically optimal and suboptimal arrangements of polar water molecules; compare also with the principle of charge and shape *complementarity* from Fig. 2.5.
Nonpolar structures do not exhibit electrostatic interactions. However, the presence of nonpolar structures in the immediate environment of polar structures may disturb the mutual saturation of the stray fields from the polar structures. In consequence, bunching together of the nonpolar structures reduces the latter disturbing effect and thus indirectly reduces W.
The *level of* ε is governed by the medium which fills V. For instance, an introduction of liquid water into crystalline table salt (NaCl) increases ε, reduces W of the stray fields between Na and Cl atoms, and thus reduces attracting forces between the latter atoms. The reduced forces may even lead to *dissociation* (separation) of the salt NaCl into its cations (Na^+) and anions (Cl^-).

The *membrane hosts* a large number of *proteins*, both attached and embedded within the membrane (Fig. 2.3). In particular, proteins on the *outer surface* of the membrane serve as *receptors*[14] (e.g., for hormones[15]), *enzymes*,[16] or even *antigens*.[17] For instance, membrane proteins linked with carbohydrates (Fig. 2.3),

---

[14]The *receptor* comprises a molecular structure, usually *protein structure*, which, on receiving

- Chemical
- Thermal or
- Mechanical stimuli

responds by changing its *molecular conformation* and thus its *molecular activity*. Synaptic transmission of neuronal signals constitutes a typical example, in which *chemical receptors* and chemical stimuli are involved (Sect. 2.1.4.2). In the synaptic cleft, biological messenger substances (e.g., acetylcholine as transmitter) may tightly and specifically *bind* to the chemical receptors in the postsynaptic cell membrane (compare charge and shape complementarity from Fig. 2.5). In consequence, the binding may *open* channel proteins affiliated with the chemical receptors, initiate *transport* of specific charged substances through the membrane (e.g., inflow of Na^+ ions into the postsynaptic cell and outflow of K^+ ions, Fig. 2.10b), and thus may change the *voltage* across the postsynaptic membrane. Besides messenger substances, chemical receptors may also recognize antibodies (Footnote 17), enzymes (Footnote 16), hormones (Footnote 15), and even foreign bacteria, cells. In the case of *thermal receptors* responding to thermal stimuli, a thermally induced expansion or constriction of channel proteins may govern the transport of charged substances through the membrane. In analogy, *mechanical receptors* respond to mechanical stimuli which, for instance, could be given by an increased intracellular pressure (compare Footnote 22). The increased pressure stretches the cell membrane, whereas channel proteins within the membrane are widened, impacting the ionic permeability of the membrane.

[15]*Hormones* are signaling molecules with an organic structure that are secreted by specific cells mainly into the bloodstream to regulate particular physiological activity of other cells. The hormones are detected by *chemical receptors* (Footnote 14) whose activation releases a specific regulatory action.

[16]*Enzymes* are proteins with a specific three-dimensional structure (produced by cells), which catalyze and regulate nearly all biochemical reactions without being altered themselves in the process. Specific reactions related to cellular metabolism and signaling are promoted by membrane enzymes. In particular, a close presence of an enzyme significantly lowers the activation energy of a particular reaction. The molecules participating in the reaction (substrates) and the enzyme show *complementary charges* and *complementary geometric shapes* (compare charge and shape complementarity from Fig. 2.5). The complementarity causes the enzyme to bind the substrates and initiate the specific reaction in between. For instance, *bonding* of the substrates may be accelerated because the enzyme precisely orientates the reactants to be bound (*binding enzyme*, Fig. 2.5). *Dividing* of a single substrate may result from additional electrostatic forces exerted on the substrate by a polar enzyme (*cutting enzyme*). Even *charge imbalance* of a substrate may be induced by an electron-donating enzyme (*reducing enzyme*), which, in turn, may lead to a different secondary structure of the substrate.

[17]An *antigen* is a substance which stimulates the immune system. The antigen can be a protein found on the outer membrane surface of a foreign cell to be removed. On the other hand, there are *antibodies*, proteins which react with specific antigens. The antibody molecules are typically Y-shaped and show three binding sites. The molecular tip structure of the antibody is very specific and can bind to a specific antigen only if the tip structure shows a *complementary match* with the structure of the antigen. Here the complementarity implies charge and shape complementarity from

known as *glycoproteins*, comprise an important class of membrane proteins, which are either integrated within the membrane or adhered to the outer membrane surface.

Glycoproteins are responsible for the *negative charge* of the outer surface so that living *cells repel* each other. In addition, the negative charge and specific spatial structure of glycoproteins facilitate *recognition* of other cells or viruses (e.g., influenza virus) on the cellular level (compare Footnote 14). That is, a *complementary match* between glycoproteins and foreign structures may trigger an appropriate *immune* response. The negative charge of the outer surface also yields an *affinity for positive ions* so that cations can be expected to play a significant role in membrane behavior (Sect. 2.1.2).

It should be noted that the membrane bilayer has a highly dynamic and *fluid structure*, as could not be inferred from Fig. 2.3; the lipid molecules move around and rotate within the single layer while the embedded proteins may also change their location in the membrane. The cell membrane is a *selectively permeable* membrane that allows certain ions or molecules to pass through. As will be described in Sect. 2.1.2, different *integral proteins* in the membrane (e.g., channel proteins) act as *gatekeepers* of cellular compartments and facilitate selective mass transfer through the membrane.

The aforementioned complementary match, i.e., *charge and shape complementarity*, deserves an extended description, since it determines, for the most part, *molecular interactions* in between polar, ional, and nonpolar structures. According to Fig. 2.5, the bonding between two (or multiple) molecules is likely to occur only if

- *Complementary charges* or complentary *polar* positions exist on both interacting sides of the molecules. The attracting electrostatic forces between the molecules favor bonding; after bonding the energy of electric stray fields is reduced because of mutual saturation of the fields (Footnote 13). Furthermore, the close proximity of *nonpolar* regions of both molecules favors their bonding, because the mutual saturation of charged regions elsewhere is not disturbed. In addition,
- *Complementary shapes* of both interacting molecular sides promote bonding in terms of their best geometrical match (Lock and Key concept); e.g., molecular tips fit into molecular notches.

Figure 2.5 also illustrates that an *enzyme* may facilitate bonding of molecules if it has a *complementary match* with both interacting molecules in terms of charge distribution and spatial structure. The enzyme helps to precisely align the molecules for their best match (Footnote 16).

Figure 2.4 demonstrates another example of *charge complementarity* between a polar lipid head and polar water molecules. The complementarity can also be observed between single water molecules, i.e., between an electronegative oxygen atom and a positively charged hydrogen atom, the resulting bond being called a

---

Fig. 2.5. After the antigen–antibody reaction, the foreign cell with the adhered antigen–antibody complex is subject to disposal by the immune system.

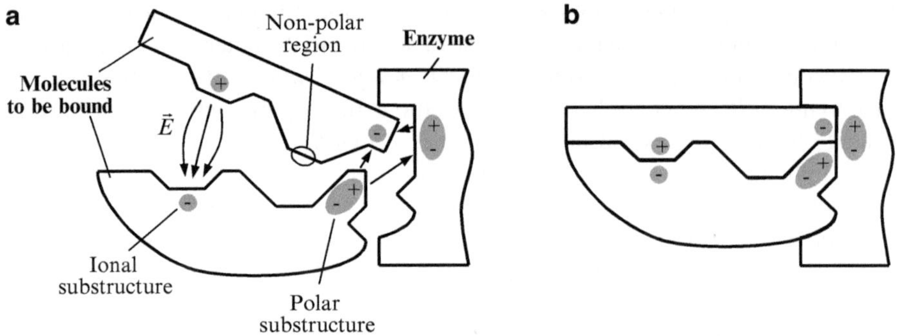

Fig. 2.5 Charge and shape complementarity in terms of the Lock and Key concept, which assists the chemical bonding of molecules. In addition, an enzyme molecule may accelerate the bonding process. (**a**) Before bonding; the vectors $\vec{E}$ denote the electric stray field. (**b**) After bonding the stray fields are minimized

hydrogen bond.[18] In addition, the complementary match is equally important for *temporal behavior of the cell membrane*, whereas electrodynamic processes around the membrane are the very origin of *biosignals*.

The *cytoplasm* constitutes a fluid environment in which organelles reside, as shown in Fig. 2.2. The cytoplasm is an aqueous saline solution with embedded organic molecules (mainly proteins); its volume amounts to about *50% of cell volume* (Silbernagl and Despopoulos 2007). The composition of the cytoplasm forms a cellular skeleton and thus provides mechanical resistance, determining the shape of the cell.

The main *cell organelles* are shown in Fig. 2.2. The *nucleus* is the largest cellular organelle with chromosomes as the carrier of genetic information and contains a dense structure termed the nucleolus. The *nucleolus* produces the ribonucleic acid needed for protein synthesis in the ribosomes. The *smooth endoplasmic reticulum* is the place where many different metabolic reactions occur, e.g., production of hormones. The surface of the *rough endoplasmic reticulum* is occupied by *ribosomes*, which act as the site of protein synthesis. The *Golgi apparatus* absorbs, modifies, and packages macromolecules into vesicles (membrane enclosed sacs) and is involved in cell secretion. The *mitochondria* are cellular power plants,

[18]*Hydrogen bonds* are of great importance for the three-dimensional form that a chain of atoms, i.e., the so-called *primary structure* of a molecule, can assume. The patterns of hydrogen bonds define the so-called *secondary structure* of the molecule. For instance, *proteins* commonly have a helix structure (= spiral conformation) as their secondary structure, while the underlying sequence of amino acids gives their primary structure. Interestingly, the helix structure is the most common protein structure crossing cell membranes and composing multifunctional channel proteins (section "Transport of Substances").

Table 2.1 Typical ional composition of intracellular and extracellular spaces of nerve cells with the respective equilibrium voltages (2.5)

Ion	Extracellular space (mMol / l)		Intracellular space (mMol / l)	Equilibrium voltage[a] (mV)
Na^+	150	➡ (thick)	15	61
K^+	5	← (thin)	150	-90
Ca^{2+}	1	➡ (thick)	< 0.0001	122
Cl^-	120	← (thin)	9	-68

Directions of the electrochemical driving force at the resting membrane voltage ($U_R = -70$ mV) are indicated by *arrows*. The *arrow's thickness* symbolizes the strength of the net driving force, without ionic permeabilities of the membrane being considered. Data has been accumulated from different sources (Silbernagl and Despopoulos 2007; Pfützner 2003; Silverthorn 2009)

[a]The voltage across the cell membrane at which there is no flux of the corresponding ions across the membrane; i.e., the electrochemical diffusion gradient is zero (Sect. 2.1.3.1)

generating adenosine triphosphate[19] out of proteins, fats, and carbohydrates, as a common source of chemical energy. The *lysosomes* are the digestive organs of the cell, which contain enzymes to decompose outdated organelles and engulfed foreign substances. Lastly, the *vacuoles* are small fluid-containing vesicles holding inorganic and organic molecules; they mainly assist in extrusion of macromolecules from the cell and in engulfment of foreign substances.

Compositional differences in the extracellular and intracellular spaces have a significant impact on biosignal origin. Namely, the composition of the *extracellular* space resembles that of aqueous 0.9% solution of salt (NaCl). According to Table 2.1, *sodium* Na^+ ions and *chloride* Cl^- ions prevail outside the cell, while other ions such as *potassium* K^+ form the minority. In the *intracellular* space, K^+ ions and *organic anions* dominate over Na^+ and Cl^-. The prevalence of K^+ ions and the minority of Na^+ ions within the cell can be attributed to an *active sodium–potassium pump* in the cell membrane, which continuously enriches the intracellular medium with K^+ ions and depletes Na^+ ions against their respective concentration gradients (section "Active Transport").

In contrast, the gradient of Cl^- ions is a subject of their *passive distribution*. That is, the minority of Cl^- ions in the cell is due to the *resting membrane potential* of the cell, which electrostatically forces negative Cl^- ions out of the cell until the

[19]*Adenosine triphosphate* (ATP, with three phosphate groups) is a source of chemical energy, which assists diverse biochemical processes that have enzymes involved (Footnote 16). For instance, ATP is used for active transport of substances across the membrane against their diffusion gradient, muscle contraction, drive of metabolic reactions, and biosynthesis. The ATP is mainly *synthesized* through the oxidation of biological molecules in mitochondria, e.g., oxidation of glucose (with oxygen) in which carbon dioxide, water, and released energy are the end products (known as *cellular respiration*). The chemical energy in ATP is *stored* as energy-rich phosphate bonds. The energy is relieved through *ATP hydrolysis* (addition of water), whereas ATP gets converted to *adenosine diphosphate* (with only two phosphate groups of lower total energy), inorganic *phosphate*, and *released chemical energy*.

opposite diffusional force affecting Cl^- balances the electrical force (Sect. 2.1.3.1). The concentration of free Ca^{2+} ions in the cell is negligibly small in comparison with that outside the cell, because Ca^{2+} is continuously stored within cell organelles or transported out of the cell by active Ca^{2+} pumps in the cell membrane. An increased intracellular concentration of Ca^{2+} usually serves as a triggering signal for multiple vital cellular functions, e.g., contraction of a muscle cell, release of neurotransmitters within the scope of synaptic propagation, or opening of specific membrane channels, as discussed in the following sections.

It should be stressed that on both sides of the membrane an approximate *charge neutrality* exists. In the *extracellular space* the charge balance is mainly given by positive Na^+ ions and negative Cl^- ions. In contrast, the *intracellular space* is balanced through K^+ ions versus anionic proteins (relatively large and immobile), negatively charged amino acids, and anionic phosphates. The cell membrane is typically *impermeable* for these large anions.

2.1.2 Cell Membrane

The cell membrane performs *vital functions* for the life of a cell (Sect. 2.1.1). Equally important is the fact that all cells exhibit an electric voltage u across their membrane or, likewise, exhibit an electric potential difference across their membrane. Per definition the *membrane voltage u* is defined as the local potential inside the cell relative to that outside. In the *resting state, u is stationary* and is usually referred to as the *resting membrane potential*, i.e., $u = U_R$ with U_R as the *resting voltage amplitude*[20] (Sect. 2.1.3.1).

While some *cells* are *not excitable*, e.g., those composing adipose and connective tissue, others are *excitable* and under electric control, e.g., nerve cells in the brain or muscle cells in the muscle. The excitability of the cell is tightly related to the *excitability of the cell membrane*. Namely, the excitation yields a *time varying* $u = u(t)$ as a short *action impulse*, usually referred to as the *action membrane potential* (Sect. 2.1.3.2). This stereotyped electric signal $u(t)$ carries manifold physiological information, e.g., about cellular receptors response to a stimulus (Footnote 14), from one nerve cell to another or from a nerve cell to a muscle cell, whereas the action potential propagates without attenuation (Sect. 2.1.4). The electric membrane behavior is equally important for biosignal generation, since it determines the *origin of all biosignals*.

---

[20]Electric voltage $u(x,t)$ stands for the voltage (or potential difference) across the cell membrane and is, in general, an arbitrary function over time t and distance x. If the voltage $u(x,t)$ changes over time, i.e., exhibits a *temporal course*, or varies along the membrane, i.e., exhibits a *spatial course*, the function $u(x,t)$ can be defined as $u(x = \text{const}, t) = U \cdot g(x = \text{const}, t) = u(t)$ or $u(x, t = \text{const}) = U \cdot g(x, t = \text{const}) = u(x)$. Here the symbol U defines the voltage amplitude while $g(x,t)$ is an arbitrary function having the physical unit of 1. In the case of a *stationary voltage* across the membrane, $g(x,t) = 1$ applies and therefore $u(t) = u(x) = U$; compare Footnote 145. In the *resting state* of the cell, $u(t) = u(x) = U_R$.

For the resting and action membrane potentials, the *passive and active behaviors* of the membrane are of crucial importance. While

- The *passive behavior* reflects sustained properties of the membrane (Sect. 2.1.2.1),
- The *active behavior* is marked by both regulatory mechanisms: (nonlinear) gated ion flow through the membrane (section "Regulatory Mechanisms"), and an active transport of substances up their chemical diffusion gradient across the membrane requiring energy consumption (section "Active Transport").

2.1.2.1 Passive Properties

The *passive behavior* of the membrane can be discussed from a chemical and electrical point of view, because biosignal information is carried by chemical and electric biosignals on the cellular level. In *chemical* terms, the transport of a (charged) substance through the membrane *along its concentration gradient* is relevant, since it determines, for instance, the ionic membrane current and membrane voltage u (Sect. 2.1.3.1). In *electrical* terms, the macroscopic (subthreshold) behavior of the membrane as a *propagation medium* is relevant. For instance, this behavior has a significant impact on the spatial propagation of both local membrane imbalance $u - U_R$ in terms of the sensing function of nerve cells and local excitation (action potential) in terms of neuronal signaling and transmission of biosignals.

Transport of Substances

The *passive transport* of substances through the cell membrane is governed by

- *Passive diffusion*
- *Facilitated diffusion*

The obvious way for a substance to enter a cell is by *crossing its membrane* through the lipid bilayer (Fig. 2.3) if there is a *concentration gradient* between the extracellular and intracellular spaces. It should be noted that most substances (vital for cellular prosperity) are usually dissolved in water; thus the substances are either charged or polar. The arising electrostatic forces between the substances and polar water molecules yield clouds of water molecules[21] around the substances.

[21]Ions in solution are enclosed by *polar water molecules* from the immediate vicinity which are electrically attracted by ions net charge q; compare Fig. 2.3. Consequently, a *hydration shell* is established. In the case of a positive ion (e.g., Na^+ ion), the electronegative oxygen atom of the water molecule is attracted to the ion; compare Fig. 2.4. A water cloud with a typical shell thickness of a few water molecules is carried by the ion as it diffuses through solution. In accordance with energetical considerations from Footnote 13, it is energetically unfavorable for an ion with the *radius* r and finite magnitude D of (radial) electric flux density on the ion's surface

Such substances are known to be hydrophilic (attracted to water) and lipophobic (repelling fats), for which the *hydrophobic tails* of lipid molecules constitute a real (energetic) *barrier* while crossing the membrane (Fig. 2.3). In addition, if the substances are charged, as given in the case of ions, their transport through the membrane produces an *ionic membrane current* (see Footnote 24).

However, *small* molecules and those that are *lipophilic* (attracted to fat) can penetrate the membrane through *passive diffusion*. For instance, water[22] (lipophobic but small in size), *oxygen*, and *carbon dioxide* can easily pass the lipid bilayer. In general, Fick's law[23] describes chemical diffusion.[24] In short, the resulting *diffusion rate* of the molecules is

---

$$D = \frac{q}{4\pi \cdot r^2}$$

to diffuse alone without the hydration shell. Interestingly, a thicker cloud of water molecules is attracted by smaller ions with smaller r, because q is more localized and D is more intense; see the above equation. For instance, although a Na^+ ion is smaller (ionic $r \approx 0.1\,\text{nm}$) than a K^+ ion ($r \approx 0.14\,\text{nm}$), the *effective radius* of a hydrated Na^+ ion is larger and its *ionic mobility* in solution—defined as the speed of ion migration (or ion's drift velocity) related to the forcing electric field in solution (Footnote 45)—is even lower. Usually, about three water molecules are completely immobilized by a single Na^+ or K^+ ion.

[22] The passive *diffusion* of *water* molecules through the cell membrane is of vital importance for so-called *osmosis*, which governs water transportation into and out of cells. That is, water molecules (solvent) move from an area of low solute concentration (e.g., outside the cell) to an area of high solute concentration (e.g., into the cell) across the membrane permeable only to the solvent. Osmotic changes may even deform the cell shape, e.g., cell swelling due to water inflow.

[23] Adolf Eugen Fick (1829–1901) was a German physiologist who introduced Fick's law of diffusion (Footnote 24) and first described a technique for measuring cardiac output by a marker substance (Sect. 3.1.3.2).

[24] In general, the *chemical diffusion* rate j_D, i.e., the amount of diffused substance per time (mol/s), along the coordinate x is given by *Fick's first law*

$$j_D = A \cdot D_F \cdot \frac{dc}{dx},$$

where A is the surface area available for diffusion (m^2), D_F is the *diffusion coefficient* (m^2/s), and dc/dx is the concentration gradient (mol/m^4) along x. The coefficient D_F basically describes the relationship between the resulting flow j_D and the stimulating diffusional "force," given by the (initial) gradient dc/dx. In fact, D_F is a function of the absolute temperature T, viscosity μ of solvent, and the effective radius r of diffusing particles. According to the Stokes–Einstein equation,

$$D_F = \frac{R \cdot T}{N_A \cdot (6\pi \cdot r \cdot \mu)}.$$

The constant N_A denotes the Avogadro constant ($N_A = 6.02 \cdot 10^{23}\,1/\text{mol}$) and R the gas constant [$R = 8.31\,\text{J}/(\text{K mol})$]. It is important to note that an ion may exhibit a greater r than given by its atomic dimensions, because of its ionic hydration shell (Footnote 21).

In electrical terms, the chemical *diffusion current density* J_D (A/m^2) for charged substances as ions with the valence z can be given according to

$$J_D = \frac{j_D}{A} \cdot z \cdot e \cdot N_A,$$

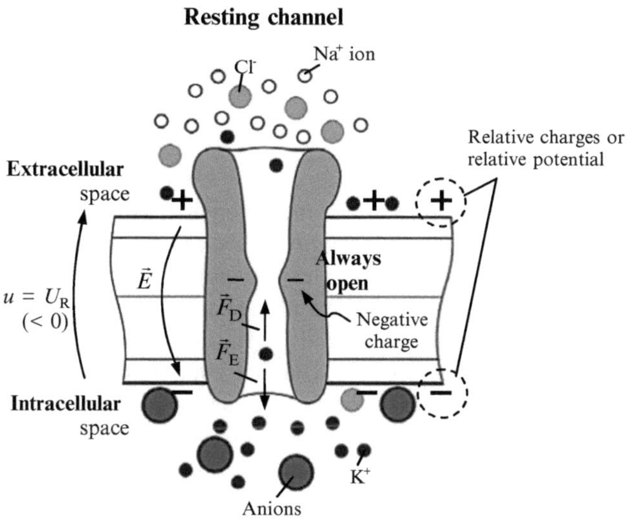

Fig. 2.6 Resting (nongated) passive channel protein for K^+ ions, transversing the cell membrane; compare Fig. 2.10. The channel is always open and governs the resting voltage amplitude U_R across the membrane in the resting state of the cell. Within the channel, the net transport of K^+ ions across the membrane is governed by an equilibrium between the diffusional force $\vec{F}_D$ and electrical force $\vec{F}_E$ (related to the electric field $\vec{E}$) acting on K^+ ions

- *Proportional* to the *surface area* of the membrane, molecule's *concentration gradient* across the membrane, and *membrane permeability* to these molecules (favoring small and lipophilic structures).
- *Inversely proportional* to the *membrane thickness*.

Because of the hydrophobic interior of the lipid bilayer, *large polar molecules* or *ions* cannot simply enter the cell. However, the membrane is *selectively permeable* for these substances because of numerous macromolecular *transport proteins* that span the lipid bilayer and act as carriers across the membrane in terms of *facilitated diffusion*. Each transport protein is specialized for a certain molecule or ion; compare transport proteins shaped as channels in Figs. 2.3 and 2.6. Here the protein-mediated mass transport of polar and charged structures still moves *along their concentration gradient*; i.e., the gradient *passively* transports the molecules from more concentrated to less concentrated regions, down the diffusion gradient.

---

where e is the elementary charge ($e = 1.6 \cdot 10^{-19}$ C). Consequently, the diffusion related *ionic membrane current* is given by the product $J_D \cdot A$ if J_D is constant over A.
In fact, if an *interface* between two *inexhaustible* substances of different c is given at $x = 0$ (c_1 for $x < 0$ and c_2 for $x > 0$ with $c_1 > c_2$), the chemical diffusion process yields a *nearly exponential* decrease from c_1 to c_2 along $x > 0$. In consequence, the rate j_D and density J_D decrease with increasing x because the effective dc/dx decreases along $x > 0$.

There are mainly two types of *transport proteins* spanning the cell membrane and facilitating diffusion of molecules and ions:

- Carrier proteins
- Channel proteins

A *carrier protein* in the membrane binds a specific molecule to be transported, in accordance with the charge and shape complementarity from Fig. 2.5. The protein is thereby induced to undergo a *conformational change*; i.e., the secondary structure of this protein (Footnote 18) changes after the molecule is bound. The reshaping process carries the molecule across the membrane and then the molecule is released on the opposite side of the membrane through another conformational change; similar to the working principle of the sodium–potassium pump (Fig. 2.11). The carrier protein then reorients to its original shape. These proteins are highly selective in binding molecules and never open to both sides of the membrane (never build a pore or channel). In addition, they have limited transport capacity in comparison with channel proteins (see below) and are usually slower[25] but transport larger substances, such as *glucose* or *amino acids*.

The *resting channel proteins*, as demonstrated in Fig. 2.6, belong to the family of transport proteins which allow *specific ions* to cross the membrane if they have specific

- *Size*, *charge*, and even
- Thickness of the *water shell* enclosing the ions (Footnote 21)

These ion channels can be pictured as *hydrophilic pores* that are always open in the middle of surrounding *membrane-spanning protein* units. The pore-forming region is usually made up of two or more protein units which have a helical elongated (secondary) structure of chained amino acids. Usually negative charges—due to polarized or charged amino acid residues—line the walls of the channel pore. The rationale for the negative charge is that it attracts positive cations to be transported. In addition, the water shell (hydration shell) of cations is sieved away at the *constriction of the channel* while the negative charge (residing in the wall) substitutes for water molecules in terms of charge complementarity.

The most important channel proteins are specialized for passing Na^+, K^+, and Cl^- ions involved in the genesis of U_R (Sect. 2.1.3.1). Figure 2.6 demonstrates a channel protein for K^+ ions, in which ions are transported out of the cell down the K^+ diffusion gradient. The channel for K^+ ions is about 10 nm in size (Malmivuo and Plonsey 1995); thus the channel length substantially exceeds the thickness of

[25]The number of molecules transported by a *carrier protein* ranges from 10^3 to 10^6 per second, whereas the number of ions flowing through an open *channel protein* is about $10 \cdot 10^6$ per second (Silverthorn 2009; Kandel et al. 2000). Thus, an ion typically stays bound in the channel protein for less than 1 µs. By comparison, an active *sodium–potassium pump* in the membrane, see section "Active Transport," is much slower and can transport at most 10^2 ions per second (Kandel et al. 2000).

the lipid bilayer (7–8 nm). The narrowest and thus rate-limiting region within the channel is only 1.2 nm in diameter (Kandel et al. 2000).

Transport of Potential Difference

The transport of *ional substances* across the cell membrane, i.e., ionic membrane current, is tightly interrelated with the resting *potential difference* or *voltage* u across the membrane; in the resting state of the cell $u = U_R$ (Fig. 2.6 and Sect. 2.1.3.1). However, if the voltage $u(x)$ varies along the membrane with $u(x) \neq U_R$, e.g., along the coordinate x of the axon of nerve cell, as shown in Fig. 2.7, equalizing *induced currents* must arise because of a relatively large *medium conductivity*[26] on both sides of the membrane. The currents aim to equalize the variations of $u(x)$ and finally evoke decremental (electrotonic) *propagation* of the local *voltage imbalance* $u - U_R$ (or of the local membrane potential). The propagation can be clearly observed, for instance, in the case of *action potential conduction* (Sect. 2.1.3.2), (natural) *stimulation of receptive fields* in a sensory nerve cell (Sect. 2.2), or even *artificial stimulation* of a nerve cell (see below).

The genesis of both propagation phenomena and the rate of change in the membrane potential are subjects of *subthreshold behavior* of excitable membranes. In fact, subthreshold behavior describes the *passive response of the membrane*, i.e., as long as the *temporal imbalance* of $u(t)$ ($\neq U_R$) or the *spatial imbalance* of $u(x)$ ($\neq U_R$) does not exceed a particular level of u ($> U_R$) to generate an action potential. Likewise, the passive response means that voltage-dependent (nonlinear) changes in the membrane conductance can be neglected. In the terminology of section "Cell Stimulation," hyperpolarizing responses are always passive, as is weak depolarization *up to a certain threshold* ($> U_R$).

The passive response yields *characteristic parameters* of the cell membrane, determining

- *Temporal and spatial extension* of imbalanced membrane regions in terms of $u \neq U_R$ (i.e., extension of hyperpolarized and weakly depolarized regions), as discussed below.
- *Propagation speed* of action potentials along the membrane, as discussed in Sect. 2.1.4.1.

[26]In general, the induced *electric current density* $\vec{J}_E$ in a medium with electrical conductivity γ and electric potential $\varphi(x,t)$ can be expressed as

$$\vec{J}_E = \gamma \cdot \vec{E} = -\gamma \cdot \operatorname{grad} \varphi(x,t) \approx -\gamma \cdot \frac{d\varphi(x,t)}{dx} \vec{e}_x .$$

It shows that $\vec{J}_E \neq \vec{0}$ only if $\varphi(x,t)$ exhibits spatial changes. The above approximation assumes changes of $\varphi(x,t)$ in x direction only. For electrolytes, see Footnote 45.

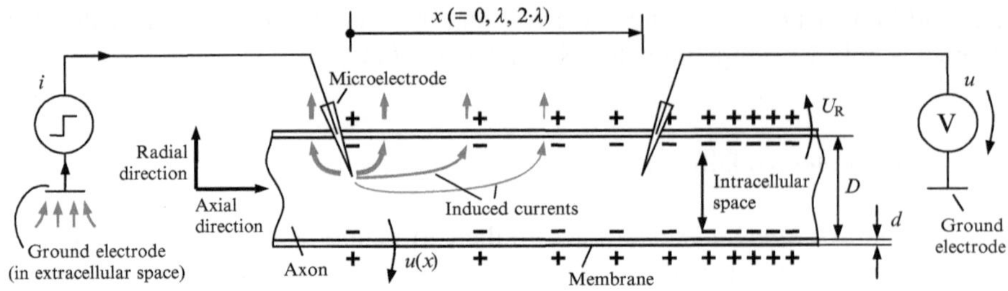

Fig. 2.7 Intracellular stimulation of the axon of a nerve cell (compare Fig. 2.21a) in order to attain characteristic data on passive (subthreshold) electrical behavior of the membrane of the axon. The step current i is applied with a simultaneous registration of the resulting membrane voltage u across the membrane at the distance x from the stimulation site. The induced currents flow from the inserted microelectrode into intracellular medium, cross the membrane outwards, enter extracellular medium, and then return to the current source. The *arrow's* thickness indicates local strength of the induced current density

Figure 2.7 demonstrates an experimental arrangement, in which a local *electrical stimulation* of the axon of a nerve cell is performed, with the axon's membrane initially at the resting membrane potential. A current[27] i is applied through an inserted *thin* microelectrode while another electrode at distance x measures the *subthreshold response* of the membrane, i.e., measures the local *deviation of* u from the resting state with U_R. It should be noted that both current application and voltage measurement are in reference to the extracellular potential; i.e., both current source and voltage meter have *large* ground electrodes at the extracellular potential (Fig. 2.7).

The *current* leaving the inserted microelectrode follows the shortest path of least resistance from the positive microelectrode to the ground electrode of the current source. Consequently, most current *crosses out of the membrane* near the site of its injection. The induced currents flow through the intracellular fluid, cross the membrane, enter the extracellular space, and then return to the current source; the currents must always complete a *closed loop*. In fact, the *currents leak out* all along the membrane; thus the current density at the membrane decreases with x; consider arrow's thickness in Fig. 2.7. In consequence, the resulting *charge density* on both sides of the membrane is relatively spare around the microelectrode (of the current source) and the density rises with x. This is because the initial charge distribution—with relatively large charge densities due to the resting U_R, see far right in Fig. 2.7—is partly compensated by the induced currents, provided that $i > 0$, $U_R < 0$, and $U_R < u < 0$. The outward currents make the intracellular side of the membrane *more positive in charge* (and more positive in potential) while the

[27]It should be noted that *current* application as an electrical *stimulus* is more comprehensible than voltage application, because all active mechanisms of the cell membrane are related to ionic transport across the membrane, which is equivalent to current inflow or outflow; see Sects. 2.1.2.2 and 2.1.3.

reverse is true for the extracellular side; the latter effects decrease with increasing x. In total, a varying u is established along x with $u(x) \neq U_R$.

If a step current is induced at $t = 0$, as depicted in Fig. 2.8a, the resulting *evolutions of the deviation* $u - U_R$ from the resting state U_R *over time t and axial distance x* (from the excitation site) are approximated in Fig. 2.8b, c. A nearly exponential increase of $u - U_R$ can be observed in the time domain,[28] whereas the *final level* depends on the distance x (Fig. 2.8b). On the other hand, the level of $u - U_R$ at a given t decreases nearly exponentially over x (Fig. 2.8c), i.e., decreases with an increasing distance between the stimulation and recording sites (Fig. 2.7). A few mutual correspondences in between the timely behavior and spatial behavior of the deviations are indicated in Fig. 2.8b, c. The thick dots connected by dashed lines show relationships for the time instant $t = \tau_R$ at $x = 0$ and, on the other hand, for the distance $x = \lambda$ and $t = \infty$.

Here τ_R is the *time constant* and λ is the *length constant* (or space constant) of the membrane, with both quantitatively describing the *exponential behavior*[29] of the membrane's response to the stimulating current. The constant τ_R describes the *temporal extension* of the response, for instance, at $t = \tau_R$ the response has reached about 63% of the final level. Similarly, the constant λ expresses the *spatial extension* of the response or, likewise, a certain distance x along the axon (Fig. 2.7), at which the response has decayed to 37% ($= 1/e$) of its level at the stimulation site $x = 0$.

The discussed temporal and spatial behavior of the membrane (*subthreshold behavior*) is the subject of an electrical model of a cylindrical cell (axon), or, more specifically, subject of an *electrical circuit model of the cell membrane*. The relevant model is called the *cable model* and is shown in Fig. 2.9. The model considers short axial elements (slices) of the membrane of length Δx from a *macroscopic* point of view. Each axial element represents all three

- Intracellular *axial* resistance

---

[28]Actually, the *temporal response of transmembrane potentials* (or the membrane voltage u) to a current stimulation pulse is usually faster than the *exponential response* if relatively small x values are considered (Fig. 2.7), i.e., smaller than the length constant λ of the membrane (2.4) (Malmivuo and Plonsey 1995). In engineering terms, the temporal response of the cable model (Fig. 2.9) is faster than the response of a simple first-order system (Fig. 2.8b), given $x/\lambda < 1$ and the same time constant (2.3). In addition, the temporal response becomes S-shaped for $x/\lambda > 1$; i.e., du/dt goes to zero for $t = 0$. Correspondingly, the *spatial response of u* deviates greatly from exponential for $x/\lambda > 1$.

[29]For instance, the *exponential decay* of $u(x) - U_R$ with the distance x and for $t = \infty$ from Fig. 2.8c is formally expressed by

$$u(x) - U_R = (u(x = 0) - U_R) \cdot e^{-x/\lambda} = c_x \cdot e^{-x/\lambda}.$$

In analogy, the *exponential growth* of $u(t) - U_R$ with the time t and for $x = 0$ from Fig. 2.8b is given by

$$u(t) - U_R = (u(t = \infty) - U_R) \cdot (1 - e^{-t/\tau_R}) = c_t \cdot (1 - e^{-t/\tau_R}),$$

with c_x, c_t being constants and having the physical unit of 1 V.

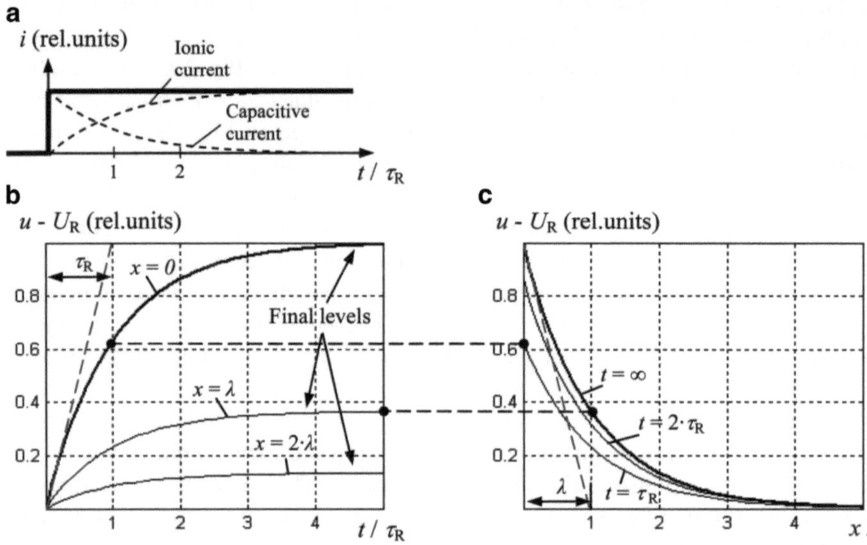

Fig. 2.8 Temporal and spatial evolutions of the membrane voltage u related to the resting amplitude U_R (compare Footnote 28). Subthreshold stimulation is given with the setup from Fig. 2.7. (**a**) Stimulating step current i is applied through the membrane with indicated ionic and capacitive currents; compare Fig. 2.9. (**b**) Temporal response of the membrane at different axial distances $x = 0, \lambda, 2\lambda$ from the stimulation site with λ as the characteristic length constant. (**c**) Spatial response of the membrane at different time instances $t = \tau_R, 2\tau_R, \infty$ with τ_R as the characteristic time constant. *Dashed lines* indicate relationships in between the temporal and spatial responses

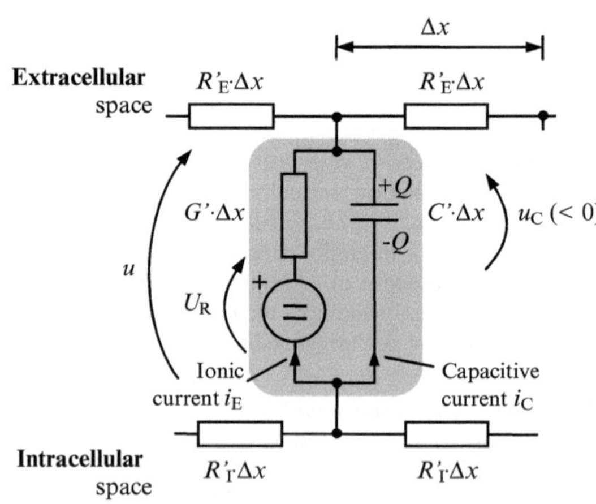

Fig. 2.9 Electrical equivalent circuit model of a cylindrical cell (axon) of small axial length Δx, referred to as the cable model. This circuit is applicable for the resting state with $u = U_R$ and for the subthreshold behavior with $u = u(t, x)$. The *gray background* refers to the membrane patch which is modeled in greater detail in Fig. 2.12

- Extracellular *axial* resistance
- *Radial* resistance and capacitance of the membrane with a voltage source to model the resting potential difference U_R.

According to Fig. 2.9, the *intracellular resistance* and *extracellular resistance* are modeled by distributed, i.e., axon length related (Fig. 2.7), electrical *resistances* R'_{I} and R'_{E}, respectively.[30] Usually the relation $R'_{\mathrm{E}} \ll R'_{\mathrm{I}}$ applies, because the total mobility of extracellular ions is higher than that of intracellular ions[31]; moreover, the intracellular cross-sectional area and therefore the number of intracellular charge carriers are limited by the axon's dimensions. For instance, the electrolytic *electrical conductivity*—directly proportional to the total *ionic mobility* (Footnote 45)—of the extracellular fluid can be estimated to be about 2 S/m while the electrical conductivity γ_{I} of the intracellular fluid amounts to only about 1 S/m (Pfützner 2003).

The *membrane* is an interface between intracellular and extracellular spaces and thus can be modeled as a distributed *leakage resistance* (or the electrical conductance G' related to axon's axial length) in parallel to the length-related *capacitance* C' (Fig. 2.9). It should be noted that G' and C' can be considered *constant* under *subthreshold stimulation* only.

In particular, the *conductance* G' reflects the fact that the membrane is a *poor conductor* (low G') and takes into account weak *ionic currents* in the membrane. In the case of the resting membrane potential, the level of G' is governed by both the *density* of the resting ion channels in the membrane and *conductance of these channels*, especially considering channel's *permeability to* K^+ *ions* and, to a lesser extent, to Na^+ and Cl^- ions (Sect. 2.1.3.1). For instance, the area-related electrical conductance G'', i.e., the membrane conductance related to membrane area, in the *radial direction* (Fig. 2.7) is given by

$$G'' = \frac{\gamma_{\mathrm{M}}}{d} = \frac{G'}{\pi \cdot D}, \tag{2.1}$$

where d is the membrane thickness and γ_{M} is the electrical conductivity of the membrane. In the resting state, the magnitude of G'' is in the order of $1\,\mathrm{mS/cm^2}$ (compare Fig. 2.14b). In contrast to γ_{I}, the size of γ_{M} is *very small* and amounts to only about 10^{-7} S/m ($\ll$1 S/m).

---

[30]The *electrical resistance* R'_{I} of a cylindrical axon with the *diameter D* (Fig. 2.7) *per unit length* in axial direction can be approximated as

$$R'_{\mathrm{I}} = \frac{1}{\gamma_{\mathrm{I}} \cdot A} = \frac{4}{\gamma_{\mathrm{I}} \cdot \pi} \cdot \frac{1}{D^2},$$

where γ_{I} is the electrical conductivity of the intracellular medium in axial direction and $A\ (=\pi \cdot D^2/4)$ is the cross-section area of the axon.

[31]The total *ionic mobility* (Footnote 45) in the extracellular space is mainly given by relatively mobile Na^+ and Cl^- ions, while the ionic mobility in the intracellular space by mobile K^+ and less mobile large anionic proteins; compare Table 2.1. Moreover, the heterogeneous structure of the intracellular medium and pronounced binding of water molecules to macromolecules within the cell facilitate collisions between ions and molecules and thus impede the total ionic mobility inside the cell. Therefore, an ion's mobility *outside the cell* is higher than *inside the cell.*

The membrane capacitance C' reflects the *inert behavior* of the temporal and spatial responses from Fig. 2.8b, c. In fact, the bilayer membrane resembles an isolator separating two good conductors, i.e., the cytoplasm and extracellular fluid. The membrane strongly resembles a parallel *plate capacitor*[32] with the plates virtually given by both conductors. As is typical for capacitors, the membrane has the ability to store and separate electric charges, which is facilitated by the *isolating behavior* of the bilayer and the *dielectric properties* of the lipid molecules comprising the bilayer (Fig. 2.3). In the case of the resting membrane, Fig. 2.6 depicts excess negative charge accumulated inside the cell and excess positive charge outside. Generally, the resultant length-related *excess charge* is given by the product $C' \cdot u$ (compare Footnote 33) and amounts to $C' \cdot U_R$ in the resting state.

Thus the *delayed response* of $u - U_R$ to a current step stimulus, as shown in Fig. 2.8, is intrinsically related to discharging the capacitor by its displacement current (or by the *capacitive current* through the membrane, Fig. 2.9). The time needed for recharging is inversely proportional to (always finite) capacitive current level and directly proportional to the size of C'.[33] Typically, the area-related capacitance C'' is given by

---

[32]The *capacitance* C' of a cylindrical axon membrane *per unit length* in axial direction can be approximated as

$$C' = \frac{\varepsilon \cdot \pi}{d} \cdot D.$$

[33]The *voltage* u_C across the *capacitor* with a constant capacity C is proportional to the time integral of the *current* i_C through the capacitor; see Fig. 2.9 for schematic illustration. The differential form of the latter relation is given by

$$i_C = \frac{dQ}{dt} = \frac{d(C \cdot u_C)}{dt} = C \cdot \frac{du_C}{dt},$$

whereas Q is the electric charge accumulated on the capacitor plates. The latter equation shows that i_C is nonzero only if u_C is changing, i.e., $du_C/dt \neq 0$. In other words, stationary cases with a constant u_C, e.g., the *resting state* with $u_C = -U_R$, require the current i_C to be zero.

A descriptive relation showing the nature of the capacitor results if its *discharge by* $u - U_R$ is considered for a given *constant* i_C, i.e., the imbalance of u is considered with respect to the (original) resting state of the membrane with $u = U_R$. Then the needed *discharge time* Δt can be estimated by

$$\Delta t = C \cdot \frac{u - U_R}{i_C}.$$

In case the voltage u_C is a *sinusoidal* signal $U_C \cdot \sin(2\pi f \cdot t)$ with the amplitude U_C and frequency f, the current i_C amounts to

$$i_C = C \cdot \frac{d(U_C \cdot \sin(2\pi f \cdot t))}{dt} = C \cdot 2\pi f \cdot U_C \cdot \cos(2\pi f \cdot t) = I_C \cdot \cos(2\pi f \cdot t).$$

It can be observed from the above equation that the oscillating amplitude I_C of i_C *increases with rising* f if U_C is given. On the contrary, if I_C is given, the amplitude U_C *decreases with*

$$C'' = \frac{\varepsilon}{d} = \frac{C'}{\pi \cdot D}. \tag{2.2}$$

In fact, the level of C'' *is nearly constant* because of the almost invariable d and a typical composition of the membrane out of lipid molecules with a particular dielectric permittivity ε (Sect. 2.1.1). The magnitude of C'' amounts to about $1\,\mu\text{F/cm}^2$, which yields the membrane's *relative dielectric permittivity*[34] of about 10 [estimated from (2.2)]. Strictly speaking, the cell membrane is a *leaky capacitor* because $G' > 0$.

If the electrical stimulation of an axon (from Fig. 2.7) is considered in the view of the cable model (Fig. 2.9), then the induced current pathways first cross R'_I and then G' and C' in parallel. The total current through the membrane is thus given as the sum of *ionic current* i_E and *capacitive current* i_C ($i = i_\text{E} + i_\text{C}$); compare Figs. 2.8a and 2.9. The *initial slope* of the temporal plots in Fig. 2.8b is mainly determined by the initial i_C ($= i$) discharging the capacitor, which yields $i_\text{E} = 0$ and the resulting initial slope $\text{d}u_\text{C}/\text{d}t = \text{d}u/\text{d}t = i/C$. The *final value* of the courses is determined by i_E ($= i$) with $i_\text{C} = 0$ and $\text{d}u/\text{d}t = 0$.

Lastly, the *voltage source* in series with G' (Fig. 2.9) models the resting voltage U_R of the membrane (Sect. 2.1.3.1). In fact, the capacitor is charged up to U_R in the steady resting state.

The aforementioned *time constant* τ_R can be derived from the cable model (Fig. 2.9) for a *current source* as the stimulus origin and *radial currents* crossing the membrane; it yields

$$\tau_\text{R} = \frac{C'}{G'}. \tag{2.3}$$

That is, the *temporal extension* of the membrane's response to the step-wise stimulating current, compare Fig. 2.8b, increases with increasing C'. Likewise, the larger is C', the longer is the time needed for the capacitor to be charged up to a particular *final voltage level*. In analogy, a decreasing level of G' increases equilibration time and thus increases the time τ_R.

---

rising f. Thus, the capacitor increasingly represents a short-circuit for the rising f, i.e., $U_\text{C} \to 0$ with $f \to \infty$.
In general, for *sinusoidal signals*, the capacitor can be seen as a *complex impedance* $\underline{Z}_\text{C} = Z_\text{C} \cdot e^{j\varphi}$ [compare (1.2)] with its magnitude Z_C and phase φ, to give

$$\underline{Z}_\text{C} = \frac{\underline{U}_\text{C}}{\underline{I}_\text{C}} = \frac{1}{j \cdot 2\pi f \cdot \text{C}} = \frac{1}{2\pi f \cdot \text{C}} \cdot e^{-j\frac{\pi}{2}};$$

compare Footnote 2. It can be observed from above that with rising f, the magnitude Z_C decreases, i.e., $Z_\text{C} \to 0$ with $f \to \infty$.

[34]The relative dielectric permittivity is given by $\varepsilon/\varepsilon_0$ with ε_0 as the permittivity of free space ($\varepsilon_0 = 8.85 \cdot 10^{-12}$ F/m).

The *length constant* λ can be given as

$$\lambda = \frac{1}{\sqrt{G' \cdot (R'_E + R'_I)}} \approx \frac{1}{\sqrt{G' \cdot R'_I}}. \tag{2.4}$$

Here, the sum $R'_E + R'_I$ can be approximated as R'_I, because R'_E is typically negligible (Malmivuo and Plonsey 1995), as already discussed. The larger the product $G' \cdot R'_I$, the smaller is the *spatial extension* of the impacted region (Fig. 2.8c). This is because *increasing* G' progressively *short-circuits the membrane* and thus inhibits further spatial propagation of both stimulating current (through the membrane) and the response $u - U_R$. In analogy, *increasing* R'_I *hinders* an efficient *current conduction* along the axon's inner core. It should be noted that the parameter C' *does not influence* λ because λ quantifies the spatial extension at $t = \infty$ or $t \gg \tau_R$ (Fig. 2.8c); so that all involved capacitors (at each Δx in Fig. 2.9) have sufficient time to be recharged and the particular size of the capacitors ($= C'$) does not play any role.

Table 2.2 illustrates *typical values* of D, τ_R and λ for unmyelinated and myelinated axons; for structural differences in between the latter axons see Sect. 2.1.4.1. It can be observed that the *size of* τ_R is usually less than 1 ms, which limits the *conduction speed* of the imbalance $u - U_R$, limits the maximum *triggering frequency* of action impulses and thus limits the maximum *rate of information flow* along axons. The *size of* λ lays within the range of a few millimeters determining *how far down the membrane is depolarized* when, for instance, an action potential is locally generated; a large λ gives rise to an (advantageous) low number of action potentials per axon length. The physiological relevance of λ can be demonstrated by the fact that an increased D lowers R'_I with $1/D^2$ (Footnote 30), increases G' with D (2.1), and thus enlarges λ with $D^{1/2}$ (2.4). Consequently, increased D improves the conduction speed of action impulse transmission.

The latter (evolutional) strategy to increase the conduction speed can be observed in unmyelinated, *giant axons* with a very large D, as described in Sect. 2.1.4.1. However, a large magnitude of λ implies a low *metabolic efficiency* through widespread equalizing currents (Fig. 2.7) and through an inefficient use of limited space in the cross section of large axons. In addition, large D increases the membrane area and thus increases C' (Footnote 32) and G' (2.1), whereas the

Table 2.2 Typical geometrical and electrical parameters of unmyelinated and myelinated axons of nerve cells

Nerve fibers	Axon diameter D (μm)	Time constant τ_R (μs)	Length constant λ (mm)	Conduction velocity v (m/s)
Unmyelinated axon	0.5–1.5	200–500	2–5	0.5–3
Myelinated axon	2–20	20–70	1–2[a]	10–100

Data has been accumulated from different sources including (Zierhofer 2001; Malmivuo and Plonsey 1995; Silbernagl and Despopoulos 2007; Pfützner 2003)
[a]Typically extends over 1–2 internodal distances

latter effects mutually compensate each other in terms of minimally varying τ_R (2.3) for the *radial currents*. In contrast, a large D has a significant impact on the extent of axial depolarization spread involving *axial currents* (Sect. 2.1.4.1). Lastly, *myelinated axons* offer another way to improve the conduction speed while maintaining a relatively small value of all three parameters, D, λ and τ_R.

2.1.2.2 Active Mechanisms

While the *passive behavior* of the cell membrane mainly involves the transport of substances down their concentration gradient (across the membrane) and the electrical subthreshold behavior of the membrane (Sect. 2.1.2.1), the *active behavior* of the membrane is mainly governed by

- *Regulatory mechanisms* in the membrane which gate channel proteins (section "Regulatory Mechanisms").
- *Active transport* of ionic substances up their diffusion gradient (section "Active Transport").

Especially in the generation of biosignals, these active mechanisms are highly relevant.

Regulatory Mechanisms

The regulatory mechanisms involve *gated channel proteins* spanning the cell membrane, in contrast to *resting channel proteins*, as depicted in Figs. 2.10 and 2.6, respectively. While the resting channels are normally *open* and are not impacted by external factors, most gated channels are *closed* when the membrane is at rest. The gated channels may open *in response to*

- Varying *voltage* u across the membrane
- Chemical *messengers* adhered to the receptor site of the channel
- Mechanical or thermal *stress* applied to the channel

During the gating of the channels, structural conformation of the channels is subject to changes over time, e.g., channel's secondary structure temporarily changes (Footnote 18). Generally, *energy* of electrical, chemical, or of mechanical origin must be supplied for a structural change, for instance, from the *closed to open* state.

In the case of the *voltage-gated channel*, as shown in Fig. 2.10a, a decreasing electric field in the membrane (Fig. 2.6) provokes a movement of a charged region residing in the channel protein and increases the probability of this channel opening. The top figure in Fig. 2.10a schematically illustrates the negatively charged region (= *activation gate* or gating charge), which occludes the channel pore for Na^+ at the resting membrane voltage U_R. An increasing u, in terms of membrane depolarization at the threshold level (Sect. 2.1.3.2), moves the charged region that rapidly opens the

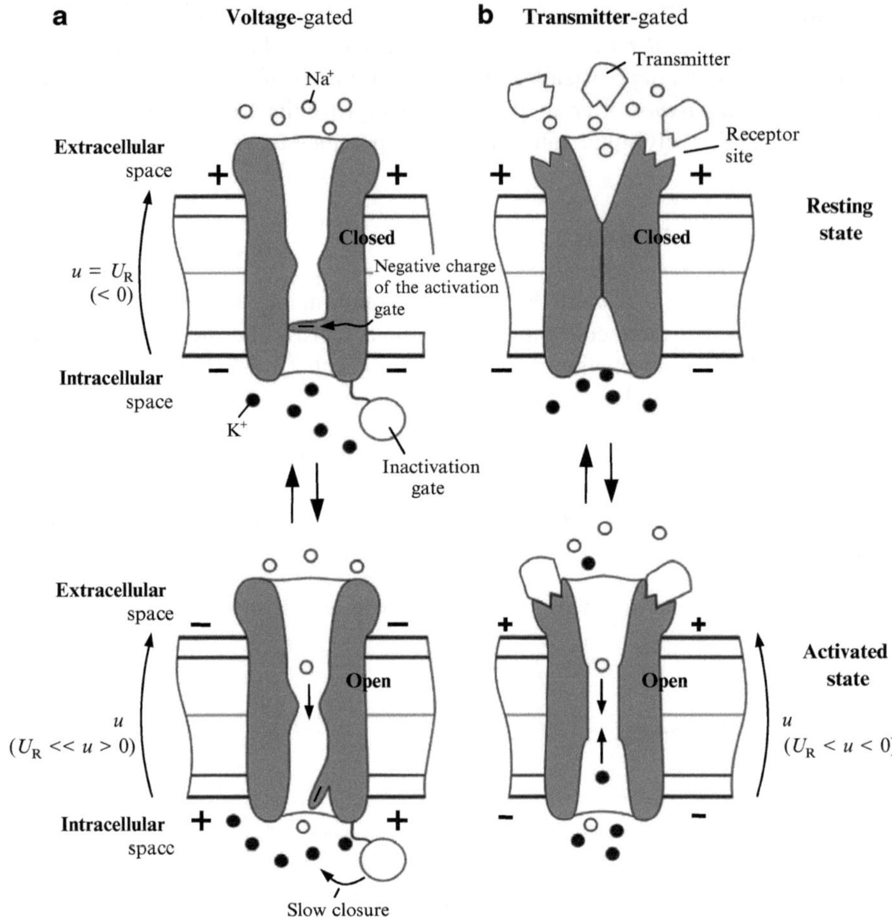

Fig. 2.10 Different types of gated channel proteins embedded into the cell membrane that govern the membrane voltage u; compare Fig. 2.6. (**a**) Voltage-gated channel for Na^+ ions, which is controlled by u. Opening of the channel yields a large increase of u from U_R towards positive values with $u > 0$. (**b**) Transmitter-gated channel for K^+ and Na^+ ions, gated directly by acetylcholine as transmitter substance. Opening of this channel yields an increase of u from U_R towards less negative values with $u < 0$

voltage-gated Na^+ channel in an *all-or-none fashion*; see bottom figure in Fig. 2.10a (Footnote 39). The channel cannot be half opened; it is either fully closed or fully open. Consequently, the inflow of Na^+ ions accelerates the increase of u toward more positive values even more.

In general, as the strength of *depolarization* increases, the *probability* of *opening* voltage-gated channels increases so that the *fraction of open channels* rises exponentially. This particularly applies to *voltage-gated* K^+ *channels* and *voltage-gated*

Na^+ channels that are involved in genesis of the action potential[35] (Sect. 2.1.3.2), whereas the Na^+ channels open about 10 times faster than the K^+ channels do (Malmivuo and Plonsey 1995). In addition, most neurons contain *voltage-gated Ca^{2+} channels* (involved in synaptic propagation, Sect. 2.1.4.2) while some neurons contain voltage-gated Cl^- channels. The *density* of the voltage-gated channels for Na^+ ions is about 5–500 per μm^2 of membrane in *nonmyelinated axons*, whereas the voltage-gated Na^+ channels in *myelinated axons* are concentrated at the nodes of Ranvier (Sect. 2.1.4.1) with a much higher density of about 1,000–2,000 per μm^2 (Malmivuo and Plonsey 1995; Kandel et al. 2000); for the different axon types see Sect. 2.1.4.1. In the membrane of muscle cells, the density amounts to about 400 per μm^2 (Hille 1978).

A *transmitter-gated channel* (Fig. 2.10b)—also referred to as ligand-gated channel—*opens* when an endogenous chemical messenger (*first messenger*) binds to a *receptor site directly on the channel*; i.e., the receptor and channel form a single macromolecule. The messenger can bind either at an extracellular site, in the case of gating *transmitters*, or at an intracellular site, in the case of gating *cytoplasmic constituents* such as Ca^{2+} ions; compare Footnote 76. This binding is facilitated by *charge and shape complementarity* between the transmitter and the receptor site (Footnote 14) so that the binding is energetically favorable (Footnote 13). Consequently, the binding yields *molecular rearrangements* of the channel proteins (twisting and bending), which are energetically favorable (Fig. 2.5a) and finally open this *directly gated channel* in the *all-or-none fashion*. Besides further alternatives[36] for an *indirect gating* (opening or closing) of a channel, a special

---

[35]It should be noted that certain *chemicals selectively block* voltage-gated Na^+ and K^+ channels, which impede action potentials vital for cellular physiology and yield paralyzing effects in humans. For instance, a substance named tetrodotoxin blocks the voltage-gated Na^+ channels, whereas tetraethylammonium blocks the voltage-gated K^+ channels. Interestingly, tetrodotoxin is found in pufferfish (or fugu fish), a Japanese traditional dish. It resides in the pufferfish's viscera which should be removed before serving this fish.

[36]A large variety of *transmitter controlled channels* are given. For instance, there is a group of channels (known as *G-protein-gated channels*), in which the *receptor site* and the *channel* itself are *spatially separated*; i.e., the receptor site and channel are built by separate transmembrane proteins (or separate macromolecules). The channel is *indirectly opened* (or closed) by a special protein (G-protein), whose *subunit*

- *Dissociates* from the receptor site inside the cell (in response to binding a transmitter molecule to the receptor site outside the cell)
- *Diffuses* along (or through) the membrane
- *Binds* to the channel, and finally *opens* the channel (Fox 2011)

It should be noted that this *indirect gating* of such channels is *slower*, but the resulting *change of u lasts longer* than in the case of *direct gating* by ligand binding (Fig. 2.10b). Concerning the direct gating, induced actions (e.g., synaptic actions) last a few milliseconds only.
In analogy, the dissociated subunit of the G-protein may bind to an enzyme which then produces *cyclic adenosine monophosphate* (cAMP). Here the substance cAMP serves as a *second messenger* which, in turn, triggers opening of special ion channels gated by cAMP. Such actions carried by second messengers are usually *very slow* and last from seconds to minutes.

ligand can indirectly (i.e., over cellular signaling cascades) activate a *covalent modification* of channel proteins through protein phosphorylation (addition of phosphate groups). Phosphorylation allows the opening of the channel which is then referred to as phosphorylation-gated channel.

As illustrated in Fig. 2.10b, an *acetylcholine-gated channel* allows the passage of both Na^+ and K^+ ions when *acetylcholine*, the transmitter molecule, binds to *two extracellular sites* on the receptor residing directly on the channel (known as *nicotinic acetylcholine receptor* on *directly gated channels*; compare Footnote 71). This, in turn, increases u from U_R toward less negative values ($>U_R$) because the inflow of Na^+ ions predominates the simultaneous outflow of K^+ ions. In fact, Na^+ ions experience a much larger driving force than K^+ ions in the resting state (Sect. 2.1.3.1); compare Fig. 2.6. For instance, at the crest of each fold of the postsynaptic membrane of a muscle cell (Sect. 2.1.4.2), the *density* of the acetylcholine-gated channels amounts to even 10,000 per μm^2 (Kandel et al. 2000).

The opening of the *stretch-gated channel*, e.g., for a (dominant) inflow of Na^+ ions and outflow of K^+ ions, is either due to direct *mechanical* stretching of the membrane or a *thermal* impact on the membrane. In the latter case, the membrane and the channels within are reshaped because of the thermal expansion of the membrane. In particular, resulting changes in the tension of the lipid bilayer or in the tension of the cytoskeleton (of the cell) gate these stress-sensitive channels, which consequently cause a disturbance in u from its resting value U_R.

It should be noted that the graded (*analogous*) opening of *gated channels* is only applicable to a *large number of channels (*typically many thousands), with the total opening level determined by the fraction of open channels. The graded opening (and closure) yields *graded u*, referred to as *graded potential* (Sects. 2.1.4.2 and 2.2). In contrast, a *single channel* opens in the *all-or-none fashion* only.

A single *gated channel* usually has three (*digital*) functional states:

- *Closed* (or resting) state
- *Open* state
- *Nonactivable* state

The *nonactivable state* (or refractory state) is particularly prominent for the *voltage-gated channels* and some *transmitter-gated channels* (Kandel et al. 2000). That is, after a channel's activation, i.e., channel opens for particular ions, the channel's *conductance is not maintained* and the channel enters the nonactivable state.[37] In this state, the channel cannot be reactivated for a few milliseconds. In the case of the voltage-gated channels, the inactivation can usually be reversed only by

[37]In the case of *voltage-gated channels*, the *inactivation mechanism* was found to be independent of the membrane's electric field (given by the ratio u/d) and to reside outside the membrane. As shown in Fig. 2.10a, a special protein (inactivation gate) is given in the shape of a *ball and chain*, dangling from the channel's cytoplasmic side (Kandel et al. 2000; Malmivuo and Plonsey 1995). What appears to be happening is that electrostatic attraction occurs in between the cytoplasmic mouth of the channel and the ball, once the channel is activated (opened). The movement of the ball, whose size exceeds that of the channel opening, plugs the opening.

repolarizing the membrane to its original resting state with the (more negative) resting voltage U_R; the repolarization allows the channel to switch to the closed (resting) state but activable state (Kandel et al. 2000). In the case of transmitter-gated channels, special *chemical substances* may deactivate the channels.[38]

This *temporary blocked* reactivation of gated channels yields *absolute and relative refractory* behaviors of the cell membrane, as described in section "Cell Response". For instance, a *voltage-gated* Na^+ *channel* (Fig. 2.10a) is nonactivable[39] for about 1–2 ms, whereas its *activation probability* decreases with increasing absolute level of U_R. For instance, the activation probability is maximal for $U_R \approx -100\,\text{mV}$ and nearly zero for $U_R > -50\,\text{mV}$ (Silbernagl and Despopoulos 2007). In contrast, a *voltage-gated* K^+ *channel* does not have a nonactivable state. To give another example, the aforementioned *acetylcholine-gated channel* has a mean opening time of only around 1 ms because of acetylcholine inactivation (Footnote 38).

Lastly, it should be noted that *gated channels* exhibit *many variants* to perform complex information processing tasks. There are channels that respond to both transmitter and voltage (including *second-messenger* systems, Footnote 36). For instance, a Ca^{2+}-activated K^+ channel is activated (opened) by intracellular Ca^{2+} ions while the channel's sensitivity to Ca^{2+} increases with depolarization of the membrane (compare Footnotes 76 and 101).

Active Transport

Besides the passive transport of substances *along their diffusion gradient*, there is a need for the means to move a solute *against its diffusion gradient* in many functional units of the cell (and organism). Thus *active transportation* powered by *chemical energy* is needed, such as from *adenosine triphosphate* (Footnote 19), or powered by *electrostatic energy*, such as from prevailing *ionic gradients* across the membrane.[40]

---

[38]For instance, in the case of *acetylcholine* as transmitter, an *inactivation* of (bound) acetylcholine molecules is attained by an enzyme (Footnote 16), known as acetylcholinesterase. The inactivation results in the closure of the relevant acetylcholine-gated channel *after about 1 ms* following channel's activation (Silbernagl and Despopoulos 2007; Kandel et al. 2000). The enzyme is present on the postsynaptic membrane and hydrolyzes acetylcholine molecules, before reaction products reenter the presynaptic axon bouton for resynthesis into acetylcholine (Sect. 2.1.4.2).

[39]Similar to the *activation gate* in the voltage-gated Na^+ channel, there is an *inactivation gate* in this channel; see Footnote 37 (Kandel et al. 2000). While the activation gate is closed in the *resting state*, the inactivation gate is open; see Fig. 2.10a. The inactivation gate closes slowly in response to *depolarization* with a delay of about 0.1–0.5 ms (Silbernagl and Despopoulos 2007; Silverthorn 2009). Therefore, the *channel can conduct* ionic current only for a short period during depolarization when *both gates* are open. In analogy, *repolarization* closes first the activation gate rapidly and then opens the inactivation gate more slowly. As soon as the channel has returned to its resting (closed) state, the channel can be opened again by a following depolarization.

[40]For instance, the prevailing Na^+ gradient across the membrane (Table 2.1) can be used to extrude Ca^{2+} ions out of the cardiac muscle cell, allowing relaxation of the cardiac muscle (compare

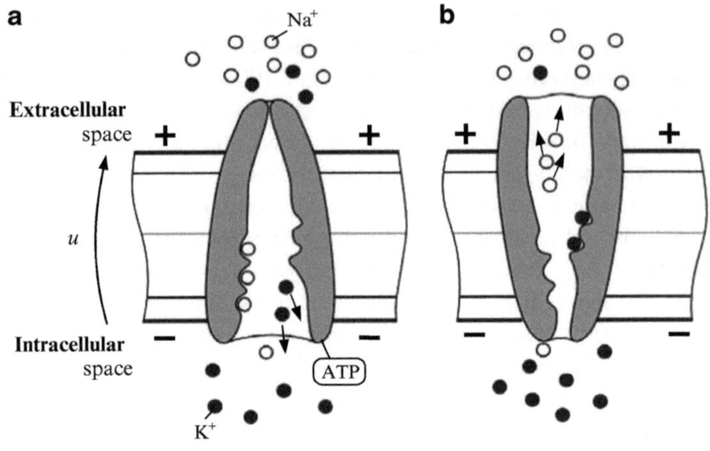

Fig. 2.11 Special membrane proteins for active transport of Na^+ and K^+ ions across the cell membrane, the so-called sodium–potassium pump. The ions are transported against their diffusion gradients, requiring hydrolysis of adenosine triphosphate (ATP) to provide the necessary energy (Footnote 19). (**a**) Binding of three Na^+ ions and release of two K^+ ions inside the cell. (**b**) Subsequent release of the three Na^+ ions and binding of two new K^+ ions outside the cell

In comparison with the channel proteins (section "Transport of Substances"), *particular characteristics* of active transportation comprise high specificity in substances (ions and molecules) being transported across the membrane, limited and saturable transport capacity, and possible inhibition of the active transportation when energy supply is disturbed.

A prominent example of active transportation is a *sodium–potassium pump* in the membrane, special membrane proteins which are universal to all cellular life. As illustrated in Fig. 2.11, the pump extrudes Na^+ ions out of the cell and moves K^+ ions into the cell, similar to the working principle of carrier proteins (section "Transport of Substances"). The membrane proteins bind *three* Na^+ *ions* on the intracellular side, in accordance with the charge and shape complementarity (Fig. 2.11a). A change of the proteins conformation follows, during which Na^+ ions are transported out of the cell; the affinity to Na^+ ions is lost and they are released into the extracellular space (Fig. 2.11b). Subsequently, the membrane proteins establish affinity to *two* K^+ *ions*, return to the previous conformational state (Fig. 2.11a), and then release both K^+ ions into the intracellular space.

The sodium–potassium pump builds and maintains gradients of K^+ and Na^+ ions across the membrane, as quantified in Table 2.1. For instance, the K^+ gradient helps to maintain the *resting membrane potential* (Sect. 2.1.3.1), while the Na^+ gradient

Footnote 103). That is, Na^+ is moved along its concentration gradient (into the cell) by special membrane proteins, which *powers* movement of Ca^{2+} against its concentration gradient (out of the cell); such proteins are known as Na^+–Ca^{2+} exchanger.

is highly relevant for rapid triggering of the *action membrane potential* through rapid Na^+ inflow, as soon as voltage-gated Na^+ channels open (Sect. 2.1.3.2).

Thus with each pump cycle, three Na^+ ions leave the cell while two K^+ ions enter. The resulting flux of ions makes the inner side of the membrane more negative, i.e., tends to *hyperpolarize the membrane*. In fact, the inner membrane side becomes even more negatively charged than would be achieved by only passive diffusion mechanisms (discussed in Sect. 2.1.3.1). For the speed of ionic transport by active sodium–potassium pumps, see Footnote 25.

2.1.3 Cell Membrane Potential

The cell membrane exhibits an imbalance of electric charges between its inner and outer side to yield the *membrane voltage u*. As already shown (e.g., in Fig. 2.6), the voltage u (or the potential difference) is defined as the local potential inside the cell (i.e., on the inner surface of the membrane) relative to that outside the cell (outer surface of the membrane), independent of the particular cell state.

- For *quiescent cells*, the *resting membrane potential* applies with a constant $u = U_\mathrm{R}$, whereas for
- *Excited cells* the *action membrane potential* applies with a varying $u = u(x,t)$[41]

Both resting and action potentials are produced by temporary changes in the *current flow across the membrane*, namely, in the ionic flow through the resting and gated channel proteins in the membrane. In particular, time and spatial courses of $u(x,t)$—significantly governed by the passive (subthreshold) propagation of u (section "Transport of Potential Difference")—carry *time-sensitive information* along cells about ongoing physiological activities. The cells may be *excitable or nonexcitable* as demonstrated, for instance, by a specialized sensory nerve cell or a secondary receptor cell, respectively (Sect. 2.2).

---

[41]In fact, the *transmembrane potentials* (strictly speaking, transmembrane voltage) can be resolved into *resting potentials* and *varying potentials* (Malmivuo and Plonsey 1995). The varying potentials typically arise in response to various *stimulating* physiological activities and are given by

- Actively propagating *action potentials* (in response to an above-threshold stimulation, Sect. 2.1.3.2)
- Passively propagating *graded potentials*, including

  receptor potentials caused by activation of membrane receptors (or membrane channels) in sensory nerve cells (Sect. 2.2) and
  synaptic potentials (postsynaptic potentials) caused by activation of the transmitter-gated channels in the postsynaptic membrane (Sect. 2.1.4.2); and

- Pacemaker *potentials*, inherently rhythmic potentials, as a result of spontaneous electrical activity of the membrane of special cells *without any external stimulation* (Sect. 2.3.2 and Footnote 111).

Action potentials are the signals through which physiological *information is conveyed* within the body and *processed* in the central nervous system. Therefore, each and every *biosignal* has its origin in a brief imbalance of the resting membrane potential, irrespective of the biosignal's type (Sect. 1.3).

2.1.3.1 Quiescent Cell

The resting state[42] of a cell, e.g., of an *excitable* nerve cell or a *nonexcitable* connective tissue cell in its natural environment, is described by the resting membrane potential. It refers to the *resting voltage* $u = U_R$ across the cell membrane, which is experimentally measurable. Figure 2.7 demonstrates the experimental assessment of u in an axon (of a nerve cell) using a voltmeter and an inserted microelectrode. The level of U_R usually amounts to about -70 mV, ranging between -50 mV and -100 mV (Silbernagl and Despopoulos 2007). Generally, the *genesis of* u and thus of U_R is mainly the result of

- The *concentrations* of diverse ions which prevail inside and outside of the cell close to the surface of the membrane.
- The selective *permeability* of the cell membrane to particular ions.

Significant ions determining the origin of u are K^+, Na^+, and Cl^- ions, whose concentrations have been discussed in Sect. 2.1.1 and with their typical values listed in Table 2.1. In short, K^+ *ions* and large immobile anions *prevail in the intracellular space* while Na^+ and Cl^- ions prevail in the extracellular space (Fig. 2.6). It should be noted that *active mechanisms* are required to maintain the above concentration gradients across the membrane; for instance, the gradients of K^+ and Na^+ ions are due to the active sodium–potassium pump (section "Active Transport").

In the *resting state*, (nongated) *channel proteins* for K^+ ions are predominantly open so that the cell membrane is mainly permeable for K^+ ions. The K^+ *ions diffuse outward* through the membrane down their chemical concentration gradient,[43] as shown in Fig. 2.6. This diffusion leaves behind a negative *excess charge* because K^+ ions are missing and a slight excess of negative ions (as Cl^- ions and other anions) results in the intracellular space. In fact, the large

[42] To start with, it should be noted that the *resting state* is not simply a passive state but actually an *active state* that is *stable* over time. A continuous maintenance of the resting state even needs metabolic energy, e.g., for a continuous operation of the sodium–potassium pump (section "Active Transport").

[43] Interestingly, the relative number of K^+ ions required to cross the membrane in order to generate the resting membrane potential is only about 0.001% (or 1 of 100,000) of all K^+ ions available in the *intracellular space* (Kandel et al. 2000; Silverthorn 2009). Since this percentage is extremely small, both intracellular and extracellular *concentrations of* K^+ ions can be considered as *constant* throughout transients and the resting state (i.e., the steady state, see text). In addition, *active transport* mechanisms continuously restore original concentration gradients; see the active sodium–potassium pump (section "Active Transport").

intracellular anions cannot follow the movement of outflowing K^+ cations because the membrane is not permeable to them. Due to the strong *electrostatic attraction* between the excess negative charge (inside the cell) and an excess positive charge of K^+ ions (outside the cell) that have already left the cell, the *K^+ ions accumulate at the outer membrane surface* and the anions (incl. Cl^- ions) collect at the inner membrane surface.

In consequence of the charge separation across the membrane, an electrical voltage u increasingly develops with a positive electric potential outside and a negative potential inside. An *electric field* arises within the membrane in proportion to the net K^+ efflux while the field is directed inward, as shown in Fig. 2.6. The increasing electric field exerts an

- Inward *electrical driving force* (with the magnitude F_E) on the permeable K^+ ions, since these ions carry a positive electric charge. This electrical force *opposes* the
- Outward *chemical driving force* (or the diffusional force with the magnitude F_D) and thus *limits the diffusional efflux* of K^+ ions; Fig. 2.6 illustrates these opposite forces[44]

An *equilibrium* is attained when the diffusional force driving K^+ out of the cell is balanced by the electrical force driving K^+ into the cell. Strictly speaking, at equilibrium the forces are equal ($F_D = F_E$) and opposite, yielding zero *electrochemical driving force* for K^+ ions; the latter force is defined as the vectorial sum of the electrical and chemical driving forces.

In fact, the diffusional force yields a certain *diffusion current density* J_D through the membrane (Footnote 24), while the electrical force yields an opposite *electric current density* J_E[45]; see also Footnote 53. At the aforementioned *equilibrium*, the

---

[44]In a first approximation, the magnitude F_E of the *electrical force* acting on a single K^+ ion (Fig. 2.6) with an excess charge e (elementary charge) depends on the *membrane voltage u*, that is

$$F_E = E \cdot e = \frac{u}{d} \cdot e.$$

The opposing force, the *diffusional force*, with the magnitude F_D is proportional to the *concentration gradient* $\mathrm{d}c/\mathrm{d}x$; see Footnote 24.

[45]In analogy with Footnote 24, the *electric diffusion* rate j_E, i.e., the amount of ions moved in an *electrolyte* by a surrounding E field per time (mol/s), is given by the *continuum form of Ohm's law* (compare Footnote 2).

$$j_E = A \cdot z \cdot c \cdot v,$$

where v is the ion's drift velocity. Accordingly, the magnitude of the ionic *electric current density* J_E (A/m^2) in the electrolyte can be expressed as

$$J_E = \gamma \cdot E = z \cdot c \cdot v \cdot e \cdot N_A = z \cdot e \cdot (c \cdot N_A) \cdot m \cdot E.$$

Here γ denotes the electrical conductivity of the electrolyte and m the *ionic mobility* (defined as the ratio v/E). In fact, the mobility m couples the electric field force to the resulting ionic flux. The net current density through the membrane must be zero; i.e., an equality of the density magnitudes $J_{\mathrm{D}} = J_{\mathrm{E}}$ *for* K^+ *ions* must apply (Malmivuo and Plonsey 1995). The equalization of both current densities is given at a particular u across the membrane, i.e., at the *equilibrium voltage* $u = U_{\mathrm{K}}$ for K^+ ions. The *Nernst equation*[46] gives

$$U_{\mathrm{K}} = \frac{R \cdot T}{z \cdot F} \cdot \ln \frac{c_{\mathrm{E}}^{\mathrm{K}}}{c_{\mathrm{I}}^{\mathrm{K}}} \cong 61\,\mathrm{mV} \cdot \log \frac{c_{\mathrm{E}}^{\mathrm{K}}}{c_{\mathrm{I}}^{\mathrm{K}}} \approx -90\,\mathrm{mV}. \tag{2.5}$$

Here F denotes Faraday's constant ($F = e \cdot N_{\mathrm{A}} = 9.65 \cdot 10^4\,\mathrm{A\,s/mol}$) and ${c_{\mathrm{E}}}^{\mathrm{K}}$, ${c_{\mathrm{I}}}^{\mathrm{K}}$ the extracellular and intracellular molar concentrations of K^+ ions, respectively. The above approximations are given for a *body temperature* of 37°C ($T = 273 + 37\,\mathrm{K}$) and concentration data from Table 2.1.

However, the *membrane* is also *slightly permeable* to other ions such as Na^+ *and* Cl^- through (resting, nongated) channel proteins in the membrane.[47] If (2.5) is adapted to Na^+ and Cl^- concentrations from Table 2.1, the following *electrochemical equilibrium* voltages result: $U_{\mathrm{Na}} = 61\,\mathrm{mV}$ for Na^+ ions and $U_{\mathrm{Cl}} = -68\,\mathrm{mV}$ for Cl^- ions. It can be observed that there is *no unique* u that would *equilibrate* electrochemical fluxes of *all ions* across the membrane (Table 2.1) because the inequality $U_{\mathrm{K}} \neq U_{\mathrm{Na}} \neq U_{\mathrm{Cl}}$ applies.

Thus, the resting voltage U_{R} represents merely the value for which a *steady state* in the ionic fluxes is achieved, but none of the involved ions is in its own equilibrium. The ionic fluxes across the membrane equalize each other so that the *total membrane current* and the net flux of ionic charges *have to be zero*. Consequently, the charge separation across the membrane (or the excess charge on the membrane, Fig. 2.6) and U_{R} remain constant. The quantitative description of the steady state is given by the *Goldman–Hodgkin–Katz equation*[48] that predicts the *equilibrium voltage* U with contributions of *each ionic species*, that is

$$U = \frac{R \cdot T}{z \cdot F} \cdot \ln \frac{p^{\mathrm{K}} \cdot c_{\mathrm{E}}^{\mathrm{K}} + p^{\mathrm{Na}} \cdot c_{\mathrm{E}}^{\mathrm{Na}} + p^{\mathrm{Cl}} \cdot c_{\mathrm{I}}^{\mathrm{Cl}}}{p^{\mathrm{K}} \cdot c_{\mathrm{I}}^{\mathrm{K}} + p^{\mathrm{Na}} \cdot c_{\mathrm{I}}^{\mathrm{Na}} + p^{\mathrm{Cl}} \cdot c_{\mathrm{E}}^{\mathrm{Cl}}}. \tag{2.6}$$

The concentrations ${c_{\mathrm{E}}}^{\mathrm{Na}}$, ${c_{\mathrm{I}}}^{\mathrm{Na}}$ and ${c_{\mathrm{E}}}^{\mathrm{Cl}}$, ${c_{\mathrm{I}}}^{\mathrm{Cl}}$ are the ionic concentrations of Na^+ and Cl^- ions, respectively. The factor p^k quantifies the respective permeability of the

---

term $c \cdot N_{\mathrm{A}}$ indicates the number concentration (or the number density) of ions. If J_{E} is constant over A then $i_{\mathrm{E}} = J_{\mathrm{E}} \cdot A$ (Fig. 2.9).

[46]Walther Hermann Nernst (1864–1941) was a German physicist and chemist who was one of the founders of physical chemistry.

[47]Of the different ion species dominating around the cell membrane only the large organic anions (in the cell) are unable to permeate the membrane (Sect. 2.1.1).

[48]David Goldman (1910–1998) was an American physiologist and biophysicist, Alan Lloyd Hodgkin (1914–1998) an English physiologist and biophysicist, and Bernard Katz (1911–2003) a German-born English physiologist.

membrane to the kth ion type; i.e., the ease with which the kth ion type crosses the membrane. In particular, the factor p^k is proportional to D_F (Footnote 24) of the respective kth ion type (Kandel et al. 2000) and inversely proportional to the membrane thickness d (Malmivuo and Plonsey 1995).

The following typical ratios of *permeabilities* for the *resting state* have been determined, being especially applicable for the squid axon (Kandel et al. 2000; Malmivuo and Plonsey 1995), to give

$$p^{\mathrm{K}} : p^{\mathrm{Na}} : p^{\mathrm{Cl}} = 1 : 0.04 : 0.45. \qquad (2.7)$$

That is, the membrane is about 25 times more permeable for K^+ ions than Na^+ ions. In addition, the permeability for Cl^- ions is also relatively high. In the *resting state*, (2.6) and (2.7) yield $U = U_R \approx -70\,\mathrm{mV}$ with the ionic concentrations from Table 2.1. Although the size of U_R seems to be relatively small, the resulting *electric field* in the thin membrane of the cell (with the field magnitude $U_{R/}d$) is very high from a technical point of view.[49]

The obvious *differences* between U_R and, on the other hand, U_K, U_{Na}, U_{Cl} indicate the *net driving force* on the particular ion type, as shown by arrows in Table 2.1. To be precise, the level of U_R determines the *electrical force* while U_K, U_{Na}, U_{Cl} the respective *electrochemical diffusional force* (Malmivuo and Plonsey 1995). The resulting fluxes of the involved ions in the *steady state* of the *resting membrane* are as follows:

- In the case of *K^+ ions*, the inequality of the driving forces yields a net *electrochemical driving force*. Namely, the difference $\Delta U = U_R - U_K = 20\,\mathrm{mV}$ applies, which means that the electrical force in the resting state is not adequate to equilibrate the outward diffusional force acting on K^+ ions. Hence, a small *outflow* of K^+ ions results because of a relatively *low driving force* (small ΔU) but *high permeability* to K^+ [large p^{K} from (2.7)].
- *Na^+ ions* show a very large $\Delta U = U_R - U_{Na} = -131\,\mathrm{mV}$, i.e., a *large driving force*. In fact, both diffusional and electrical forces acting on Na^+ ions are directed into the cell. However, the *very small permeability* (p^{Na}) strongly limits the net *inflow* of Na^+ ions.
- Very few *Cl^- ions* will also leave the cell at the resting membrane potential (Silbernagl and Despopoulos 2007). Their *outflow* is very weak because of a very *low driving force*, i.e., a very small $\Delta U = U_R - U_{Cl} = -2\,\mathrm{mV}$ but still *considerable permeability* (p^{Cl}). Here the diffusional force is a bit smaller than the electrical force ($|U_{Cl}| < |U_R|$).

---

[49]The *strength* of the *electric field* in the membrane (Fig. 2.6) should be illustratively put into perspective with electric fields common in technological applications. In fact, the field within the membrane attains impressive levels of about $100\,\mathrm{kV/cm}(= U_R/d = 70\,\mathrm{mV}/7\,\mathrm{nm})$ in the resting state. This demonstrates an extremely strong and durable structure of the bilayer membrane from an electromechanical point of view. For comparison, the level of the breakdown electric field in air is only about 30 kV/cm while in oil about 150 kV/cm.

That is, K^+ ions flow to the outside of the membrane in synchrony with Na^+ inflow and Cl^- outflow. As already noted, the above *inflow and outflow* of the different ions[50] *compensate and equalize* each other in electric terms, considering ional charges and ional flow directions; so that the total membrane current becomes zero.

It is important to observe that the weak *Na^+ inflow* and *Cl^- outflow* in the resting state tend to *increase u* (due to $U_R \ll U_{Na}$ and $U_R \leq U_{Cl}$) or, likewise, tend to *depolarize* the cell membrane (Sect "Cell Stimulation"). In analogy, the weak *K^+ outflow* tends to *decrease u* ($U_R > U_K$) or, likewise, tends to *hyperpolarize* the membrane. However, the resting level of U_R *is principally given* by the electrochemical diffusion of only K^+ and Na^+ ions because the intracellular *concentrations of K^+ and Na^+ ions are fixed* by the active sodium–potassium pump (section "Active Transport"). The diminished Cl^- concentration in the cell (Sect. 2.1.1) originates from the resulting electrochemical driving forces on Cl^- ions (at $u = U_R$) which float around the membrane relatively freely (2.7).

The mutual interrelations of the different ionic fluxes can be demonstrated if, for instance, additional positive Na^+ ions are assumed to be inside the cell. It would increase the *intracellular potential* and thus the absolute level of u (from U_R) toward less negative values. In consequence, the electrical force acting on K^+ ions in the membrane channels would decline while the diffusional force on K^+ ions would become dominant and would accelerate the (compensating) outflow of K^+ ions (Fig. 2.6). In analogy, removal of negative Cl^- ions or additional positive Ca^{2+} ions (Footnote 50) inside the cell, i.e., *outflow of anions* or *inflow of cations*, would raise the intracellular potential and thus reinforce the *outflow of K^+ cations.*

A potential imbalance of the involved ionic concentrations in the cell (Table 2.1), e.g., due to a slow *depletion of ional gradients* across the cell membrane in the course of the *steady state* (Footnote 43), is prohibited by diverse *active*[51] *transport mechanisms* in the membrane. These mechanisms maintain the ional gradients: K^+ and Cl^- ions are pumped back into the cell while Na^+ ions back out of the cell. Thereby, *sodium–potassium pumps* (section "Active Transport"), sodium–calcium exchangers as membrane proteins, and sodium–chloride carrier proteins are involved (Silbernagl and Despopoulos 2007).

If the electrical *cable model* from Fig. 2.9 is considered from the perspective of the resting membrane potential, a membrane patch can be represented by an electrical *parallel conductance model.* This model provides an intuitive and quantitative description of the ional movements across the membrane. Figure 2.12a

---

[50]It should be noted that very few *Ca^{2+} ions* enter the cell as well in the resting state (Table 2.1) because of a very high *electrochemical gradient* ($\Delta U = U_R - U_{Ca} = -70\,\text{mV} - 122\,\text{mV} = -192\,\text{mV}$ with $z = 2$ (2.5)] but a very small membrane *permeability* for Ca^{2+} ions ($p^{Ca} \ll p^{Na}$; compare (2.7)]. Like in the case of Na^+ ions, both diffusional and electrical forces acting on Ca^{2+} ions are directed into the cell.

[51]Interestingly, active transport mechanisms use different *energy sources*. The sources comprise not only chemical energy stored in adenosine triphosphate (Footnote 19) but also energy stored in the preexisting electrochemical gradients across the cell membrane, e.g., energy stored in the gradient of Na^+ ions across the membrane (section "Active Transport").

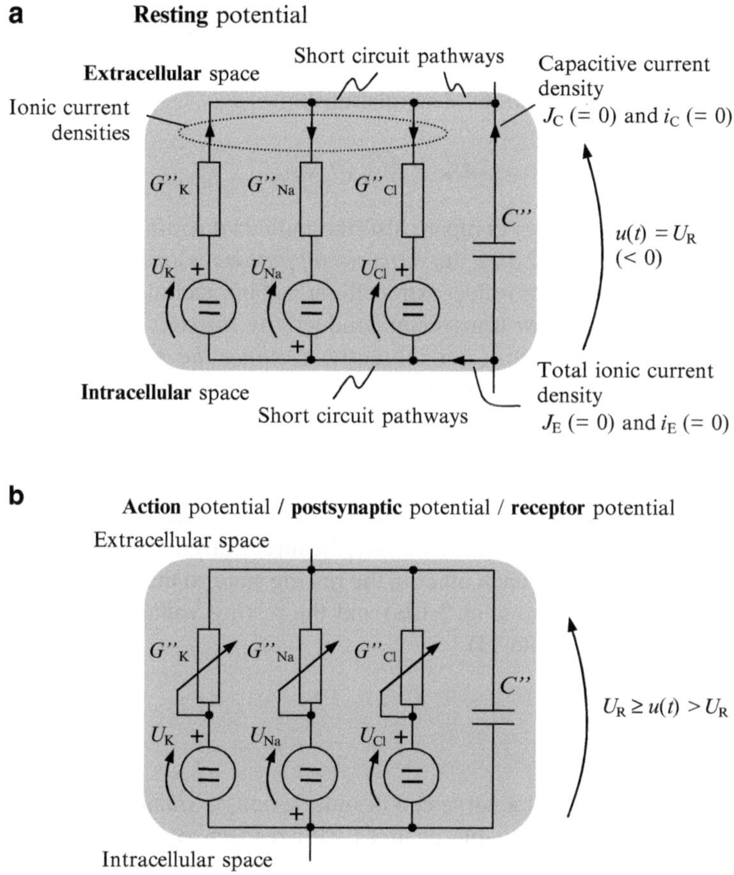

Fig. 2.12 Electrical equivalent circuit model of a membrane patch; compare with Fig. 2.9. Different channel proteins are considered (**a**) at the resting state and (**b**) beyond the resting state. The equivalent circuit is referred to as the Hodgkin–Huxley model (Malmivuo and Plonsey 1995)

depicts three *ion-conducting branches* in parallel to a *capacitor*, which models the ability of the membrane to accumulate electric charges (section "Transport of Potential Difference"). The ion-conducting branches comprise voltage sources with the equilibrium voltages U_K, U_{Na}, U_{Cl} in series with the area-related electrical conductances G''_K, G''_{Na}, G''_{Cl}; the sources and conductances describe the electrochemical behavior of K^+, Na^+, and Cl^- ions in the membrane. The *voltage sources*[52] represent chemical forces driving the relevant ions across the membrane

[52]An even more accurate electrical equivalent circuit of a membrane patch can be made by adding two *current sources* as two additional parallel branches to the membrane capacitor (Fig. 2.12). The continuous currents would model the active but unequal K^+ and Na^+ fluxes driven by the *sodium–potassium pump* (section "Active Transport").

because of the prevailing gradients of *ionic concentrations* (2.5). The *conductances* describe how readily particular ions cross the membrane. In fact, the conductances are related to both *resting permeability* of the corresponding ion channels (2.7) and gradients of ionic concentrations. The relationship

$$G''_{\mathrm{K}} > G''_{\mathrm{Cl}} \gg G''_{\mathrm{Na}} \tag{2.8}$$

applies for the *resting state*; compare also the indicated resting levels of G''_{K} and G''_{Na} in Fig. 2.14b. In Fig. 2.12a, the *short circuit pathways* located above and below the ion-conducting branches reflect extracellular and intracellular fluid, respectively. In comparison with the low transversal conductivity *through the membrane*, both fluids are excellent conductors for the currents *along the membrane* because of (relatively) large cross-sectional areas of fluids (Footnote 30) and many ions available for charge transport.

The *steady state* is characterized by *small currents* in the ion-conducting branches, as shown in Fig. 2.12a. These currents model ion flow through the membrane; e.g., the (plotted) outward current in the K^+ branch corresponds to the flow of K^+ ions[53] out of the cell. Obviously, the three currents in the ion-conducting branches must compensate each other in the resting state so that the *current through the capacitor becomes zero* (Fig. 2.12a) and the resting voltage $u = U_{\mathrm{R}}$ remains constant over time (Footnote 33).

2.1.3.2 Excited Cell

If an *excitable cell*, such as a nerve cell or muscle cell, is *stimulated*, its transmembrane voltage u shows inevitable changes with respect to the resting state (with $u = U_{\mathrm{R}}$).

- As long as the resulting u does not reach a specific value ($U_{\mathrm{R}} < u < 0$), i.e., u does not exceed the *stimulation threshold*, the *subthreshold stimulation* is given. As described in section "Transport of potential difference," the membrane responds *passively* to this stimulation.
- Provided that the threshold is reached, the *above-threshold stimulation* occurs and the membrane responds fundamentally different, that is, it responds *actively*. The stimulation triggers the excitable membrane to generate an *action membrane potential* in terms of a time-dependent electrical *action impulse* $u(t)$. As an approximation, the voltage $u(t)$ rapidly rises from the resting U_{R} (< 0) to positive values ($u > 0$) and then slowly recovers back to U_{R}.

---

[53]The *direction of electric current flow* is conventionally defined as the direction of net *movement of positive charges*. Given a directional electric field in an ionic solution (Footnote 26), *cations* move in the direction of the electric current while *anions* in the opposite direction.

The *difference* between both types of membrane responses could be illustrated by the fact that the generated action impulse $u(t)$ *does not become attenuated* as it moves down an excitable membrane because the impulse is actively regenerated. In contrast, the passive response of the membrane is a subject of a *strong attenuation* along the membrane. On the other hand, when the subthreshold *stimulation ends*, the level of u returns exponentially to the resting (original) level U_R. This is in contrast to the above-threshold stimulation, in which, once the threshold is exceeded, an impulse $u(t)$ is generated, whose duration is independent of the stimulus duration.

In this context, the *action membrane potential* or *action impulse* refers to an (experimentally measurable) electrical *nerve impulse* or *muscle impulse* $u(t)$ with the nerve cell or muscle cell being *excited*, respectively. In contrast, the *resting membrane potential* refers to the resting state of all cells, excitable and nonexcitable (Sect. 2.1.3.1).

Cell Stimulation

In general, the stimulation of the cell membrane is based upon a net *flow of ions* into or out of the cell. At the site of ion flow, the charge separation across the membrane gets disturbed which alters the polarization of the membrane (compare Fig. 2.10). Consequently, the local level of u changes. The membrane is said to be

- *Hyperpolarized*, if the charge separation across the membrane is increased as related to the resting state. For instance, an inflow of anions (e.g., Cl^-) or outflow of cations (K^+) increases the excess charge and thus the charge separation. Correspondingly, the voltage u *decreases* below its resting level U_R with the potential inside the cell becoming more negative than in the resting state; see Fig. 2.13c. In contrast, the membrane is said to be
- *Depolarized* (or hypopolarized), if the resting charge separation is reduced (still $u < 0$) or even reversed ($u > 0$), e.g., due to an inflow of cations (Na^+). The voltage u *increases* from U_R toward more positive values (Fig. 2.13c).

In an analogous way, if a stimulating *current* (Fig. 2.8a) is artificially *injected* into the axon of a nerve cell (Fig. 2.7), the *outflowing currents*—outward through the membrane—establish a specific distribution of u along the axis of the axon (Fig. 2.8b, c). The resulting u deviates from the resting U_R, particularly in the vicinity of the stimulating current electrode. Figure 2.13a illustrates the impact of the stimulating current in more details:

- If *positive current* pulses are applied (Fig. 2.13b), *excitatory stimulation* occurs and a *depolarization* is produced at *membrane sites*, where the induced *current outflows* the axon. The membrane initially behaves as a combination of resistors, capacitors, and voltage sources, corresponding to the equivalent electrical circuit model (zoomed region in Fig. 2.13a). However, this model is valid only over a limited range of $u(t)$ in terms of the *subthreshold stimulation* (section "Transport of Potential Difference"). That is, as long as $u(t)$ does not reach the membrane threshold, the membrane response is basically *exponential* (Fig. 2.8b); see the exponential course of u in Fig. 2.13c for the constant excitatory stimulus at $t < 0$. During subthreshold stimulation, the response duration depends on the stimulus duration. In case the *depolarization exceeds the threshold* (above-threshold stimulation) at a certain membrane location (Fig. 2.13c), an *action*

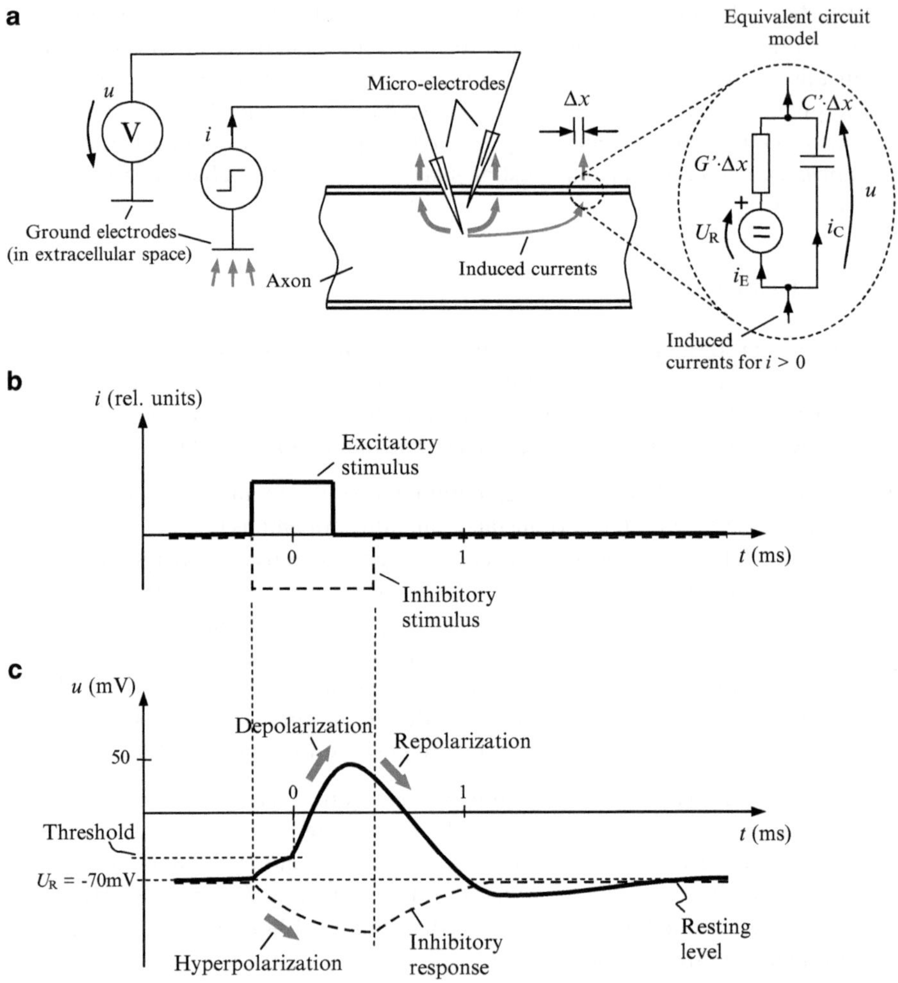

Fig. 2.13 Stimulation of an excitable cell. (**a**) Injection of the stimulating current i into the axon of a nerve cell with a synchronous registration of the resulting voltage u across the cell membrane; compare Figs. 2.7 and 2.9. In the equivalent circuit model for the passive membrane response (*right subfigure*), the current i_E denotes the electric ionic current and the current i_C the capacitive current. (**b**) Rectangular waveforms of the excitatory and inhibitory current stimuli. (**c**) Membrane voltage u in the course of the excitatory (depolarizing, above-threshold) stimulus that yields an active membrane response, i.e., an action membrane potential. In contrast, the inhibitory (hyperpolarizing, subthreshold) stimulus yields a passive membrane response in terms of an exponential deflection of u

impulse is elicited at that location, the duration of which is independent of the stimulus duration (section "Cell Response").

- Likewise, when *negative current* pulses (or negative charge) are injected into the cell (Fig. 2.13b), *inhibitory stimulation* occurs and a temporal hyperpolarization is produced. The membrane responds *passively and exponentially* (section "Transport of Potential Difference"). The *duration* of hyperpolarization is tightly related to the stimulus duration (Fig. 2.13c).

The equivalent electrical circuit model from Fig. 2.13a (right subfigure) illustrates the flow of induced outward currents in the membrane in the case of a positive stimulating current (compare Fig. 2.9). The resulting ionic current yields a proportional voltage drop across the resistor while the capacitive current discharges the capacitor (Footnote 33) so that u increases and the membrane depolarizes.

Therefore, the *origin of an action potential* is a local excitatory stimuli, in the course of which local u begins to increase and the membrane depolarizes above threshold. The excitatory *stimuli* may be of a different origin:

- *Artificial excitation*, as given by an injected positive current crossing the membrane from inside to outside (Fig. 2.13a)[54]
- *Outward local currents* that originate in the already excited proximal areas; for instance, spatial propagation of action potentials is based on such local currents (as discussed later, Fig. 2.18a) or even
- *Opening of gated channels* in the membrane that induces an inflow of positive cations into the cell, i.e., induces an inward current (Fig. 2.10). For instance, receptors in the sensory nerve cell respond to a specific physical stimulus by activation of specific gated channels in the cell membrane (Sect. 2.2). In consequence, *outward currents* are induced in proximal areas close to the location of the gated channels (compare Fig. 2.19)

Cell Response

To be clear, if the threshold of u is not reached during *depolarizing stimulation*, the passive *subthreshold* behavior of the membrane takes place (section "Transport of Potential Difference"). The *genesis of the threshold* as such is of particular relevance (as described below) since the threshold determines if the membrane's response remains passive or becomes active.

In particular, the conductance of the *voltage-gated Na^+ channels* increases in a graded manner as *depolarization* level increases in the course of a *depolarizing*

[54]From a practical point of view, the artificial stimulation can be performed not only by *inserting a microelectrode* into the axon of a nerve cell (Fig. 2.13a), but also by *applying two electrodes* to the skin, below which the axon is located. The stimulating current may flow from the positive electrode (*anode*) into the nerve cell and then leave the cell toward the negative electrode (*cathode*). In analogy with Fig. 2.13a, the axon's membrane would be depolarized and could be excited only *below the cathode* because here the induced current outflows the axon.

stimulus (section "Regulatory Mechanisms"). This is because each increment in depolarization level (although small) increases the total number of the channels that have switched from the closed (resting) state to the open state (Kandel et al. 2000). However, the ongoing *depolarization*—still within subthreshold levels—augments not only the Na^+ inflow through the voltage-gated channels but also the outflow of K^+ ions and inflow of Cl^- ions through the *resting channel proteins*. The electrochemical driving forces on K^+ and Cl^- ions increase with increasing depolarization level (compare Sect. 2.1.3.1 and Footnote 70). In addition, the ongoing depolarization opens additional *voltage-gated* K^+ *channels*, fortifying the outflow of K^+ ions even further. During the depolarizing stimulus, voltage-gated Na^+ channels become inactive and are not allowed to return to the closed state (i.e., activable and ready for opening state). The threshold level even increases with slow depolarization because numerous voltage-gated Na^+ channels have already been inactivated; in fact, if most of these channels are inactive, no action potentials can be initiated at all.

In total, the depolarizing inflow of Na^+ ions (comprising inward current, Footnote 53) is (over) compensated by the hyperpolarizing outflow of K^+ ions and inflow of Cl^- ions (outward currents). This *equilibration of inward and outward currents* prevents a further acceleration of the membrane depolarization, resists the depolarizing action of the voltage-gated Na^+ inflow, and hence corresponds to the *subthreshold behavior* of the membrane and to *accommodating mechanisms* from section "Response to Different Stimuli."

However, a strong and fast depolarization may yield an *excess in the inward current* because of a *large voltage sensitivity* and *rapid opening of the voltage-gated* Na^+ *channels*. The voltage-gated K^+ channels are slower in their response (section "Regulatory Mechanisms"), which delays the compensatory outward current. The net inward current depolarizes the membrane even more to yield an *avalanche-like opening* of additional voltage-gated Na^+ channels. Consequently, an *action potential is triggered*, as described below. Therefore, the *threshold* of the membrane excitation is given by a specific value of u at which the *net ionic current* has just changed from outward to inward; i.e., the *membrane response* diverts *from passive to active*.

As soon as the *threshold is reached* (above-threshold stimulation), i.e., the level of u *rises* usually by about 20 mV from $U_R = -70$ mV to the *threshold level* of about -50 mV, the activation of the voltage-gated Na^+ channels begins rapidly. The Na^+ ions flow inward because of the large Na^+ gradient across the membrane (Table 2.1), with the inflow exceeding both resting and compensatory K^+ outflow (Sect. 2.1.3.1). Figure. 2.14a demonstrates the course of u that exceeds the threshold level, while Fig. 2.14b illustrates an immediately following rapid increase in the membrane *conductance to* Na^+ ions. In fact, the conductance and permeability of the membrane to Na^+ ions increase more than to either K^+ or Cl^- ions. According to Fig. 2.12b, the following relationship applies [compare (2.8)]:

$$G''_{Na} > G''_K > G''_{Cl}. \tag{2.9}$$

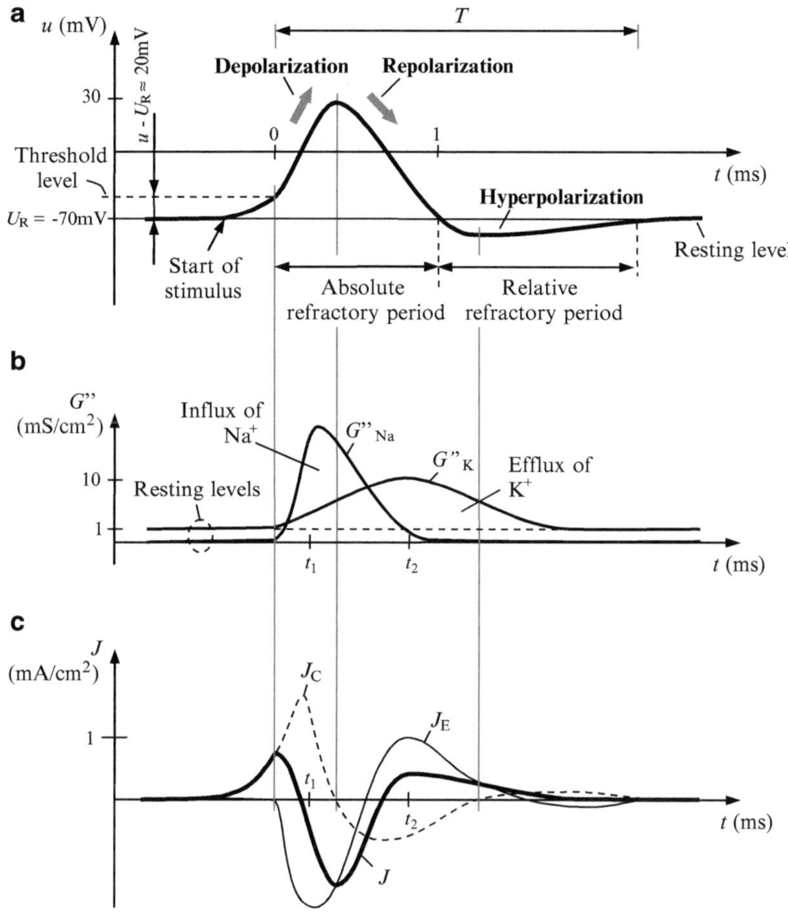

Fig. 2.14 Action membrane potential. (**a**) Voltage u across the cell membrane as an action impulse; compare Fig. 3.44a. (**b**) The corresponding behavior of the area-related membrane conductance G''_{K} and G''_{Na} for K^+ and Na^+ ions, respectively; see the equivalent circuit model in Fig. 2.15. (**c**) The corresponding ionic component J_E and capacitive component J_C of the total current density J $(= J_E + J_C)$ through the membrane. Typical absolute values of u, G'', and J are given in each case with the data being partially taken from (Malmivuo and Plonsey 1995; Silbernagl and Despopoulos 2007)

A net influx of positive charge is created, which reduces the resting charge separation across the membrane, discharges the membrane as capacitor, speeds up the membrane's u, and thus causes further depolarization of the membrane. That is, the ongoing depolarization causes more voltage-gated Na^+ channels to open, which *accelerates the depolarization* even more. This feedback (i.e., *positive feedback cycle*) drives the voltage u toward the positive *equilibrium voltage* U_{Na} (= 61 mV, Sect. 2.1.3.1) for Na^+ ions, i.e., toward the equilibrium voltage of the ions to which the membrane is most permeable. As shown in Fig. 2.14a, the level of u even becomes positive ($u > 0$); just as in the resting state u approaches the negative equilibrium voltage U_K [$= -90\,\text{mV}$ (2.5)] with the membrane being most permeable to K^+ ions.

As depolarization continues, the voltage-gated Na^+ channels gradually close through *temporal inactivation* after about 0.5 ms (Footnote 39). In parallel, the *voltage-gated K^+ channels slowly begin to open* and accelerate K^+ efflux. Actually, the increase in permeability to K^+ ions is slower than the increase in permeability to Na^+ ions. As illustrated in Fig. 2.14b, the K^+ conductance rises from its resting level. The delayed increase in the K^+ efflux starts to compensate (already declining) Na^+ influx and then produces even a *net efflux of positive charges.*

The membrane voltage *u reaches its maximum* of about 20 mV shortly after the peak in the Na^+ conductance and during the ongoing increase in the K^+ conductance, as can be observed in Fig. 2.14a, b. At the peak of u, the *permeabilities* of ion channels in the membrane are about (Kandel et al. 2000)

$$p^{K} : p^{Na} : p^{Cl} = 1 : 20 : 0.45. \tag{2.10}$$

In comparison with the permeabilities at the resting state (2.7), the level of p^{Na} is increased by a factor of 500. The estimated level of U [under the assumption of the steady state according to (2.6)] approaches U_{Na}, given the permeabilities from (2.10). However, the nonzero values of p^K and p^{Cl}, and, on the other hand, the quickly following inactivation of the voltage-gated Na^+ channels, prevent U from reaching U_{Na}; i.e., the K^+ efflux and Cl^- influx oppose the temporally limited Na^+ influx.

Then the level of *u starts to decline* because of the net efflux of positive charges, i.e., the K^+ efflux exceeds the Na^+ influx, and the so-called *repolarization* begins (Fig. 2.14a, b). The K^+ conductance is elevated above its normal values, which accelerates the repolarization phase even more (in terms of a *negative feedback cycle*). Therefore, while the first half of the action potential is governed by the Na^+ conductance, the second half is governed by the K^+ conductance.

When the declining part of u crosses the resting level at U_R (Fig. 2.14a), the K^+ conductance is still increased (Fig. 2.14b) because of the relatively inert voltage-gated K^+ channels. Therefore, it causes the voltage u to approach U_K, i.e., to become even more negative than U_R ($U_K < U_R$). Hence the so-called *hyperpolarization* (also known as after-potential) may arise. Finally, the K^+ conductance reaches its resting level after a few milliseconds and the voltage u levels off at U_R with a certain time delay quantified by the time constant of the membrane [from (2.3)]; the *action potential is now over.*

It is important to stress that an initiation of the action potential in the course of the *avalanche-like* opening of the voltage-gated Na^+ channels is a subject of the *all-or-none law*.[55] That is, the vigor of the cell response (evoked by a

[55] At first glance, the *all-or-none law* may seem *contradictory* to the *graded increase* in the conductance of the voltage-gated Na^+ channels as the depolarization level increases (section "Regulatory Mechanisms"). However, as already discussed, a large voltage sensitivity and a rapid opening of

depolarizing above-threshold stimulus) and the resulting shape of $u(t)$ (Fig. 2.14a) are *independent of the strength and duration* of the stimulus. Basically, the resulting *duration of the action potential* amounts to about 1 ms and is limited (fixed) by both the *gradual inactivation* of the voltage-gated Na^+ channels and *delayed opening* of the voltage-gated K^+ channels (Fig. 2.14a, b). Likewise, the *maximum amplitude of the action potential* is nearly constant because the concentration gradient of Na^+ ions across the membrane (Table 2.1) is relatively constant.

The beginning of the action potential is followed by a period of *diminished excitability*. Namely, there is

- An *absolute refractory period*, see Fig. 2.14a, during which another depolarizing stimulus—no matter how great the stimulus is—can not trigger another (subsequent) action potential in the *already excited area*. In fact, the necessary condition for the active membrane response is not given; i.e., the net inflow of Na^+ ions cannot exceed the outflow of K^+ ions because the *voltage-gated* Na^+ *channels* are in the *nonactivable state* (section "Regulatory Mechanisms").
- A *relative refractory period* immediately follows the absolute refractory period, during which a *larger stimulus*—i.e., larger than normally required to reach the threshold—is needed to initiate another action potential. However, if the subsequent action potential is initiated, it is lower in both ascending slope and peak amplitude (Silbernagl and Despopoulos 2007). The longer the pause in between two consecutive stimuli, the greater will be an increase in G''_{Na} during the second stimulus. This is because a greater fraction of the voltage-gated Na^+ *channels* will have *recovered* from the nonactivable state and these channels can be opened again by the second stimulus. Actually, a *reduced* magnitude of the total Na^+ *inflow* flattens the ascending slope of $u(t)$. In addition, the elevated (delayed and hyperpolarizing) outflow of K^+ *ions* in the second half of the ongoing action potential period partly *compensates the second depolarizing stimulus*, impedes a prompt release of a subsequent action potential, and thus *reduces the peak amplitude* of the subsequent action potential. The relative refractory period ends when $u(t)$ reaches its resting level U_R after the hyperpolarization phase (Fig. 2.14a).

Usually, the absolute *refractory period lasts* about 1 ms while the relative refractory period lasts approximately another 2–4 ms. In fact, the refractoriness[56] of the

---

the *voltage-gated* Na^+ *channels boost prevailing depolarization* and override slow compensatory repolarizing effects (dominated by the K^+ outflow). The strengthened depolarization yields an *avalanche-like* opening of additional voltage-gated Na^+ channels that actually *irrevocably triggers* the action potential.

[56]The importance of the refractory period can be demonstrated by the fact that the *parasympathetic system* (Footnote 189) affects the duration of this period. That is, an increased *activity of the parasympathetic vagus nerve* (controlling the sinoatrial node, Sect. 2.4) decreases *atrial refractory period* while increasing *ventricular refractory period.* In consequence, ventricular tachycardia and ventricular arrhythmias can be cancelled by an accelerated vagal activity or heightened parasympathetic activity (Zemaityte 1997).

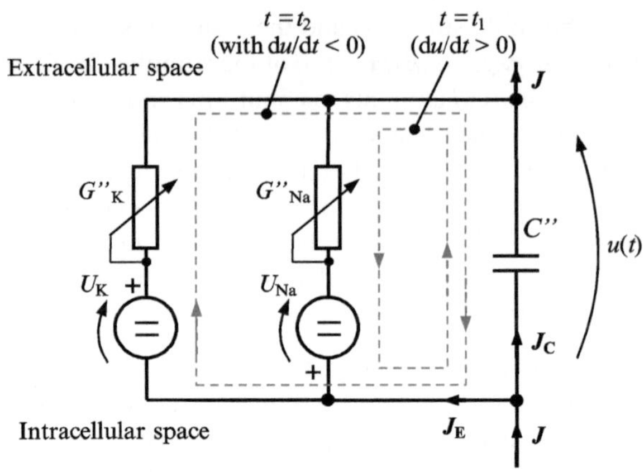

Fig. 2.15 Reduced electrical equivalent circuit model (after Fig. 2.12b) demonstrating the genesis of the action membrane potential. The model considers ion-conducting branches for K^+ and Na^+ ions only. Current pathways are indicated for two time instants $t = t_1$ (dominating Na^+ conductance G''_{Na}) and $t = t_2$ (dominating K^+ conductance G''_K). Both time instants are explicitly denoted in Figs. 2.14b, c considering the courses of G''_{Na}, G''_K, and the membrane current density J

membrane *prevents* temporal *overlap* of action potentials (or action impulses) and *excludes* the possibility of *backpropagation* of action potentials along an excitable membrane (Sect. 2.1.4.1).

The electrical equivalent circuit model from Fig. 2.12b can be used to determine *dynamic changes* of $u(t)$, as shown in Fig. 2.14a, in response to discussed changes of ional permeabilities p^k or ional conductances G'' during the action potential.[57] A *reduced equivalent circuit model* in Fig. 2.15 considers only *ion-conducting branches for* K^+ *and* Na^+ *ions*, which practically determine the *genesis of the action potential*. The resulting course of $u(t)$ is tightly related to the behaviors of the *ionic current* density J_E and the *capacitive current* density J_C through the membrane (Fig. 2.15).

To begin with the interpretation of the courses of J_E and J_C, as depicted in Fig. 2.14c, the exciting depolarizing stimulus should be considered first. The onset of the dynamic stimulus yields mainly capacitive currents with $J_C \neq 0$ while the total J_E is approximately zero due to the low conductance of gated channels; recall the previously discussed depolarization within subthreshold levels. As soon as the threshold is reached, the current J_E rapidly becomes negative because of the Na^+ inflow. Figure 2.15 illustrates the corresponding *current pathway*, given for the time instant $t = t_1$ from Fig. 2.14c. Later, the delayed and slow outflow of K^+ ions causes J_E to reverse; compare another current pathway for the time

[57]Here it should be recalled that (2.6) (from Sect. 2.1.3.1) is applicable only for a constant level of $u(t)$ (i.e., $u(t) = U$, Footnote 20), but not for dynamic and temporal changes of $u(t)$.

instant $t = t_2$. At the peak of $u(t)$, the current J_C becomes zero because at this point $du(t)/dt = 0$; see Footnote 33. After the peak, the direction of the current J_C also reverses, since $u(t)$ begins to decline, i.e., $du(t)/dt < 0$. In the terminal phase of the action potential, the current J_E even causes the membrane to become hyperpolarized. As the voltage $u(t)$ approaches the resting level U_R at the end of the hyperpolarization phase, both J_E and J_C converge toward zero while the membrane becomes discharged. In total, the net current density J $(= J_E + J_C)$ through the membrane yields a *biphasic waveform* (as an approximation) in the course of the action potential, with a peak value at the peak of $u(t)$.

It should be stressed that changes in the membrane voltage $u(t)$ are caused by substantial changes in the membrane's *permeabilities* p^k to different ions (dominating around the membrane) but not by changes in the concentrations of the ions (Table 2.1). The *bulk concentrations* of Na^+, K^+, Cl^- ions inside and outside the cell remain *constant* under most physiological conditions and also during the action potential. This is because

- The *relative number* of ions crossing the membrane is very small and amounts to only about 0.001% of all intracellular ions (compare Footnote 43).
- Original resting ionic gradients across the membrane are restored and maintained by diverse *active mechanisms* in the cell membrane. For instance, the active *sodium–potassium pump* (section "Active Transport") is involved along with other transport mechanisms for Cl^- and Ca^{2+} ions (Sect. 2.1.3.1).

Response to Different Stimuli

The aforementioned *threshold* of u determines whether an action potential is generated or not. In general, an *inhibitory* (*hyperpolarizing*) *stimulus* increases the amount of a simultaneous *excitatory* (*depolarizing*) *stimulus* needed to reach the threshold and to generate an action potential. Particularly notable is the fact that a *dynamic stimulus* is needed to provoke an active response of the cell membrane, as could be inferred from the discussed subthreshold behavior given a *depolarizing stimuli* (section "Cell Response"). In other words, a *constant stimulus*, e.g., a constant current continuously passing an excitable membrane, has the least impact on the excitable cell.

In a first approximation, the excitation threshold largely depends on the *strength and duration* of the *stimulus*. Strictly speaking, the threshold level is a function of a particular *waveform* of the stimulus (Pfützner 2003). Figure 2.16 illustrates a few of the most common waveforms of *currents* used as *depolarizing stimuli*, such as

- *Rectangular* pulse
- *Sawtooth* pulse
- *Sinusoidal* oscillation

The respective amplitude of the current i stimulus is denoted as I and the impulse duration (or period duration) as T. Figure 2.13a shows the relevant experimental

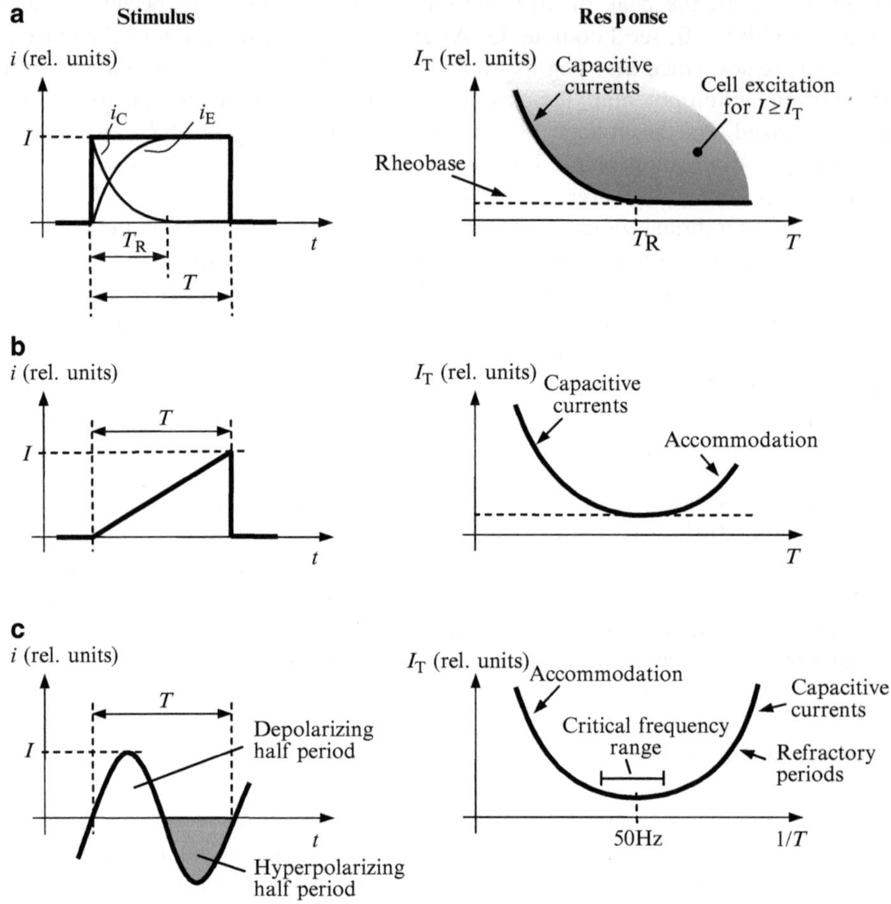

Fig. 2.16 Various waveforms of the current stimuli i (left subfigure) injected into an excitable cell according to Fig. 2.13a. The corresponding cell response is shown (*right subfigure*), quantified as the threshold current amplitude I_T which is now sufficient to excite the membrane; i.e., the current amplitude $I \geq I_T$ leads to triggering of an action membrane potential. (**a**) Rectangular current stimulus with the impulse duration T. Arising current components in the membrane are indicated within the rectangular shape of i, such as the electric ionic current i_E and the capacitive current i_C; in accordance with the equivalent circuit model from Fig. 2.13a, (**b**) Sawtooth current stimulus with the ascending slope I/T. (**c**) Sinusoidal current stimulus with the oscillating period T

setup with i being injected into an excitable cell. In Fig. 2.16, the corresponding *threshold current* I_T is given as a function of T; i.e., as soon as $I \geq I_T$ for a given T, an action potential is generated.

For the *rectangular pulse stimulus* (Fig. 2.16a), the threshold may be reached either by a short, strong stimulus or by a long, weak stimulus. That is, the threshold I_T increases for decreasing T. This results from the behavior of the electric ionic current i_E and the capacitive current i_C in the membrane, given the rectangular shape of the stimulus i $(= i_E + i_C)$; compare with the equivalent circuit model from Fig. 2.13a. As shown in Fig. 2.16a, the level of i_C declines while that of i_E rises during the rectangular current impulse (compare Fig. 2.8a). As soon as i_E *has reached a certain level* at which $i_E/(G' \cdot \Delta x) \approx 20\,\text{mV}$ (Fig. 2.13a), the membrane is activated. Here, the voltage drop of about 20 mV across the membrane conductance $G' \cdot \Delta x$ corresponds to the difference $u - U_R$ at the *threshold level* (Fig. 2.14a). In analogy, if the electric current density J_E (proportional to i_E, Footnote 45) is considered, the stimulation threshold is reached when $J_E/G'' = J_E \cdot d/\gamma_M \approx 20\,\text{mV}$ (2.1). This means that with increasing I the depolarization amount of about 20 mV is reached faster with shorter T; in fact, it implies increasing I_T for decreasing T. It can also be observed in Fig. 2.16a that i_E levels off at I after the time T_R while i_C becomes substantially zero. As an approximation, T_R amounts to about $5 \cdot \tau_R$ (2.3). Now the constant level of i_E causes I_T to stop declining for $T > T_R$. Actually, the minimum value of I_T is called *rheobasic current* which is adequate for increasing the value of u up to the excitation threshold provided that the stimulus is relatively long ($>T_R$).

For the *sawtooth pulse stimulus* (Fig. 2.16b), the course of I_T over T shows a peculiarity of increasing I_T for large values of T. That is, a certain minimum slope I/T is needed to depolarize the cell and to trigger an action potential, whereas an increasing slope (increasing dynamics of the stimulus) potentiates the probability of the active cell response. This can be attributed to *accommodating mechanisms* in the cell membrane, which compensate depolarizing effects of slowly changing stimuli. Similar compensating effects have been discussed in section "Cell Response" within the scope of depolarization within subthreshold levels.

For the *sinusoidal oscillation stimulus* (Fig. 2.16c), the level of I_T increases for a long oscillating period T (or a small oscillating frequency $1/T$), which is due to the *accommodating mechanisms* described above. In the range from 30 Hz up to 100 Hz the sinusoidal stimulus is *sufficiently dynamic* with a depolarizing *positive half period* which has a *sufficient duration* for the excitation of the cell. It is notable that the most commonly used technical frequencies of 50 Hz and 60 Hz in power lines fall into this *critical*[58] *frequency range* with a low I_T. In contrast, the *negative half period* hyperpolarizes the membrane, given the stimulation setup from Fig. 2.13a; compare with the inhibitory stimulus and inhibitory response from Fig. 2.13b, c. With increasing frequency $1/T$, the level of I_T increases again, because the depolarizing half period begins to overlap with the *relative refractory period* (section "Cell Response") from a previous action potential. The association between the stimulus and the corresponding action potential begins to disappear. For frequencies higher than about 500 Hz, the depolarizing half period starts to overlap with the *absolute refractory period*, impeding the generation of new action potentials. In the kHz range, mainly *capacitive currents* bridge the membrane; likewise, the component $i_C(\approx i)$ dominates while i_E (relevant for reaching the

[58]In addition, stimulation frequencies *above 50 Hz* are critical in terms of *unwanted muscle stimulation* resulting in sustained muscle contraction; see Footnote 105.

threshold) approaches zero (Footnote 33). Practically, no action potentials can be triggered for frequencies higher than 30 kHz (Pfützner 2003). It should be noted that the previously discussed behavior of I_T plays a crucial role in the establishment of induced electric biosignals for diagnostic aims (Sect. 4).

2.1.4 Propagation of Excitation

Cell excitation is related to temporal changes of $u(t)$ across the cell membrane, as discussed in Sect. 2.1.3.2. The action potentials (or action impulses) that arise carry, for instance, *sensory or actuating information*, originating in sensory neurons or directed towards muscle cells as body actuators, respectively. Thus the transient electric signal $u(t)$ *travels along nerve and muscle cells* to carry time-sensitive information throughout the human body.

In the following section, the focus is put on the propagation of the excitation along nerve cells, whereas the propagation along excitable muscle cells is based on similar phenomena (Sect. 2.3.2). To anticipate the evolutionary design of *nerve cells* (Sect. 2.2), it should be noted that two competing needs determine their design:

- *Rapid conduction* of action potentials to facilitate a high reaction speed
- *Small axon* size, allowing large numbers of them to fit into limited biological space

2.1.4.1 Axon Propagation

Figure 2.17b demonstrates a typical interconnection of nerve cells, which begins with the sensorial ending of a sensory neuron (acting as *input*) and ends with the terminal region of a motor neuron, i.e., a nerve–muscle synapse (as *output*). At the sensorial ending—a stretch-sensitive region with stretch-gated channels in the membrane (section "Regulatory Mechanisms")—a depolarizing imbalance $u - U_R\ (> 0)$ across the membrane is induced in response to mechanical stretch of this ending. The imbalance *spreads passively*—in terms of the *subthreshold behavior* of the membrane, section "Transport of Potential Difference"—in the axial[59] direction of the sensorial ending towards the trigger zone and *experiences* a particular *attenuation* over distance. Arriving at the trigger zone, an action potential can be released if the graded local imbalance $u - U_R$ exceeds the local level of the membrane threshold (section "Cell Response"). Then the action potential *actively propagates* without any attenuation along a *myelinated axon* of the sensory neuron, as described in the following. Between the terminal region of the sensory neuron and, on the opposite side, the input region of the motor neuron a *nerve–nerve*

[59]When considering a *propagating* imbalance $u - U_R$ or a propagating action potential instead of a *nonpropagating* local action potential (Sect. 2.1.3.2), *axial currents* along the surface of cell membrane must be considered in addition to the transmembrane *radial currents*; compare Fig. 2.7 and Footnote 63.

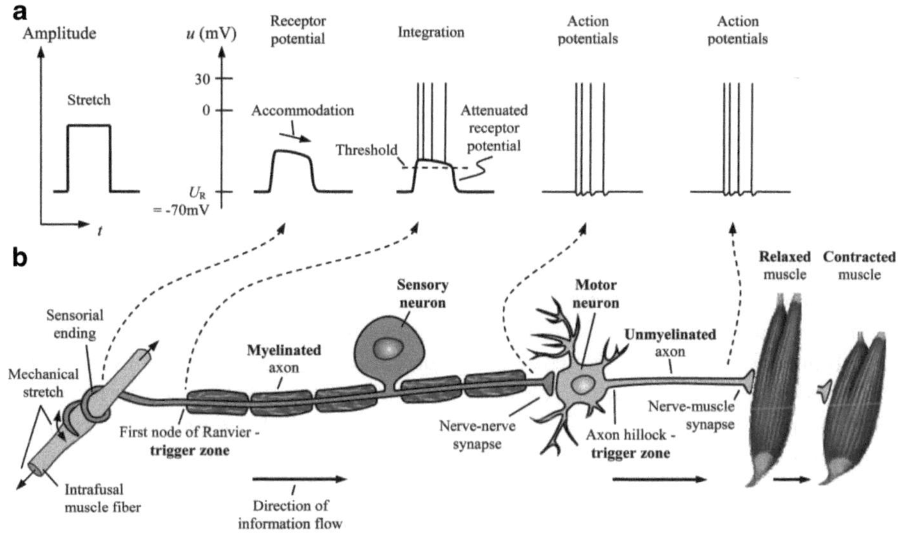

Fig. 2.17 The sequence of a mechanical stimulus, a sensory neuron (with its sensorial ending in the muscle spindle), a motor neuron, and an innervated muscle that constitutes a reflex loop. (**a**) Mechanical stretch stimulus has a rectangular shape, whereas the corresponding membrane voltage u at different sites of both neurons shows specific waveforms. (**b**) The stretch-sensitive region of the sensory neuron continues into the trigger zone where action potentials are generated. After saltatory conduction of the action potentials along the myelinated axon, these potentials are passed to the motor neuron over the neuronal synapse. The motor neuron controls the contraction of muscle cells over the neuromuscular synapse (compare Fig. 2.24)

synapse resides. The synapse transmits excitatory information towards the motor neuron with the use of chemical substances (Sect. 2.1.4.2). In analogy, the action potentials propagate *without failure* and *without attenuation* along an *unmyelinated axon* of the motor neuron towards the *nerve–muscle synapse*.

The passive spread of a local action potential along an excitable membrane *without reaching the membrane threshold* or, more generally, the *passive and exponential spread* of any imbalance $u - U_R$ is spatially restricted by the size of λ (2.4). In fact, this imbalance decays with increasing distance from the site of origin and effectively *disappears after* 5λ; compare Fig. 2.8c. According to Table 2.2, the size of λ is in the range of a few millimeters, while typical spatial distances for information processing (and propagation) in humans exceed a meter. In other words, there must be *active approaches* in axon propagation.

There are basically two ways for the action potential to *propagate without being attenuated*,

- First along *unmyelinated axons*
- Second along *myelinated axons*

In the case of the *unmyelinated axon*, every patch of its membrane contains voltage-gated Na^+ and K^+ channels. Thus *action potentials* can be produced along the *entire length* of the axon. To discuss conduction in the unmyelinated axon, it should be assumed that an already excited region exists. Here, it does not matter if this excitation was artificially or naturally created (section "Cell Stimulation"). Figure 2.18a demonstrates such an excited region of the axon. On the intracellular surface of the membrane, an excess of *positively charged* Na^+ ions occurs in the depolarized region (with $u > 0$) while an excess of *negative charge* occurs in the adjacent regions that are in the resting state ($u = U_R < 0$). Consequently, (closed-loop) *equalizing ionic currents* form along the inner and outer surfaces of the membrane down the axon because of the prevailing potential gradient along the membrane (Footnote 26). The equalizing currents are demonstrated in Fig. 2.18a to the right of the excited region. These *currents* in *adjacent regions* cross the membrane from inside to outside so that the currents serve as the *depolarizing stimulus* and the local value of u ($> U_R$) gradually tends to reach the threshold; compare Fig. 2.13. As shown in section "Transport of Potential Difference," the membrane can be approximated as a leakage resistance and capacitance in parallel. In fact, the level of u within the distance limits of λ—to the right in Fig. 2.18a—reaches the *threshold*. Consequently, the relevant patch of the membrane experiences an *accelerated depolarization* and generates another action potential. This *action potential*, in turn, serves as a stimulating source for another action potential even further to the right in Fig. 2.18a. It is important to recognize that the action potentials are *regenerated* along the entire length of the unmyelinated axon with the *same amplitude and duration* of the action potential (section "Cell Response"). Ion channels have to be opened along the entire length of the axon in the course of the regeneration. It is as if a *voltage wave was traveling* from the excited region to the right in Fig. 2.18b.

In principle, the passive spread of the depolarization along the axon is relatively fast compared to the time it takes to open voltage-gated channels and to generate an action potential or an additional action potential. This observation is supported by the fact that the time constant τ_R of the membrane (up to 0.5 ms, Table 2.2) is less than the temporal extension of the action potential (about 2 ms, Fig. 2.14a). In other words, for a higher conduction speed *fewer action potentials* need to be produced per axon length. In addition, the very thin membrane of the unmyelinated axon shows a quite high capacitance C' (Footnote 32) which results in a large τ_R (2.3), slowing down the conduction of changes in u.

The above issues are addressed by *myelinated axons*, in which the axon is covered with an *insulating layer*. The layer is formed out of tightly wrapped Schwann[60] cells (in the peripheral nervous system) forming multilamellar sheets of insulating membrane, known as the *myelin sheath* (Fig. 2.18a). The myelin sheath is

[60]Theodor Schwann (1810–1882) was a German physiologist who discovered cells producing myelin sheath and contributed to the development of the modern cell theory, defining the cell as the basic unit of live.

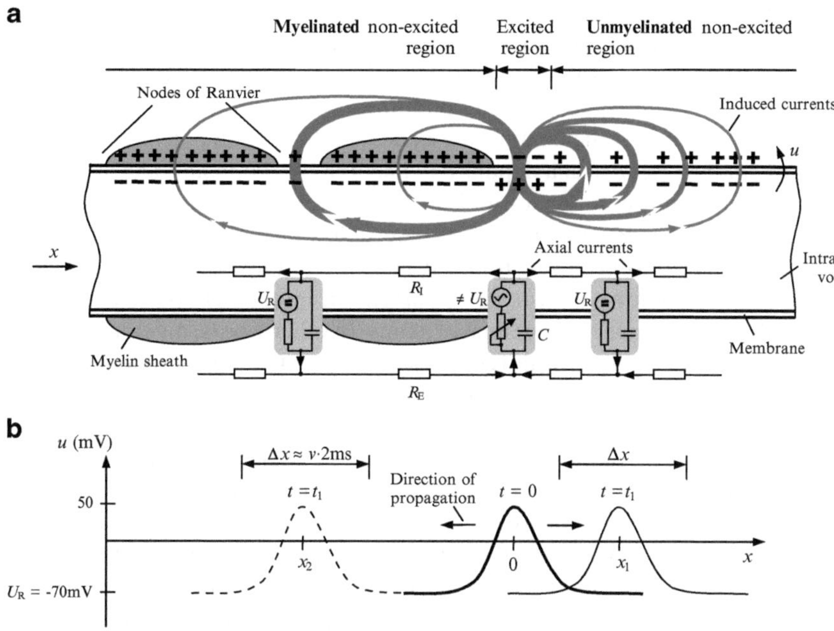

Fig. 2.18 Conduction of excitation along unmyelinated and myelinated axon regions. (**a**) The excited (depolarized) region at the location $x = 0$ yields an action potential which is conducted in both directions. Conduction along the unmyelinated region of the axon is illustrated to the right of the excited region (compare with Fig. 2.7) while that along the myelinated region is illustrated to the *left*. The *arrow's* thickness indicates local strength of induced current densities. The cable model below illustrates the corresponding current flow from the excited region to adjacent (resting) regions on each side. (**b**) The spatial course of the voltage *u* (across the membrane) has propagated more along the myelinated region than along the unmyelinated region ($|x_2| > x_1$) after time t_1 because of the varying conduction speeds ($|x_2/t_1| > x_1/t_1$). The spatial extension Δx is estimated according to (2.21) (with $\Delta x = \lambda$ and an assumed impulse duration of 2 ms (= T, Fig. 2.14a) of the action potential). The spatial courses of action potentials depicted do not account for the hyperpolarization phase (Fig. 2.14a)

not continuous but regularly and sharply interrupted every 1–2 mm by bare patches of axon membrane about 1–2 μm in length; these gaps are known as the *nodes of Ranvier.*[61] Because of the isolation preventing (depolarizing) current flow across the membrane, action potentials can be triggered only at the nodes.

As shown in Fig. 2.18a to the left of the excited region (active node), the inward currents, which originate at the active node, *flow axially*. The *major part* of the induced net current transverses the membrane only at the *next node* because the isolation increases the transverse membrane resistance and strongly reduces the

[61]Louis-Antoine Ranvier (1835–1922) was a French histologist and pathologist who discovered the nodes (or gaps) in the myelin sheath of myelinated axons.

current through the internodal surface.[62] In other words, a *decreased* G' is observed in *myelinated regions* while the *current* from the active node is *less attenuated over distance* and more of it is available for *discharging the membrane* at distant nodes (i.e., depolarization implies discharging due to $|u \cdot C'| < |U_R \cdot C'|$). As soon as the *threshold is reached* at the next distant node, another action potential is elicited there, which, in turn, serves as a depolarizing source for the membrane at the next node but one, even further to the left in Fig. 2.18a. Consequently, the *spread* of the action potentials *slows down* as it crosses bare nodes of high C' (see below) and action potentials have to be *regenerated.* It is as if the impulses jump from node to node in a process known as *saltatory conduction*, in contrast to the continuous conduction that occurs in the unmyelinated axons. In other words, myelinated axons offer a more cable-like spread with fewer sites for an active regeneration of action potentials, which speeds up the propagation of action potentials.

It should be noted that the *axial current* introduced above is also available to discharge C' of the adjacent myelinated regions ahead of the current, as indicated in Fig. 2.18a by the *minor* current paths through the myelin sheaths. Advantageously, the *myelination* significantly *decreases local* C' because the axonal membrane (or the isolating layer) is about 100 times thicker; compare Footnote 32. The decreased C' reduces the time needed[63] to discharge the myelinated axonal membrane and thus *speeds up internode discharging* for a given (minor) transmembrane current. In other words, the decreased C' of the myelinated regions reduces the current needed for a given discharge time; compare Footnote 33.

At the *nodes*, the exposed membrane is relatively thin and has a *reduced axial extension*. Consequently, the reduced area of the membrane tends to *reduce C of the nodal area*, whereas the *nodal* C' *is still relatively large* because of the membrane's thinness; compare Footnote 32. The reduced C *speeds up nodal discharging* for a given nodal current that allows the membrane threshold to be reached more quickly in order to regenerate an action potential further down the myelinated axon.

---

[62]The current always takes the *pathway of the least resistance* so that the transmembrane current density is highest in the nodal regions.

[63]In fact, the *charging speed* of the membrane—or the rate of passive spread of the depolarization—*varies inversely with* τ_R $(=C'/G')$ (2.3). However, in the case of the *depolarization spread* in the *axial direction*, the equivalent circuit model from Fig. 2.18a should be considered. At the already excited region (or at the active node) a *voltage source* with a variable voltage ($\neq U_R$) and a variable membrane conductance are connected in series; compare with Fig. 2.13a. Since *axial currents* are involved in the depolarization spread toward nonexcited regions with $u = U_R$ (Fig. 2.18a), axial resistances R_I and R_E ($\ll R_I$) are critical, whereas the membrane capacitance C must be discharged during the local depolarization. Thus, the effective time constant τ_A for a *voltage source* as stimulus origin (located at the excited region) and involved *axial currents* amounts to

$$\tau_A = C \cdot (R_I + R_E) \approx C \cdot R_I.$$

It should be noted that the above derivation is in contrary to the assumptions of a current source as stimulation origin and transmembrane radial currents, which were made for the derivation of τ_R. That is, the *myelination* reduces all three C, τ_R, and τ_A.

Thus, the *propagation* of action impulses is *accelerated* in comparison with unmyelinated axons.

On the basis of what has been discussed, it is interesting to note that the *internodal distance* is in the range of λ, for the passive spread of depolarization is spatially limited to values similar to λ. Furthermore, the *nodal area* is *very rich in voltage-gated Na^+ channels*[64]; typical values of the channel density are given in section "Regulatory Mechanisms". In fact, a high channel density contributes to the generation of an intense (depolarizing) inward Na^+ current—in response to the passive spread of depolarization—that periodically boosts the amplitude of the action potential. However, the *mean density of the voltage-gated Na^+ channels* over the axon's length is much lower than for unmyelinated axons, which highlights the *metabolic benefits* of myelinated axons. In addition, less current is needed to facilitate propagation of excitation, which results in a *smaller ionic imbalance* to restore, e.g., by the active sodium–potassium pumps (section "Active Transport"). Consequently, the restoration of only small concentration gradients requires *lower energy* expenditure by the active pumps.

The *propagation speed* v of the action potential along unmyelinated and myelinated axons deserves an extended discussion. Interestingly, a high speed v and a small size of axons constitute two competing needs, as will be shown below. Obviously, a *higher* v increases the speed of information circulation in the nervous system, enhances processing capacity and temporal precision of the central nervous system, and provides shorter delays in reflex loops (Fig. 2.17). On the other hand, a *small diameter* D of axons offers advantages in required biological space and metabolic efficiency.

In the case of *unmyelinated axons*, the *speed* v of the depolarization spread is a *function of* τ_A, as shown in Footnote 63. Strictly speaking, the smaller the product $C \cdot R_I$, the sooner the nonexcited regions reach their excitation threshold and the faster the action potential travels. That is, the value of v can be increased by decreasing R_I or by decreasing C (R_E is predefined by the extracellular environment, Footnote 63). *Increasing the interior D*

- Lowers R_I with D^2, as can be derived from Footnote 30
- Raises C and C' with D [(2.2) and Footnote 32]
- Raises G' with D (2.1)

Thus, the net effect is a *decrease in* $\tau_A \approx C \cdot R_I$ with increasing D (Kandel et al. 2000) and consequently an *increase in* v (Hodgkin 1954), according to

$$v \propto \sqrt{D}. \tag{2.11}$$

---

[64]At the *nodes of Ranvier*, the mechanism of membrane *repolarization* is mainly a fast *inactivation* of the voltage-gated Na^+ channels (section "Regulatory Mechanisms"), combined with a large outward leakage current. It seems that the voltage-gated K^+ channels do not play a significant role in the nodal repolarization (Chiu et al. 1979; Rattay and Aberham 1993; Kandel et al. 2000). This is in contrast to the repolarization phase of a typical action potential (section "Cell Response"), e.g., in an unmyelinated axon.

In addition, the *length constant* λ (2.4) *tends to rise* with rising D that consequently enlarges the axial depth of (passive) membrane depolarization, reduces the number of action potentials per axon length, and thus increases v of action impulse propagation. This developmental strategy for increasing v led to *giant axons*, e.g., the squid axon with a huge D of about 1 mm. However, large axons take up a relatively large amount of space in the organism.

In the case of *myelinated axons*, an *increasing* D increases the number of myelin layers and thus reduces C' of the myelinated regions even more. The increased speed of the internode discharging in combination with the aforementioned decrease of τ_A (without considering myelination) yields an *increase in* v (Hartline and Colman 2007), according to

$$v \propto D. \tag{2.12}$$

That is, a strong first power dependence results between v and D in myelinated axons rather than a weak square root dependence in unmyelinated axons (2.11). In other words, myelination results in a proportionately greater decrease in τ_A than does the same increase in the total D (Kandel et al. 2000). Lastly, it should be noted that the greater the *nodal density of voltage-gated* Na^+ *channels*, the greater is the *level of* v. This is because more channels (per area) allow for more ions (larger current) to enter the axon at the excited region (or active node, Fig. 2.18a). In consequence, a *greater difference in electric potential* is established between excited and resting regions that increases equalizing currents and thus facilitates a more rapid discharge of C in neighboring resting nodes.

Nervous systems have developed two mechanisms for *large* v (Hartline and Colman 2007):

- *Giant axons*
- *Axons encased* by the myelin sheath

As listed in Table 2.2, the *typical values for* v are in the range from 0.5 to 100 m/s for both types of axons and, in general, *increase with* D [(2.11), (2.12)]. For instance, a typical distance in the human body of 1 m has a pass-through time of about 20 ms (at an assumed $v = 50\,\mathrm{m/s}$), favoring *prompt reflex actions*. For thin axons with D of a few μm, *myelin sheath speeds* up the impulse propagation by a factor of more than 10 compared to unmyelinated axons; see Footnote 247. A schematic example is given in Fig. 2.18b, in which a voltage wave propagates along the myelinated region (to the left from the excited region) with a higher v in comparison with the propagation along the unmyelinated region (to the right). Conversely, for a given value of v, myelinated axons are smaller, require less biological space, and are more energy efficient[65] than their unmyelinated counterparts. To give an *example*,

[65]Brain activity (or nervous system) accounts for about 20% of the total energy budget in resting humans (Silbernagl and Despopoulos 2007), which illustrates the need for an efficient use of energy resources.

a myelinated frog axon with D of 10 μm is as fast as an unmyelinated squid axon with D of 500 μm (Silverthorn 2009).

The *spatial extension* Δx of an action potential in the axial direction of an axon at a given *time instant* (Fig. 2.18b) can be estimated from (2.21). Provided that T is the impulse duration of the action potential (according to Fig. 2.14a), (2.21) can be rewritten as

$$\Delta x \approx v \cdot T. \tag{2.13}$$

For unmyelinated and myelinated axons, Δx typically amounts to a few centimeters, e.g., $\Delta x = 10\,\text{cm}$ for $v = 50\,\text{m/s}$ and $T = 2\,\text{ms}$. Here it is important to stress that a *single action potential* spans over *numerous nodes* of a myelinated axon; about 50 neighboring nodes are usually located within the action potential. Likewise, numerous nodes are excited at the same time.

It should be stressed that the propagation of action impulses goes only in one direction. *Backpropagation is impossible*[66] because *refractory periods* prevent an immediate re-excitation of adjacent axonal regions (nodes). As described in section "Cell Response," this is due to temporally inactivated (depolarizing) voltage-gated Na^+ channels and (hyperpolarizing) voltage-gated K^+ channels that are still open *backward* in the propagation direction.

2.1.4.2 Synaptic Propagation

Synapses serve as the *electrochemical* elements for *switching* of nongraded action potentials and *amplifying* of graded potentials in human physiology. As already demonstrated in Fig. 2.17b, the synapse is a *communicational junction* between

- Two nerve cells or between
- A motor nerve cell and a muscle cell (or gland)

In the former case, the synapse works as a *mediating* junction with *chemical transmission*[67] so that *electric* action potentials can propagate from one nerve

[66]In the case of *artificial excitation* of a resting axon (section "Cell Stimulation"), the propagation of an action impulse may be induced in both directions, forward and backward along the axon; compare Fig. 2.18a. However, the *backward propagation ceases* once the impulse arrives at the next *unidirectional synapse* (Sect. 2.1.4.2).

[67]While chemical transmission seems to have evolved for a unidirectional propagation of action potentials over relatively long distances (still <100 nm), *electrical transmission* between cells does exist for *very short distances*, *rapid* and *bidirectional transmission* of cell depolarization. The electrically coupled cells are very close to each other [separated by only 2 nm (Fox 2011)] and are joined by contact areas of low electrical resistance, called *gap junctions* or *electrical synapses*. These junctions consist of *ion-conducting channels* with a pore diameter of about 1.5 nm (Kandel et al. 2000), built out of proteins. From a functional point of view, the induced *currents* from an action potential in the presynaptic cell cross the gap junction, depolarize the membrane of the postsynaptic cell up to its threshold, and induce another action potential there; compare Fig. 2.18a.

cell (presynaptic neuron, Sect. 2.2) to another nerve cell (postsynaptic neuron). Figure 2.17b illustrates a *nerve–nerve synapse* between a sensory neuron and a motor neuron. This synapse is known as a *neuronal synapse* which is typically about 1 μm in size.

On the other hand, the junction allows for innervation of a *skeletal muscle cell* (Sect. 2.3) by a motor neuron over an intermediate chemical step (see also Footnote 99). Figures 2.17b and 2.24 demonstrate a *nerve–muscle synapse* connecting the motor neuron with the muscle cell. The latter synapse is called a *neuromuscular synapse* with a size of about 10 μm.

As depicted in Fig. 2.19, the *synapse* basically consists of

- A presynaptic terminal (bouton) at the end of the *axon*, which ends with a *presynaptic membrane*
- A very narrow gap (10–50 nm in width in neuronal synapses and 100 nm in neuromuscular synapses), known as the *synaptic cleft*
- An adjacent postsynaptic terminal, which begins with a *postsynaptic membrane* of the receiving *nerve cell* or *muscle cell.* In the case of the nerve cell, the postsynaptic terminal is usually located on its *dendrites* or its *cell body*[68] (compare Fig. 2.21a)

The presynaptic ending contains *synaptic vesicles* (about 50 nm in size) which *hold neurotransmitter* molecules (e.g., acetylcholine) encapsulated in a membrane for chemical transmission across the cleft (Fig. 2.19). Numerous vesicles are already *docked* to the internal face of the presynaptic membrane (active zones). After an *action potential* arrives at the swelling of the axon terminal, the presynaptic membrane becomes depolarized. Depolarization opens *voltage-gated* Ca^{2+} *channels* which are abundant in this membrane. A rapid Ca^{2+} *inflow* follows because of a high electrochemical driving force (Table 2.1). The entry of Ca^{2+} ions—as much as a thousandfold increase in the Ca^{2+} concentration at the active zones—commands vesicles to merge with the presynaptic membrane. In the course of the fusion, *transmitter molecules* are *released* into the synaptic cleft. The remaining vesicle membrane is recycled back into the cell.

For instance, gap junctions are present between *muscle cells in the heart*, allowing the excitation to spread from cell to cell (Sect. 2.3). Such junctions are also found in some *smooth muscle* cells so that a stronger contraction can be produced. Some gap junctions are even found between *neurons in the brain* to synchronize their firing. There are indications that gap junctions can even be *gated by neurotransmitters* (section "Regulatory Mechanisms"). Some neurotransmitters cause the junctions to close or open through activation of *second messengers* such as Ca^{2+} or cyclic adenosine monophosphate (Fox 2011); compare Footnotes 76 and 36.

[68] A *synapse* can also be formed on an *axon terminal*, which then selectively controls this particular axon terminal. Usually, the inflow of Ca^{2+} ions into the controlled axon terminal is synaptically governed, either enhancing or depressing transmitter release from this axon terminal (known as presynaptic facilitation or presynaptic inhibition). Alternatively, transmitter-gated Cl^- channels in the presynaptic membrane can be activated, which decreases the final amplitude of the action potential in this presynaptic axon terminal; in consequence, the presynaptic depolarization is diminished and fewer voltage-gated Ca^{2+} channels are opened.

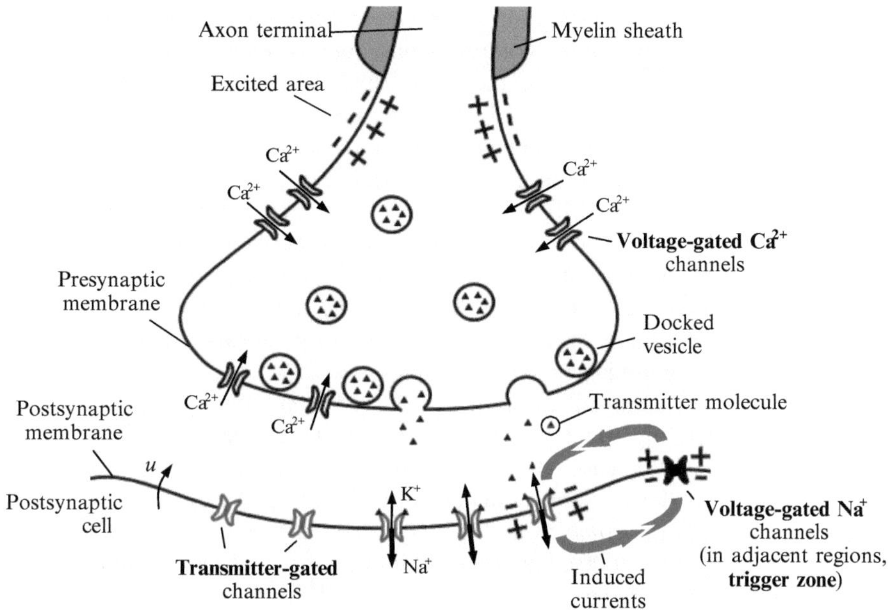

Fig. 2.19 Excitatory electrochemical synapse. After action potentials reach the axon terminal bouton, the resultant membrane depolarization opens voltage-gated Ca^{2+} channels. An increased number of intracellular Ca^{2+} ions activates vesicle fusion with the presynaptic membrane. The fusion releases transmitter molecules into the synaptic cleft. Arriving at the postsynaptic membrane, the molecules open transmitter-gated channels, which, for instance, generate a strong inflow of Na^+ ions and a weak outflow of K^+ ions. Consequently, the postsynaptic membrane becomes depolarized and equalizing currents arise toward adjacent nondepolarized regions of the membrane

Neurotransmitter molecules diffuse through the synaptic cleft, arrive at the postsynaptic membrane, and *bind* to highly specific receptors, namely, receptor sites of *transmitter-gated channels* (section "Regulatory Mechanisms"). Once the transmitter molecules are bound, the channels *open* (Fig. 2.19). For instance, in the case of *acetylcholine* as neurotransmitter, two acetylcholine molecules bind to acetylcholine receptor sites (two sites per single channel, Fig. 2.10b). The opened channel allows *Na^+ ions to enter* the postsynaptic cell and *K^+ ions* to leave it *at the same time*. Actually, *Na^+ inflow is stronger* than K^+ outflow because of a larger electrochemical driving force for Na^+ ions at the resting state of the membrane; compare Sect. 2.1.3.1. As a consequence, the intracellular side of the postsynaptic membrane becomes less negative; i.e., the membrane gets *depolarized* in terms of a *graded change* of *u* (graded potential)[69]; compare Fig. 2.23c. This local

[69]The *initial voltage* U_I of the *membrane voltage u* strongly influences the *expected change* Δu during *depolarization* of the membrane (by the *excitatory synapse*) if opened channels allow Na^+ ions to enter the postsynaptic cell and K^+ to leave it at the same time. In an approximation, the

depolarization is called the *excitatory postsynaptic potential* and the corresponding synapse is called the *excitatory synapse*. Consequently, *equalizing currents* arise along the inner and outer surfaces of the postsynaptic membrane toward neighboring nondepolarized membrane regions. In fact, the *depolarization spreads passively and decrementally*. The prevailing gradient of the electric potential (along the membrane) forces currents to originate, as shown in Fig. 2.19, similar to the passive spread of depolarization (sections "Transport of Potential Difference" and "Axon Propagation"). These induced currents cross from inside the membrane to outside in the adjacent regions so that the currents serve as *depolarizing stimuli* for the (still) closed *voltage-gated* Na^+ *channels* residing in the adjacent regions (Fig. 2.19). There the local u gradually tends to reach the *threshold* at this *trigger zone*, after which an *action potential* is generated; compare Fig. 2.23c, d.

Interestingly, the chemical transmitters involved in *neuronal synapses* may not only have a depolarizing effect but also a hyperpolarizing effect on the *postsynaptic membrane*. In the latter case, the synapse is called an *inhibitory synapse*. The particular effect of the transmitter depends on the receptors in the postsynaptic membrane, whereas the *same transmitter* may have *different effects* on different receptors (Kandel et al. 2000). For instance, in the case of *glycine* (amino acid) as neurotransmitter, the corresponding transmitter-gated channels open and initiate an *inflow of* Cl^-.[70] In analogy, the aforementioned *acetylcholine* triggers an *outflow of* K^+ when it binds to special channel receptors.[71] The inflow of Cl^- or the outflow of

final value of the *local* u at the site of depolarization converges to the *average equilibrium voltage* $(U_{Na} + U_K)/2$, i.e., the *target voltage* given by the average of U_{Na} and U_K; compare Fig. 2.12b with the high conductivities G''_K and G''_{Na}. For *muscle cells*, this local target voltage is close to 0 mV and—in view of the equalizing currents toward neighboring membrane regions with the resting $U_R = -90$ mV—effectively amounts to about -15 mV (Malmivuo and Plonsey 1995; Kandel et al. 2000). Thus, if U_I is equal to U_R (< 0) then $\Delta u > 0$ and the resulting final value u ($= U_R + \Delta u$) increases, i.e., the membrane gets depolarized. Otherwise, if $U_I > 0$ then $\Delta u < 0$ and the membrane gets hyperpolarized because Na^+ inflow is even less than K^+ outflow. In other words, at the average equilibrium voltage the inward Na^+ flux is balanced by outward K^+ flux so that the net current through the membrane is zero and $\Delta u = 0$ for $U_I = (U_{Na} + U_K)/2$.

[70]In fact, *Cl^- ions can diffuse* into the cell only as long as $u > U_{Cl}$; i.e., the cell membrane is depolarized more strongly than given at $u = U_{Cl}$ (Sect. 2.1.3.1). This may not be the case in the *resting state* with $U_R \leq U_{Cl}$, as indicated in Table 2.1. However, if the membrane is *depolarized* ($u > U_{Cl}$), the Cl^- inflow and the corresponding hyperpolarizing effects become prominent. In other words, the local membrane depolarization, e.g., due to propagating excitatory postsynaptic potentials that originate in neighbouring regions of the membrane, can be partly *compensated* by Cl^- inflow.

[71]A prominent example is given by the *vagus nerve* which *synapses* with *pacemaker cells* in the heart (Footnote 111). Activation of this parasympathetic nerve slows down the heart rate (Footnote 189). This *inhibitory effect* of vagus nerve activation is based on the indirectly gated *G-protein-gated* K^+ *channels* (Footnote 36) with the so-called *muscarinic acetylcholine receptors* in the postsynaptic membrane. An indirect stimulation of these channels by the transmitter *acetylcholine* opens them and leads to an *outflow of* K^+ *ions*, in the course of which the pacemaker cells become hyperpolarized. The cell's spontaneous *self-excitation slows down* or, likewise, the time required for the diastolic self-depolarization to reach the threshold is increased (Footnote 111).

K^+ makes the inside of the postsynaptic membrane more negative. The membrane becomes *hyperpolarized* with the local u moving *farther away from the threshold*[72]; compare negative values of U_{Cl} and U_K (from Table 2.1) with a typical threshold level of about −50 mV (section "Cell Response"). This membrane hyperpolarization is called *inhibitory postsynaptic potential*; compare Fig. 2.23c.

While excitatory postsynaptic potentials stimulate the postsynaptic *nerve cell* to generate action potentials, the *inhibitory postsynaptic potentials antagonize* this stimulation. In other words, the *hyperpolarized region* of the postsynaptic membrane (at inhibitory synapses) *short-circuits* the passive spread of the depolarization which has originated in the neighboring *depolarized regions* of the membrane (at neighboring excitatory synapses). The shunting of equalizing currents—heading toward the trigger zone—inhibits the current's depolarizing effect on the trigger zone, as will be shown in Sect. 2.2.2 (Fig. 2.23). In ionic terms, inhibitory Cl^- inflow or K^+ outflow renders excitatory Na^+ inflow less effective in reaching the threshold at the trigger zone.

In *excitatory neuronal synapses* which usually contact *dendrites* (see top synapse in Fig. 2.21a), the *excitatory postsynaptic potential* exhibits a graded and depolarizing change Δu (>0). Provided that a *single action potential* has arrived in the presynaptic terminal, Δu amounts to less than 1 *mV* toward the threshold; the maximal Δu is about 20 mV (Silbernagl and Despopoulos 2007). In other words, a single excitatory postsynaptic potential is, by far, *not sufficient* to trigger an action potential (distally at the trigger zone) or even reach the local *target voltage* $u = -15$ mV (Footnote 69). If U_I is the *initial voltage* across the postsynaptic membrane, then the resulting u can be written as

$$u = U_I + \Delta u. \tag{2.14}$$

For depolarization of the membrane, the inequality $u > U_I$ applies (2.14). If the resting state is given postsynaptically before depolarization, then $U_I = U_R$. However, U_I can differ significantly from U_R (e.g., $U_I > U_R$) because of *previous*, still not decayed postsynaptic voltage contributions. Namely, U_I can be predetermined by still present excitatory or inhibitory postsynaptic potentials coming from neighboring synapses or even from a previous firing of the same synapse. Thus all these additive contributions—compare the *spatial and temporal summation* of

[72]In analogy with Footnote 69, the final value of u during *hyperpolarization* of the membrane converges to a local *target voltage* of about −70 mV (Kandel et al. 2000) [or about −80 mV after (Pfützner 2003)]. The target voltage is mainly given by negative U_{Cl} or even more negative U_K (Table 2.1) depending on the ions involved. Therefore, the opening of K^+ channels has a stronger hyperpolarizing effect than the opening of Cl^- channels ($U_K < U_{Cl}$); compare Fig. 2.12b with either a high conductivity G''_K or a high G''_{Cl}. Thus, a hyperpolarizing $\Delta u < 0$ can be synaptically induced only if $U_I > -70$ mV (most usual case). Likewise, no change $\Delta u = 0$ is given for $U_I = -70$ mV; in this particular case, the *inhibitory synapse* is without effect. Lastly, $\Delta u > 0$ applies for $U_I < -70$ mV, in which the inhibitory synaptical transmission increases u and thus has even a depolarizing effect.

postsynaptic potentials from Sect. 2.2.2—determine if the threshold will be distally achieved for the generation of an action potential.

It is important to note that the excitatory postsynaptic potential *cannot reverse postsynaptic u* (though $U_{\mathrm{Na}} > 0$, Table 2.1) due to a simultaneous[73] outward diffusion of K^+ ions ($U_{\mathrm{K}} < 0$). If the electrical equivalent circuit of the membrane (Fig. 2.12b) is considered, the following relationship applies for ional *conductances* within the *postsynaptic* membrane:

$$G''_{\mathrm{Na}}, G''_{\mathrm{K}} > G''_{\mathrm{Cl}}. \tag{2.15}$$

In *inhibitory neuronal synapses* which usually contact the *cell body* (see bottom synapse in Fig. 2.21a), the *inhibitory postsynaptic potential* exhibits a graded and hyperpolarizing change Δu (< 0) of a relatively small amplitude; the maximal Δu is only about 4 mV (Silbernagl and Despopoulos 2007). Typically, the inequality $u < U_{\mathrm{I}}$ (2.14) is valid for the inhibitory synapse. The *target voltage* is about -70 mV (Footnote 72); the level of U_{I} can also differ significantly from U_{R}. In analogy with (2.15), the *inhibitory synapses* yield

$$G''_{\mathrm{K}} > G''_{\mathrm{Na}}, G''_{\mathrm{Cl}} \quad \text{or} \quad G''_{\mathrm{Cl}} > G''_{\mathrm{Na}}, G''_{\mathrm{K}}; \tag{2.16}$$

compare (2.8) and (2.9).

In *neuromuscular synapses*, only *excitatory postsynaptic potentials*, known as endplate potentials, are generated. Here *acetylcholine*[74] acts as the sole neurotransmitter and only a single type of *directly gated channels* is involved (Fig. 2.10b). In contrast to neuronal synapses, the graded depolarizing change Δu (> 0) is much greater and amounts to *about 70 mV* for a *single action potential* arriving in the presynaptic terminal. The estimated number of channels opened per action potential is around 200,000 (Kandel et al. 2000). Typically, the postsynaptic membrane of a muscle cell with $U_{\mathrm{R}} \approx -90$ mV becomes depolarized (discharged) nearly up to the *target voltage* of -15 mV (Footnote 69). Thus, a single presynaptic action potential (from a motor neuron) is *sufficient to trigger* postsynaptically an action potential, leading to the contraction of a muscle cell.

The behavior described above is based largely on the fact that the *neuromuscular synapse* is larger in size and has a wider gap than the neuronal synapse (see above). As shown in Fig. 2.26, the postsynaptic membrane of the skeletal muscle cell has *junctional folds*, i.e., deep depressions in the surface with a high density of *acetylcholine-gated channels* (with nicotinic acetylcholine receptors) at the crest

---

[73]If the *outflow of K^+ ions is delayed* (nonsimultaneous) with respect to Na^+ inflow, as given in the case of the *action potential* (Fig. 2.14b), the absolute level of *u can even reverse* (Fig. 2.14a).

[74]For instance, the *drug curare*, used on blowgun darts by Indians for hunting, *competes with acetylcholine* (as a neurotransmitter) for attachment to receptor sites of the transmitter-gated channels but *does not open* these channels. Consequently, curare diminishes the depolarization strength of the postsynaptic membrane and produces flaccid *paralysis* in hunted animals.

of each fold and *voltage-gated Na^+ channels* located below the crest extending into the fold toward its vale. The folds enlarge the effective *synaptic area*. Thus *more vesicles* can be emptied per single action potential forcing more acetylcholine-gated channels to open. Consequently, the neuromuscular synapse yields a large Δu across the postsynaptic membrane and a strong depolarization spread toward adjacent membrane regions exhibiting voltage-gated Na^+ channels (in the folds). *Action potentials* are generated in the junctional folds, which then propagate along the membrane of the muscle cell, *triggering its contraction* (Sect. 2.3).

Generally, the *amount of transmitter* substance entering the synaptic cleft (or the number of fused vesicles) is directly proportional to the *number of action potentials* entering the presynaptic terminal.[75] On the other hand, the *amplitude of the local postsynaptic potential* depends on the amount of transmitter released, whereas the *duration* of this postsynaptic potential depends on how long the transmitter is active. To illustrate this *quantitatively*, about 100–150 and 1–10 vesicles are emptied per single action potential in the case of the neuromuscular synapse and neuronal synapse, respectively. About 5,000–7,000 acetylcholine molecules reside in a single vesicle and the minimal depolarizing response from a single emptied vesicle is about 0.5 mV (quantal unit of the postsynaptic potential because of *all-or-none emptying* of a vesicle) (Kandel et al. 2000; Silbernagl and Despopoulos 2007). Obviously, the change of the postsynaptic potential is not instantaneous in response to an action potential entering the presynaptic terminal. The *onset* of the *postsynaptic potential is delayed* by about 0.5 ms due to a relatively inert release of the transmitter and its diffusion across the synaptic cleft (Silbernagl and Despopoulos 2007). The *duration* of the postsynaptic potential, i.e., the duration of its excitatory or inhibitory action based on equalizing currents, ranges from milliseconds (due to a rapid response of directly gated transmitter-gated channels) to minutes (slow response of indirectly gated transmitter-gated channels, Footnote 36).

In order to maintain proper neuronal control and to avoid refractory behavior of the synapses, the *stimulatory effect* of the *transmitter molecules* on the receptor sites has to be quickly *interrupted*. In fact, the transmitter molecules are either *returned* back into the presynaptic axon terminal (mediated by active transport mechanisms in the membrane, section "Active Transport"), or *absorbed* by neighboring cells, or diffused away, or even broken down and temporally *inactivated* (Footnote 38).

It should be noted that physiological *information*, carried by the action potentials, *proceeds unidirectionally* from the presynaptic membrane of the axon terminal, over the synaptic cleft, up to the postsynaptic membrane, in which graded postsynaptic potentials are induced. The synapses serve not only for information gating but also for physical *amplification* (Kandel et al. 2000). That is, a *small* presynaptic

---

[75] A *prolonged synaptic firing* may *diminish the amount of transmitter released* (per action potential) because local transmitter reserves in the axon terminal are limited and a new transmitter supply needs a certain time to replenish. On the other hand, a temporally *increased frequency* of the synaptic firing *accumulates* (entered) Ca^{2+} within the axon terminal and thus increases the amount of released transmitter in the *short term* (Footnote 67).

terminal can depolarize a *large* postsynaptic cell over the synaptic amplification. This is because the *chemical transmission* involved is an *active process* in which many thousands of postsynaptic transmitter-gated channels are opened in response to presynaptic depolarization.

As discussed with respect to the resting membrane potential (Sect. 2.1.3.1) and action membrane potential (Sect. 2.1.3.2), *intracellular concentrations* of ions involved, as a rule, do not change over time. This is because of a relatively low number of ions involved (transversing the membrane) and active pumps (transporting ions back to their site of origin). However, the amount of *Ca^{2+} ions* is a *notable exception* to the above rule if *synaptic propagation* is considered. Actually, the intracellular concentration of Ca^{2+} ions is very low (Table 2.1). The concentration of Ca^{2+} in the *presynaptic terminal* can significantly and temporarily increase as a result of Ca^{2+} inflow through the *voltage-gated Ca^{2+} channels* during depolarization[76] of the presynaptic membrane.

Interestingly, *neuronal synapses* may breakdown and reform within hours, even in the mature central nervous system (Fox 2011), a phenomenon known as *synaptic plasticity*. For instance, a repeated activation of a particular synapse supports its growth and ease of the electrochemical transmission within the synapse. This plasticity may play a crucial role in the ability to *learn* and *memorize* in humans. Namely, frequently used neural pathways (interconnected neurons) are established and favored, which include frequently used and easy-to-pass synapses.[77] The plasticity also seems to be a consequence of slow but *long-lasting effects* of indirectly gated transmitter-controlled channels (Footnote 36). In addition, there are retrograde chemical messengers that diffuse from the postsynaptic neuron back to the presynaptic neuron to regulate its transmitter release. Such messengers provide a *chemical feedback* and may also contribute to the long-lasting effects.

---

[76] In fact, a *transient increase in Ca^{2+}* concentration in the cell has *several effects*, besides an additional depolarization of the membrane (Ca^{2+} ions carry positive charge into the cell). Namely, the *Ca^{2+}-activated K^+ channels*, with Ca^{2+} as a *second messenger* (Footnote 26), open to increase the outward ionic current (composed of outflowing K^+ ions) and cause the cell to repolarize. In addition, there are some *voltage-gated Ca^{2+} channels* which are *cross-sensitive* themselves to the level of intracellular Ca^{2+}. These Ca^{2+} channels close (become inactivated) when Ca^{2+} ions (now excessive in the cytoplasm) bind to their intracellular receptive surface. Both effects from above oppose further Ca^{2+} inflow. In other words, the *depolarizing influx of Ca^{2+} ions* through the voltage-gated Ca^{2+} channels in the presynaptic membrane is self-limited and aids *repolarization* of the membrane (Kandel et al. 2000). This repolarization also seems to contribute to *adaptation effects*; see Sect. 2.2.2.

[77] To give a more tangible example for *synaptic memory* persisting for minutes, synaptic effectiveness can be considered in view of its *intense activity*. A high-frequency train of action impulses leads to saturation and *temporal excess of Ca^{2+} level* in the presynaptic bouton. In consequence, *more vesicles* will be emptied per action impulse because of the increased resting level of Ca^{2+} ions. Successively *larger postsynaptic potentials* will be produced per single action impulse, with this effect known as *potentiation* due to tetanic stimulation (compare Sect. 2.3.2). When the frequency of action impulses is then reduced, the postsynaptic potentials remain enhanced for several minutes because of the previously *accumulated Ca^{2+}*. This effect is known as *short-term posttetanic potentiation*.

Lastly, the *differences* between *action potentials* and *postsynaptic potentials* (excitatory or inhibitory) should be highlighted:

- Unlike action potentials with their *all-or-none behavior* (typical Δu of about 100 mV with impulse duration of about 2 ms), postsynaptic potentials are *graded* in their nature [Δu of about 0.1–10 mV with duration from 5 ms up to 20 min in neuronal synapses (Kandel et al. 2000)].
- Unlike action potentials triggered at a certain *threshold* level, postsynaptic potentials have *no threshold.*
- Unlike action potentials which propagate *actively without failure* and *without attenuation* over long distances (up to 1–2 m in the human body), postsynaptic potentials spread *passively and decrementally* only over short distances (up to 1–2 mm).
- Unlike action potentials showing *absolute and relative refractory periods*, postsynaptic potentials have *no refractory period.*
- Unlike action potentials with *no overlap*, postsynaptic potentials are capable of *summation*.

Obviously, the differences above are tightly interrelated with the differences between the *voltage-gated channels* (e.g., for Na^+ ions) involved in the genesis of *action potentials* and *transmitter-gated channels* (e.g., acetylcholine-gated channels) involved in the genesis of *postsynaptic potentials*. It should be stressed that an action potential can only be triggered by voltage-gated Na^+ channels because these channels are *regenerative*. It means that a progressing depolarization of the membrane caused by Na^+ influx opens even more voltage-gated Na^+ channels, which *accelerates this depolarization* (section "Cell Response"). In contrast, a depolarization produced by a dominant Na^+ influx through the acetylcholine-gated channels does not lead to the opening of more acetylcholine-gated channels.

2.2 Neurons and Receptors

Nerve cells, known as *neurons*, comprise the basis of the nervous system and play a crucial role in the *genesis of diverse biosignals*. In particular, the neurons serve as

- *Signaling units* of the nervous system to *relay action potentials* which carry time-sensitive information
- Cellular *receptors* for sensing diverse stimuli or
- *Functional units* to realize miscellaneous *regulative and controlling functions*

On the other hand, *receptor cells*, known as *receptors*, act as sensor transducers which sense external stimuli of different physical origin and deliver an appropriate electrical response. A receptor cell can be given by

- A specialized *sensory neuron* which expresses a *receptive field* in its membrane (primary receptor) or by

- A specialized *nonneuronal cell* (secondary receptor) which is synaptically connected to an afferent sensory neuron

Generally, the stimulus strength is translated into the *frequency* of action potentials while the stimulus *duration* is translated into the *number* of released action potentials. Nearly all receptors adapt to a constant stimulus to some extent so that contrast in time and space is predominantly detected.

2.2.1 Structure

The composition of the excitable nerve cell is similar to that of all other cells (Sect. 2.1.1); however, there are some morphological and functional peculiarities. A *typical neuron*, as illustrated in Fig. 2.20, has four principal regions in its morphological structure[78]; see Fig. 2.21a:

- Numerous short cellular extensions, known as *dendrites*, which branch out in a tree-like fashion and provide a *receptive area*. The receptive area serves as a *cellular input* for *graded inputs* from other neurons (over neuronal synapses), from a physical stimulus (direct impact), or even from a nonneuronal receptor cell (over synapses). A membrane imbalance $u - U_R$, known as the (electric) *graded potential*, is generated in this receptive area in response to any of the cellular inputs.
- The *cell body*, known as soma, serves primarily as the metabolic center of the cell.
- A single long tubular extension (from 0.1 mm up to 3 m in length), known as the *axon*, conducts *nongraded action potentials* (all-or-none events, Sect. 2.1.4.1).
- At the end of the axon, a presynaptic terminal resides which releases a chemical transmitter in a *graded* manner (though in quantal steps, Sect. 2.1.4.2). The transmitter release serves as a *cellular output* toward another neuron or a muscle cell.

The uninsulated origin of the axon serves typically as a *trigger zone* located at the axon hillock or the first node of Ranvier (Fig. 2.17b). Numerous voltage-gated Na^+ channels reside in the trigger zone where action potentials originate. Typically, the *axon branches out* and builds numerous collaterals that form communication sites (synapses) with other neurons or muscle cells. The *terminal region of an axonal collateral*, known as axon bouton, is usually thickened in comparison with the axonal diameter. These swellings of the axon terminals serve as *presynaptic terminals*. A *bundle of axons* located outside the central nervous system is known as a *nerve*.

---

[78]In fact, there are *different morphological shapes* of neurons which are typical for humans. The illustrated *sensory neuron* in Fig. 2.21a is known as a pseudo-unipolar cell, in which a *single extension* basically leaves the cell body. In contrast, both association neurons to the right in Fig. 2.21a and the *motor neuron* in Fig. 2.17b are known as multipolar cells, in which *multiple extensions* emerge from the cell body.

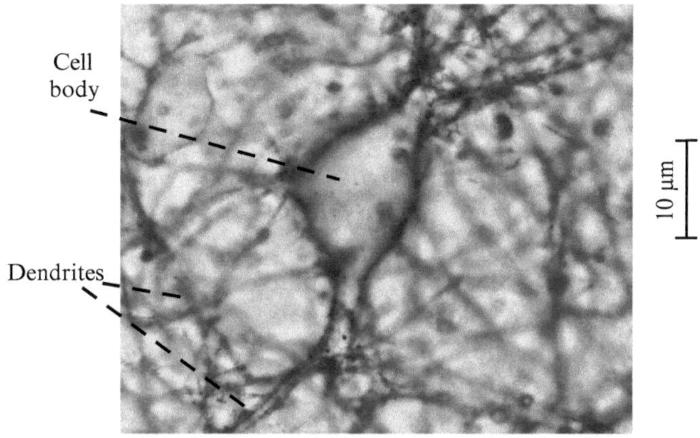

Fig. 2.20 Nerve cell (neuron) from the hippocampus, major region of the brain. A voluminous cell body can be recognized with branches called dendrites; compare Fig. 2.21a. The microscopic structure is visualized by immunoprecipitation with fluorescent quantum dots (Dodt 2010)

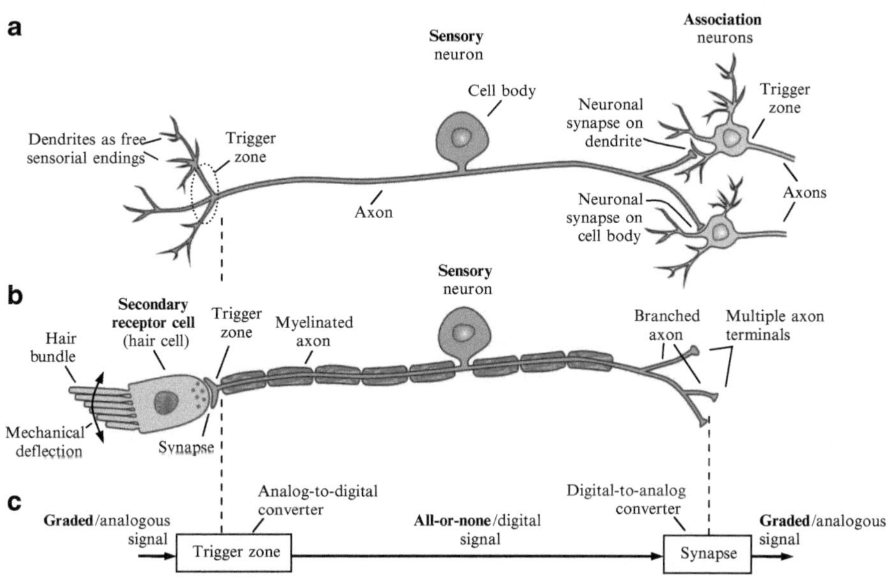

Fig. 2.21 (**a**) Sensory neuron as a primary receptor cell. Bare dendrites (inputs) act as free sensorial endings to sense temperature, mechanical tension, or chemical substances. Branches of the axon (outputs) transmit frequency modulated action potentials to downstream association neurons. (**b**) Secondary receptor cell (nonneuronal cell) synaptically connected with an afferent sensory neuron. The specific receptor cell facilitates high sensitivity and specificity to an ambient physical stimulus. In the case shown, the hair cell allows for registration of miniature deflections of a hair bundle in atomic dimensions. (**c**) Signal conversions in the sensory neuron from an engineering point of view. A graded input signal (receptor potential or postsynaptic potential) is encoded into all-or-none signals at the trigger zone (series of action potentials), whereas the all-or-none signals are re-encoded into a graded output signal in the synapse (amount of transmitter released)

From a structural and functional point of view, three *types of neurons* can be distinguished:

- *Sensory neurons*, i.e., afferent neurons, include (or are interconnected with) sensory receptors located in the body or body's periphery. The sensory neurons convey sensory information toward the central nervous system. Namely, *sensory inputs* such as light, sound, pressure, chemicals, or heat activate the corresponding *receptors* on the cellular level, e.g., activate gated channels for ions in the membrane of the dendrites (Footnote 14). The resulting graded output of a receptor is converted into all-or-none action potentials which then propagate along the axon toward the *neuronal synapses* (as communication units with other neurons). Arriving in the central nervous system, an appropriate response is provoked, e.g., as a conscious perception or an involuntary reflex action. A typical morphological structure of a sensory neuron is shown in Fig. 2.21a; compare Footnote 78.
- *Association neurons* (interneurons) interconnect other neurons via synapses and ensure functional integrity of the central nervous system; compare Fig. 2.21a. These neurons comprise by far the largest class of neurons.
- *Motor neurons*, i.e., efferent neurons, conduct action potentials from the central nervous system toward *effector organs* as muscle cells or glands (compare Fig. 2.24). Namely, series of action potentials are converted into a graded release of a neurotransmitter (acetylcholine) in the neuromuscular synapses that triggers muscular contraction. A typical morphological structure of a motor neuron is illustrated in Fig. 2.17b.

Neurons build specific (but not random) synaptic connections to other neurons, forming densely interconnected *networks* for functional processing, e.g., of incoming sensory information from the sensory neurons. The particular *function* of an embedded neuron in the network is greatly determined by its *anatomical relationships* to other neurons. On the *input side*, the neuron receives inputs either *from sensory receptors* (residing in the same neuron or in a receptor cell synaptically connected to this neuron, see below) or *from other neurons* (up to *many thousands*) via neuronal synapses located on dendrites and the cell body[79] (Fig. 2.21a). On the *output side*, the axon branches out and builds collaterals (up to *thousand*) with the respective synapses in their terminal regions. These collaterals output a train of action potentials from the relevant neuron to numerous other neurons.[80] In fact, the neuron is confronted with the *task of decision-making* based on prevailing *inputs*, to

[79]When numerous neurons have synapses with a single neuron downstream, it is known as the *convergence* of neural pathways. The convergence is typical for the motor neurons in the *output stage* of the nervous system because the motor neurons receive input from different neurons and integrate this input (Fig. 2.23a).

[80]When one neuron has synapses with numerous other neurons downstream, it is known as the *divergence* of neural pathways. The divergence is typical for the sensory neurons in the *input stage* of the nervous system because the sensory neurons distribute sensory information to many target neurons (Fig. 2.21a, b).

fire or not to fire action potentials on the *output* side—the very task of the nervous system.

As shown in Fig. 2.21, the *sensory unit* (or sensory organ) can be conceptually divided into

- Receptive field (typically electrically nonexcitable) yielding a graded response to an external stimulus
- Trigger zone for generation of all-or-none action potentials
- Conducting region for propagation of action potentials
- Synaptic region for intercellular communication

In general, the *receptor region* acts as a transducer and responds primarily to one particular type of the sensory stimulus (the type known as modality of stimulus, Footnote 88). Depending on the anatomical and functional formation of the sensory organ, *two types of receptors cells* can be distinguished:

- Primary receptors
- Secondary receptors

The *primary receptor* cell is a *specialized sensory neuron* which exposes specific membrane regions in its *bare dendrites* (Fig. 2.21a), i.e., exposes *receptive fields* with *stimulus-gated channels* in the membrane (section "Regulatory Mechanisms"). The free *sensorial endings* respond to a particular external stimulus by a corresponding graded change in the local u (depolarization or hyperpolarization), known as *receptor potential.* The stimuli can be of thermal, mechanical, or chemical origin. The imbalance of u ($\neq U_{\mathrm{R}}$) passively spreads toward the trigger zone of the sensory neuron and if the threshold is reached there, all-or-none action potentials are generated. From an energetic point of view, stimulus energy is converted into electrical energy. Usually the receptive field in the dendrites is surrounded by a *specialized end organ* (nonneuronal mechanical structure) to shape the dynamic response and dynamic adaptability of the receptive field.[81]

To give a few examples, *primary receptors* include olfactory sensory neurons for the sense of *smell.* Thin cilia from dendrites of these neurons extend into the

---

[81]An obvious example is the case of mechanoreceptors in the skin for the sense of *touch.* There is a rapidly adapting *primary mechanoreceptor*, known as the *Pacinian corpuscle*. A specialized *end organ* around an *unmyelinated dendrite*—built out of concentrically arranged fluid filled lamellae of connective tissue that form a *capsule* (resembling an onion, about 0.5 mm in size)—has to be deformed in a *dynamic* way to activate stretch-gated channels in the dendrite membrane (Kandel et al. 2000). Responding to *pressure application* on the mechanoreceptor, the opening of the stretch-gated channels increases the conductances of Na^{+} and K^{+} ions. The *receptive field* of the membrane becomes depolarized, i.e., a *receptor potential* arises. In contrast to the application of a dynamic pressure, during *steady* pressure the applied mechanical load is absorbed by the end organ, i.e., by the outer lamellae of the capsule. The static pressure is prevented from being transmitted to the inner core of the capsule and thus to the stretch-gated channels, which actually establishes a dynamically *adaptive structure*. When the *pressure is removed*, the capsule resumes its original shape and the inner core is mechanically stimulated again yielding another depolarizing receptor potential; compare Fig. 2.22d.

nasal cavity (sensing site), with the odorant receptive field[82] residing in the cilia. The axons of the sensory neurons project to the olfactory bulb of the brain (output site). Besides the Pacinian corpuscle from Footnote 81, bare nerve endings in the skin serve as primary receptors for the sense of *touch*, *pain*, and *thermal sensation*.

A *secondary receptor* cell is a separate *nonneuronal receptor cell* which is synaptically connected to an afferent sensory neuron (Fig. 2.21b). The receptor cell is usually highly specialized and converts the sensory stimulus (e.g., of acoustical, optical, or gustatory origin) into a graded change in its membrane u, i.e., into its *receptor potential*. The receptor potential then activates synaptical transmission. Consequently, the induced postsynaptic potential generates a few action potentials at the trigger zone of the sensory neuron downstream. The action potentials can be interpreted as an *indirect response* to the sensory stimulus which acts directly on the nonneuronal receptor cell.

To give a few examples, *secondary receptors* include hair cells connected to afferent sensory neurons via synapses for the sense of *hearing*, as illustrated in Fig. 2.21b. The cell's hair bundle is deflected by acoustical (mechanical) pressure waves, which mechanically open sensitive ion channels in the bundle.[83] Consequently, the receptor potential of the hair cell changes to impact the synaptic transmission. Another example involves taste cells clustered in taste buds on the tongue for the sense of *taste*. The taste cells expose miniature cell extensions (microvilli) to the oral cavity, in which sensory transduction takes place. Then the sensory information is synaptically transmitted to the gustatory sensory neuron.

Typically, the secondary receptor cells are more *specific* in their sensing capabilities and their *activation energy* is much lower in comparison with the primary sensing cells. The physical stimulus acting on a secondary receptor cell is only a *triggering event*. The stimulus energy is not required to change the membrane's local u to such a large extent that an action potential could be distally released. In addition, the nonneuronal receptor cells may build numerous *stages*, i.e., signaling cascades in series where chain reactions take place. For instance, a *single first messenger* generates *numerous secondary messengers* (compare Footnote 36) the aggregate of which is able to govern conductance of many ion channels and to provoke a reasonable response of u. Such numerous stages strongly increase the *total amplification* of the original physical stimulus under the active consumption of

---

[82]The *odorant molecules* bind to receptor proteins on the cell membrane to activate G-proteins linked with the receptor proteins (Footnote 36). The activated G-proteins release *second messengers* (such as cyclic adenosine monophosphate) which then mediate the opening of ion channels (at remote sites) and membrane depolarization.

[83]The hair cell contains a *hair bundle* (Fig. 2.21b), in which elastic structures reside, known as *gating springs*, that are involved in gating the channels for K^+ ions (Kandel et al. 2000). The springs are connected with molecular gates in the channels. A *rapid deflection* of the hair bundle (in response to an acoustical/mechanical wave) increases the tension of the springs that directly opens the channels with which the springs are connected. During a *prolonged deflection*, the tension of the springs decreases and the channels become closed. Consequently, the channel's response to a continuous stimulation is diminished which actually comprises *adaptation effects* of the hair cell.

chemical energy (e.g., from ATP, see Footnote 19). The amplification can be quite large so that mechanical movements of mechanoreceptor's extensions as small as atomic dimensions or even single light photons[84] hitting special photoreceptors (both stimuli of ultra *low energy*) yield action potentials (of relatively *high energy*) as the response at the output of an appropriate sensory organ.

It is obvious that the discussed *morphology* of the neurons and receptors is tightly related to their particular *function* and determines the cell response. Proportions of the *different channel proteins* at different cell regions play a decisive role here. In particular, the distribution of the various *voltage-gated channels* is important. That is, the *dendrites* of the neurons have some voltage-gated Ca^{2+}, K^{+}, and Na^{+} channels to modify the passive conduction of postsynaptic potentials.[85] In some neurons the latter channels can even conduct action potentials from the trigger zone back into the dendrites, thereby actively influencing synaptic sites in the dendrites (Kandel et al. 2000). The *trigger zone* has an exceptionally high density of the voltage-gated Na^{+} channels yielding a low threshold for the generation of action potentials. The *axon* contains the voltage-gated K^{+} and Na^{+} channels for the active conduction of action potentials. The *presynaptic* terminal at the end of the axon has a high density of voltage-gated Ca^{2+} channels to trigger transmitter release. The *transmitter-gated channels* in the neurons are mainly found in the *dendrites and cell body* where synapses adhere.

[84] *Interconnected stages* for the amplification and processing of a sensory stimulus exist, for instance, in the retina of the human eye. There are *photoreceptors* (as secondary receptor cells), namely, cones for colored day vision and rods for highly sensitive night vision. These *photoreceptors* are synaptically connected with *ganglion cells* over intermediate *bipolar cells*. These ganglion cells, in turn, induce action potentials and project this visual information to the brain along the optic nerve. When the light photons hit the photoreceptors, the concentration of specific *second messengers* in the cytoplasm decreases (compare Footnote 82) and the *receptor potentials* arise which causes changes in the transmitter release toward the bipolar cells. In turn, the bipolar cells generate other *receptor potentials* impacting the transmitter release onto the ganglion cells. Finally, the induced postsynaptic potentials in the *ganglion cells* initiate *action potentials*. In other words, at least two receptor potentials are in between the optical stimulus and the action potential as a neuronal response, providing a *huge amplification* of the optical stimulus.

[85] There are indications that *dendrites* of most neurons contain some *voltage-gated* Na^{+}, K^{+}, *and* Ca^{2+} *channels* for a *local amplification* of weak excitatory postsynaptic potentials (Kandel et al. 2000). The region with the voltage-gated channels from above (*local trigger zone*) sums the local excitatory and inhibitory postsynaptic potentials and, if the net depolarization is above the local threshold level, an action potential is generated in the local trigger zone. The resulting depolarization front from this (intermediate) action potential *passively spreads* from this local trigger zone toward the axon hillock (*global trigger zone*). It should be noted that a regeneration of this intermediate action potential along dendrites (like in the unmyelinated axons, Sect. 2.1.4.1) is not possible due to a relatively *low density* of *voltage-gated channels in dendrites*.

2.2.2 Function

The strategic function of neurons and receptors is demonstrated in Fig. 2.17b. In terms of the *knee-jerk reflex* loop,[86] the *sensorial ending* of a sensory neuron (primary receptor) wraps around the specialized muscle cells[87] located within the fleshy part of the *skeletal muscle*. These muscle cells with sensorial endings, packed within a connective tissue sheath, comprise the so-called *muscle spindle*, i.e., a sensory organ to record muscular length and rate of change in its length.

The stimulus in the reflex loop is provided by a *mechanical stimulation* of the sensorial ending. Namely, the ending is stretched by tapping the kneecap with a hammer, pulling the tendon of the (quadriceps) muscle, and *stretching this muscle* with an embedded muscle spindle. This stimulation activates stretch-sensitive *receptive fields* in the sensorial ending; i.e., the stretch-gated channels in the membrane are opened. In consequence, a net influx of positive ions results into the sensorial ending (section "Regulatory Mechanisms"). As shown in Fig. 2.17a, the local u drives toward more positive values, yielding a *local depolarization* of the membrane (or an *imbalance of* u related to the resting state with $u = U_R$). The arising local difference $u - U_R$, i.e., the *graded receptor potential*, is proportional to the *intensity of the stretch*. In fact, the stronger and longer is the mechanical stretch of the receptive field, the *larger and longer* is the resulting *receptor potential* in the sensorial ending; compare Fig. 2.22b, c.

The *local receptor potential* induces axial currents inside and outside the sensorial ending flowing toward and from the resting membrane regions, respectively; compare Fig. 2.7 and currents in the unmyelinated region in Fig. 2.18a. That is, the imbalance $u - U_R$ *spreads passively* along the sensorial ending and *attenuates* with increasing distance from the receptive field (site of origin). Actually, the imbalance cannot be conveyed much further than 1–2 mm (*short-range* conduction). Arriving at the *trigger zone* of the sensory neuron, namely, at the *first node of Ranvier* (Fig. 2.17b), action potentials can now be generated here provided that the

[86]The *reflex loop* comprises the anatomical route which typically connects

- The *receptive field* or receptor cell responding to a particular stimulus (e.g., bare nerve endings in the finger tip as thermal receptors perceiving heat or cold).
- The *sensory neuron* transmitting afferent impulses to the *central nervous system* (spinal cord or brain).
- The *processing units* in the central nervous system.
- The *motor neuron* transmitting efferent impulses from the central nervous system toward periphery.
- The *effector* organ (hand or arm muscles) which responds with a specific motor response (recoil of hand when fingers are burned).

[87]These specialized and relatively thin muscle cells, known as *intrafusal muscle cells*, are only partly contractile. Like the ordinary skeletal muscle cells (Sect. 2.3.1), the intrafusal muscle cells insert into tendons on each end of the skeletal muscle. Therefore, stretching a muscle causes the intrafusal muscle cells to stretch which mechanically stimulates the sensorial endings surrounding these intrafusal muscle cells.

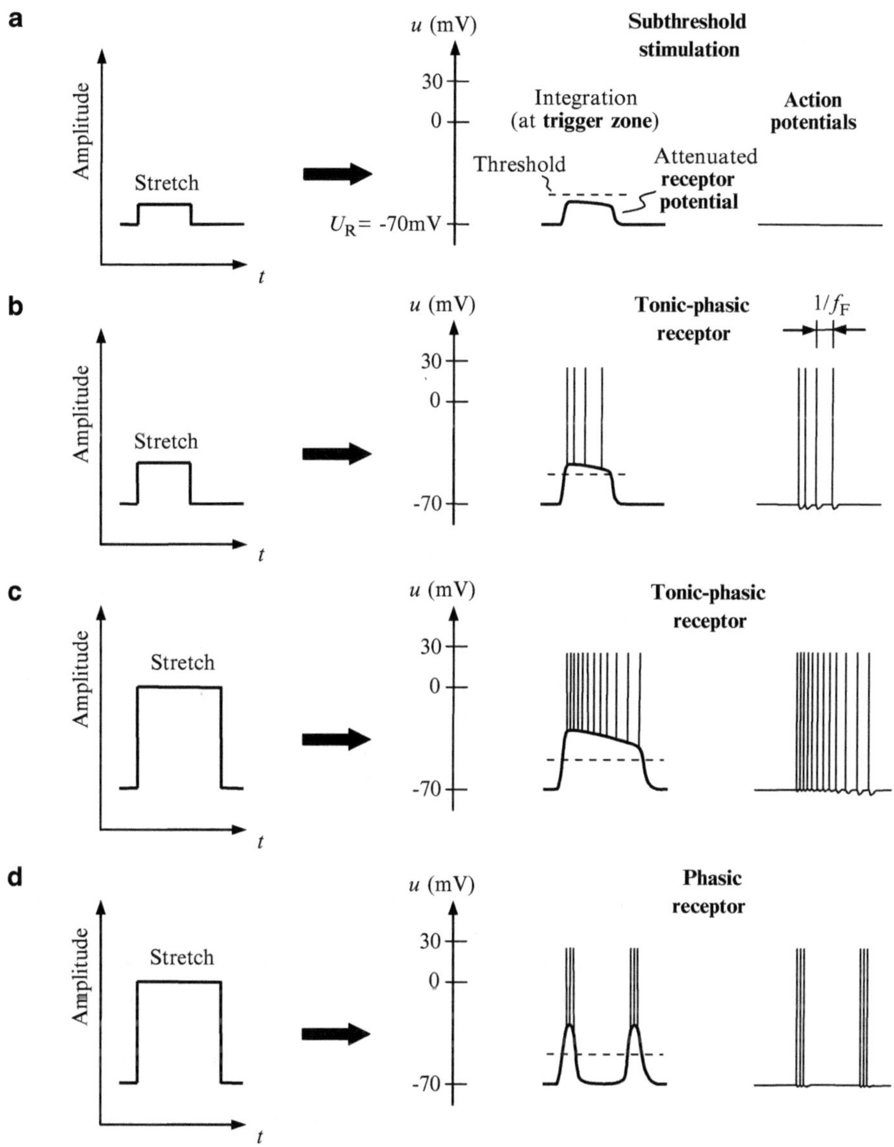

Fig. 2.22 Responses of the membrane voltage u (*right subfigures*) to different stimuli (*left subfigures*) considering different types of primary receptors (sensory neurons with an embedded receptive field). (**a**) Subthreshold stimulus yields no action potentials at the trigger zone because the corresponding deviation of u from its resting level U_R (attenuated receptor potential) does not reach the threshold level (compare Fig. 2.17a). (**b**) The tonic-phasic receptor under weak stimulation outputs a few action potentials with a monotonically decreasing firing rate f_F, i.e., decreasing instantaneous frequency of the action potentials. A slow adaptation of the receptor to a constant stimulus lowers f_F. (**c**) Strong stimulation of the tonic-phasic receptor yields a high f_F decreasing over time. (**d**) The phasic receptor rapidly adapts to a constant stimulus and thus its response is limited to periods, during which the stimulus intensity changes

induced local depolarization (graded difference $u - U_R$) exceeds the *local membrane threshold*; compare Fig. 2.22a, b.

Then the action potentials *actively propagate* without any attenuation or waveform change along the axon of the sensory neuron (*long-range* conduction). The action potentials reach the axon's terminal region, a *nerve–nerve synapse* (located in the spinal cord in the case of the knee-jerk reflex), in which a chemical neurotransmitter is released into the synaptic cleft as the information carrier. The transmitter molecules interact with (e.g., open) the transmitter-gated channels in the postsynaptic membrane of the downstream *motor neuron*.

Again, the local postsynaptical u becomes imbalanced; i.e., a *graded postsynaptic potential* arises which spreads passively and does not reach beyond the *trigger zone* of the *motor neuron*, namely, beyond the *axon hillock* (Fig. 2.17b). Here, another action potential can be triggered if the local membrane threshold is exceeded. Afterward, the action potential propagates actively along the axon toward the *nerve–muscle synapse*. Likewise, a *graded postsynaptic potential* is induced in the membrane of the muscle cell. The latter potential spreads passively along this membrane and, in turn, generates action potentials in the muscle cell. Lastly, the action potentials cause the muscle cell to contract (Sect. 2.3.2).

Thus the extensor (quadriceps) *muscle actively contracts* in response to its *passive stretch* induced by tapping the kneecap. In addition, motor neurons innervating the opposing flexor (hamstring) muscle are synaptically inhibited (Footnote 96) via the following neuronal pathway: sensory neurons, inhibitory interneurons, and motor neurons. The inhibition prevents a simultaneous contraction of the latter *antagonistic muscles* and increases the *stability* of the knee-jerk reflex.

Beginning with the mechanical stretch stimulus, as illustrated in Fig. 2.17a, the *graded input stimulus* first affects

- *The receptive field* of the sensory neuron and becomes encoded by
- *The graded receptor potential* which spreads in a *lossy* way over a *s hort-range* toward
- *The trigger zone*, in which graded changes of the receptor potential are transformed into a digital train of action potentials (analog-to-digital converter, Fig. 2.21c).
- Afterward, *the all-or-none action potentials* regenerate themselves along the axon, spread in a *lossless* way over a *long-range*, and arrive at
- *The synapses*, the secretory components, in which the digital train is transformed again into a graded signal, i.e., into a *graded release of neurotransmitter* (digital-to-analog converter, Fig. 2.21c)

It should be stressed that all electric signals from above are *encoded by* $u - U_R$, i.e., the imbalance of the actual membrane voltage u from its resting level U_R. Such electric signals in neurons usually *propagate only in one direction* (Sect. 2.1.4.1).

Despite the *diversity of human sensations*,[88] all sensory systems convey four elementary types of information: *strength*, *timing*, *type*, and *location* of the stimulus, as described below. To begin with, it is striking to see how a *graded sensory stimulus* (of various physical origins) is ubiquitously transformed into a sequence of *stereotyped action potentials* (of electrical origin), i.e., into a digital pulse code. This transformation codes physiological information conveyed by the *stimulus intensity* and its *time course*, irrespective of the receptor type and stimulus type involved.

In fact, two features of the sequence of action potentials communicate *physiological information*[89] hidden in the sensory stimulus toward the central nervous system:

- *Instantaneous frequency* of action potentials (frequency coding).
- *Number* of action potentials.

A physical *graded stimulus* generates a *receptor potential* (depolarizing or hyperpolarizing) proportionally *graded* in its *amplitude and duration*. As shown in Fig. 2.22b, c, a rectangular stimulus yields a nearly rectangular receptor potential in a first approximation (discussed later). Thereby the sensory channels in the receptive field are opened (or closed) in response to the stimulus, whereas the channels are (typically) electrically nonexcitable and less specific to the ion type.[90] The receptor potential has properties *similar to* those of the excitatory *postsynaptic potential*, as summarized in Sect. 2.1.4.2; the *typical amplitude* of the receptor potential is about 0.1–10 mV with a *duration* in the range of 5–100 ms (Kandel et al. 2000).

The receptor potential arrives at the *trigger zone* with a *reduced amplitude* because of its lossy passive spread (within only 1–2 mm), as shown in Fig. 2.17a. Then if the local level of u reaches the *threshold*, simply speaking, an *action potential* is generated. However, the particular generation and timing of action

---

[88]Major *sensory modalities* can be recognized: seeing (Footnote 84), hearing, tasting, smelling, touching (Footnote 81), balancing, and senses of pain and temperature; see Sect. 2.2.1. Interestingly, only four basic *classes of receptors* serve the needs of the above modalities and respond to (primarily) only one form of physical energy (Kandel et al. 2000):

- Mechanical receptors (e.g., for hearing, touching, and balancing)
- Thermal receptors (e.g., sense of temperature, regulation of the core body temperature)
- Photoreceptors (e.g., seeing)
- Chemical receptors (e.g., tasting, smelling, sense of pain)

[89]As *vividly summarized* by a British electrophysiologist Edgar Douglas Adrian (1889–1977): "... all impulses are very much alike, whether the message is destined to arouse the sensation of light, of touch, or of pain; if they are crowded together the sensation is intense, if they are separated by long intervals the sensation is correspondingly feeble...."

[90]It seems that *ions* furthest from their electrochemical equilibrium (Sect. 2.1.3.1) and greatest in their concentration (Table 2.1) contribute to the genesis of the *receptor potential*, as soon as *less specific gated channels* open. Typically, K^+ and Na^+ ions contribute to the ion fluxes through the membrane while the influx of Na^+ ions dominates and evokes depolarization of the membrane (section "Regulatory Mechanisms").

potentials depend not only on the amplitude of u (waveform of the receptor potential) at the trigger zone but also on the *actual threshold* level.

Interestingly enough, the actual threshold level depends on the preceding firing. Immediately after release of an action potential, there is an *absolute refractory period* (Fig. 2.14a) lasting for about 1 ms, during which another (subsequent) action potential cannot be generated (section "Cell Response"). Consequently, there is *no overlap* of action potentials in the time domain and a *minimum time interval* exists between successive action potentials. The absolute *upper limit* for the instantaneous *frequency* of action impulses—or maximum firing frequency (Fig. 2.22b)—results to less than 1 kHz ($= 1/1$ ms).

If the attenuated receptor potential at the trigger zone is only slightly larger than the (resting) threshold level—provided a relatively *weak physical stimulus* is present—the *following action potential* is generated only after the *relative refractory period.* In contrast, a *strong stimulus* yields a large amplitude of the receptor potential, which helps to overcome the *elevated threshold* during the relative refractory period. That is, the large amplitude of the receptor potential fires the *following action potential* earlier in comparison with the small amplitude given the weak stimulus. Thus *weak physical stimuli* tend to yield weak receptor potentials and a relatively *low firing frequency*; in contrast, *strong stimuli* generate strong receptor potentials and a *high firing frequency.*[91] Typically, there is a *linear relationship* between the amplitude of the receptor potential and the firing frequency. In other words, strong stimuli generate a greater number[92] of action potentials in a given time frame (Fig. 2.22b, c).

Likewise, the *duration* of a physical stimulus determines the duration of the corresponding graded potential and, provided that the receptor potential exceeds the threshold, determines

- The total *number of action potentials* being generated
- The *period* over which action potentials are generated
- The total *amount of transmitter* released in the synapses

[91]In addition to the discussed *frequency coding* of stimuli strength, stronger stimuli activate a greater number of receptors (sensory neurons), referred to as *population coding* (or recruitment). This is because the individual receptors differ in their sensory thresholds. For weak stimuli only *low threshold receptors* are recruited while for strong stimuli all *low and high threshold receptors* are recruited.

[92]Obviously, a certain strength S of the stimulus is needed to generate action potentials and to convey a sensory message to the brain. Elevation of the *sensory threshold* S_T typically signals malfunction in sensory receptors, such as loss of hair cells in the ear.
The minimal *difference* ΔS *in the stimulus strength* that can be discriminated depends on the absolute stimulus strength, that is $\Delta S \propto S$ (Kandel et al. 2000). Likewise, the stronger the stimuli, the larger should be the difference in their magnitudes in order to be perceived as separate stimuli. On the other hand, the *intensity I of the sensation* experienced by a subject is usually a logarithmic function of S:

$$I \propto \log(S/S_T),$$

whereas a linear relationship exists between I and S for some stimulus types.

In summary, the four elementary types of information about a physical stimulus are coded as follows:

- The stimulus *strength* (and temporal changes of the strength) is coded by the *frequency of action potentials.*
- The stimulus *duration* is coded by the *number of action potentials.*
- The stimulus *type* (e.g., mechanical, chemical, or optical) is coded by *distinct receptors* responding to only a particular stimulus type and by distinct *neural pathways* that carry the relevant action potentials towards the central nervous system.
- The stimulus *location* (and its spatial dimensions), provided by topographically distributed receptors, is also coded by *distinct receptors* and distinct *neural pathways.*

It should be stressed that informative characteristics of the stimulus can not be coded by the amplitude or waveform of a single action potential itself (Fig. 2.14a) because the *action potential is an all-or-none* event (section "Cell Response"). Likewise, the *pattern of action potentials* does not code the stimulus type and location. For instance, tactile information from fingertip mechanoreceptors to the brain takes a different sensory pathway than pain information from fingertip thermosensitive receptors.

An interested reader will have noted in Fig. 2.17a (and Fig. 2.22c) that the *receptor potential weakens slightly* and the instantaneous frequency of action potentials decreases in response to a *prolonged and constant* excitatory *stimulus* (above the threshold). That is, the receptor *adapts* (or *accommodates*) to this persisting *continuous stimulus* and the sensory neuron reduces its firing rate. The adaptation is equivalent with a rise in the excitation threshold. Obviously, the firing stops when the stimulus ends.

Generally, the *adaptation effects* of the receptor cells and sensory neurons are due to

- *Electrochemical mechanisms* at the ion channel level, which take place in the membrane of the receptors and neurons.
- *Mechanical* structure of the receptor cells.

In the case of *electrochemical mechanisms,voltage-gated K^+ channels* are slowly activated in response to a persisting depolarization. The induced outflow of K^+ ions shifts the membrane's u to more negative levels, offsets the latter depolarization, and thus slows down the train of action potentials or even prevents further action potentials.[93] Another contributing mechanism is the inactivation of *voltage-gated Na^+ channels* by the depolarizing receptor potential, which excludes long-lasting

[93]For instance, if an initial *depolarization* of the membrane is *just above the threshold*, the sensory neuron fires only a few action potentials. However, a delayed activation of the voltage-gated K^+ channels by the depolarizing receptor potential induces a hyperpolarizing outward current (across the membrane) which now offsets the depolarizing inward current (compare section "Cell Response"). The membrane repolarizes toward U_R and the sensory neuron *stops firing*.

and constant receptor potentials. In addition, numerous other electrochemical mechanisms contribute to the adaptation effects that are related to Ca^{2+} and K^{+} channels in the presynaptic terminal and the transmitter-gated channels in the postsynaptic terminal.[94]

The *mechanical structure* of the receptor cells usually has a strong impact not only on the sensitivity and specificity of the receptor but also on its adaptation capabilities. Usually the mechanical structure around the receptive field *filters out the steady component* of the stimulus. A static stimulus (typically of large amplitude) deforms the receptor without affecting the receptive field. In contrast, a transient stimulus (typically of small amplitude) is transmitted directly to the receptive field. Prominent examples are the Pacinian corpuscle as a mechanoreceptor in the skin (Footnote 81) and the hair cell as an acoustical receptor (Footnote 83).

Nearly all *receptors* or, more generally, sensory organs *adapt to a constant stimulation*. The receptors monitor only the rate (or velocity) at which the strength of the stimulus changes by rapidly changing their firing rate. Sensorial contrasts are detected in time and space; i.e., the receptors signal *time and space derivatives* of the stimuli; e.g., the Pacinian corpuscle monitors the speed of skin indentation (Footnote 81). The beginning and end of a steep stimulus typically yield dominant receptor responses about a *changing sensory environment* (Fig. 2.22d). The receptors can be subdivided into *three types*, according to their *adaptation speed* to a constant excitatory stimulus:

- *Phasic receptor*, the *rapidly adapting* receptor which is silent when a constant stimulus occurs. That is, if the stimulation persists its perception gradually fades from consciousness. The receptor fires only when the stimulus intensity increases or decreases, as illustrated in Fig. 2.22d. The corresponding receptor potentials last only for a short time, independent of the stimulus duration. Typical examples are olfactory sensory neurons which detect only the change in scents and mechanoreceptors in the skin which respond to the establishment of skin contact (or to skin vibrations) rather than to a continuous skin contact (Footnote 81).
- *Tonic-phasic receptor*, the *slowly adapting* receptor which yields a train of decelerating action potentials at a constant stimulus (Fig. 2.22b, c). The corresponding receptor potentials weaken slightly during the stimulus. An example is given by the discussed muscle spindle to record the actual muscle length (Fig. 2.17b).
- *Tonic receptor*, the *very slowly adapting* receptor which yields a continuous train of action potentials with a constant instantaneous frequency at a constant stimulus. This receptor gives information on the stimulus level throughout

[94]Besides the discussed role of voltage-gated channels, *adaptation effects* seem to be related to the opening of *Ca^{2+}-activated K^{+} channels* and the closure (or inactivation) of *cross-sensitive voltage-gated Ca^{2+} channels* by an increased intracellular Ca^{2+} concentration in the presynaptic terminal, as described in Footnote 76. In addition, the *refractory state* of transmitter-gated channels supports adaptation effects. That is, receptor sites in the postsynaptic membrane (e.g., acetylcholine receptors) are progressively inactivated given a continuous presence of the transmitter (acetylcholine); compare Footnote 38 (Kandel et al. 2000).

sustained and prolonged stimulation. For instance, baroreceptors in certain arterial vessels continuously monitor the blood pressure and its changes (Sect. 3.2.2.1 and Footnote 247).

Figure 2.22 illustrates the *adaptive behavior* of both tonic-phasic and phasic receptors. Obviously, the basic prerequisite for neuronal response is that the (attenuated) receptor potential at the trigger zone exceeds the local *threshold* level, as demonstrated in Fig. 2.22a. The firing frequency of the *tonic-phasic receptor* is proportional to the actual level of the receptor potential, while the frequency monotonically decreases throughout the stimulus because of adaptation (Fig. 2.22b, c). In the case of the phasic receptor, only the stimulus change is coded by a train of action potentials (Fig. 2.22d).

The *trigger zone* of the neuron deserves an extended functional description, for it not only serves as the place of *origin of action potentials*, but it also *integrates inputs* from other neurons into a *single neuronal response*. Typically up to ten thousands of axon terminals converge via synapses to the relevant neuron, with some of the synapses being excitatory and others inhibitory.[95] The trigger zone has a high density of voltage-gated Na^+ channels that account for a relatively *low threshold* (at about −50 mV) if compared with the thresholds in other regions of the neuron; e.g., at the cell body the threshold is much higher at about −35 mV. Therefore, a passively spreading depolarization is more likely to generate an action potential at the trigger zone with the lowest threshold (or discharge first the trigger zone to the threshold level) in comparison with other regions of the neuron (Footnote 85). The trigger zone is typically located at the *axon hillock* in motor neurons and association neurons, whereas in sensory neurons this zone is located at the *first node of Ranvier* (of myelinated axons) or just beyond the receptive field; compare Figs. 2.17b and 2.21.

The *integrative effect* of the trigger zone becomes obvious if *numerous inputs* are given, originating either from spatially extended receptive fields or from other neurons. Likewise,

- Decrementally *spreading receptor potentials* are integrated at the trigger zone in sensory neurons, whereas
- Decrementally *spreading postsynaptic potentials* (excitatory and inhibitory) are integrated at the trigger zone in motor and association neurons.

If the *sum level* of all potentials, i.e., the total depolarization at the trigger zone, is at or above the threshold level by the time the potentials reach the trigger zone, an action potential is generated. Any further increase in the sum level increases the frequency of action potentials, according to the aforementioned *frequency coding*. It should be noted that the *translation* of graded *postsynaptic potentials* into a pulse

[95] *Inhibitory synapses* (coming from inhibitory interneurons) comprise an important tool to enhance *sensorial contrasts in time and space*. For instance, inhibition allows the most active afferents to reduce the output of less active (neighboring) afferents. The inhibition facilitates an expression of *only one response* to the central nervous system out of multiple (competing) responses; compare Footnote 96.

sequence of *all-or-none action potentials* is analogous to the discussed translation of receptor potentials into action potentials.

In particular, Fig. 2.23 illustrates spatial and temporal summation of excitatory and inhibitory postsynaptic potentials within the trigger zone. In fact, *spatial locations of synapses* are critical for their effectiveness. While

- *Excitatory synapses* are usually located on dendrites.
- *Inhibitory synapses* are usually found on the cell body near the axon hillock.

As soon as *excitatory synapses* are activated, e.g., the synapses A, C, and D in Fig. 2.23a, equalizing currents are *induced* flowing from the corresponding (distant) excited regions (with $u > U_R$) toward the resting regions ($u = U_R$); compare section "Transport of Potential Difference" and Fig. 2.18a. In other words, the depolarization spreads toward the trigger zone while the intracellular currents must *pass through the cell body*. The latter *currents sum up* in dendrites and the cell body. Arriving at the trigger zone, the net current crosses the local membrane patch outward and thus depolarizes this patch. It should be noted that inward currents—at receptive fields or postsynaptic membranes—are concomitant with the outward current at the trigger zone.

In the case where an *inhibitory synapse* located close to the cell body is activated, i.e., the synapse B in Fig. 2.23a, this synapse acts as a *current shunt* (or short-circuit) for the above currents from the excitatory synapses. Consequently, the excitatory currents spreading toward the trigger zone are reduced and thus their depolarizing influence on the trigger zone is inhibited.[96] Fig. 2.23a depicts a relatively large nonshunted current density from the excitatory synapse C in comparison with a relatively low current density from the excitatory synapse A shunted by the inhibitory synapse B.

As shown in Fig. 2.23d, when the *total depolarization* at the trigger zone exceeds the membrane threshold, an action potential is generated. In the given example, the excitatory postsynaptic potential from synapse A weakly affects the response at the trigger zone because the inhibitory synapse B is active at the same time and offsets the excitatory currents leaving synapse A (Fig. 2.23a, c, d). Later after time Δt, when both synapses C and D become active, the total depolarization at the trigger zone is sufficient to reach the threshold level. In addition, Fig. 2.23d also depicts the case in which synapse D would be missing; the resulting total depolarization would be insufficient to reach the threshold and trigger an action potential.

This *integrative behavior* is the quintessential action of the neuron which *weights* the different *input* information and then responds appropriately at the *output* side (as a triggering component). For a single synapse,

---

[96]Evidence that the *integrative effects* at the trigger zone are reasonable can be illustrated within the scope of the discussed *knee-jerk reflex* (Fig. 2.17b). That is, during the contraction of the extensor muscle, a possible simultaneous contraction of the flexor muscle is inhibited. Inhibitory interneurons *activate inhibitory synapses on motor neurons* that govern the flexor. These integrative effects prevent simultaneous contractions of antagonistic muscles and increase the stability of the knee-jerk reflex.

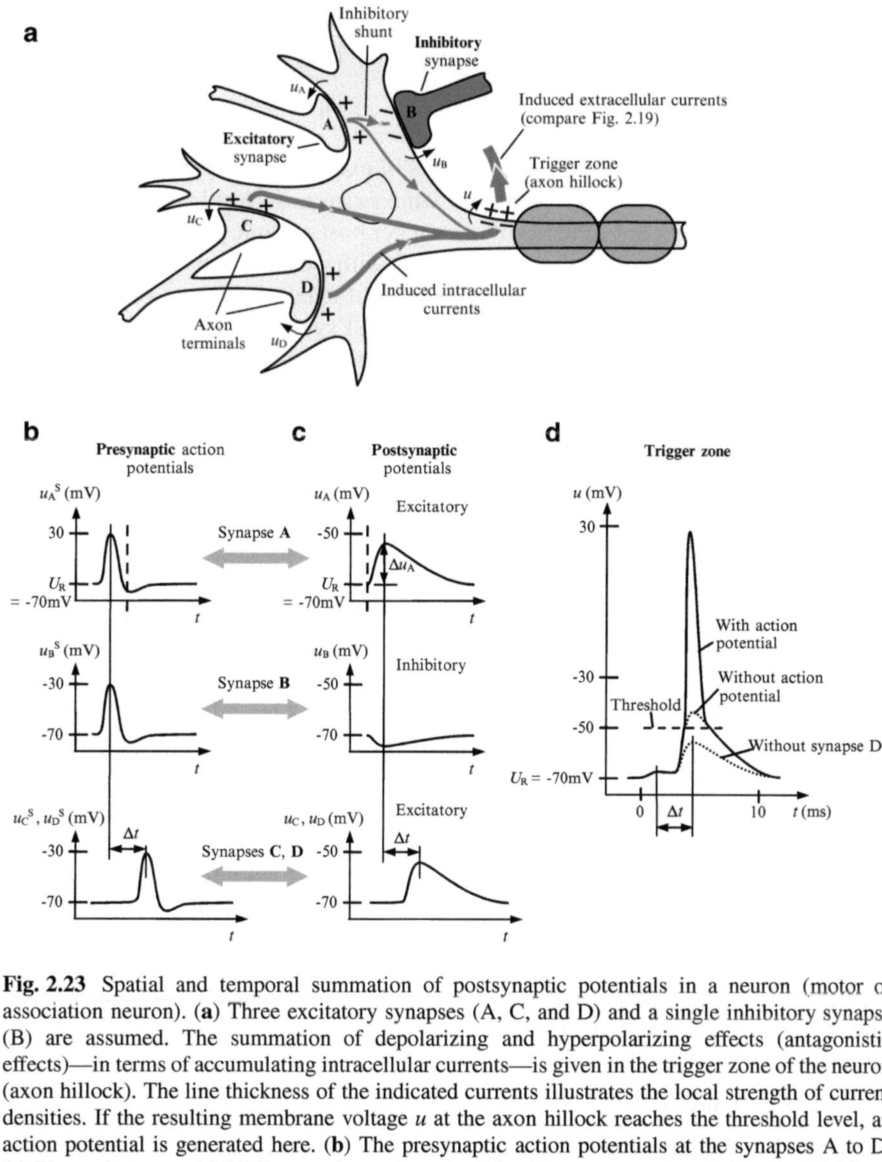

Fig. 2.23 Spatial and temporal summation of postsynaptic potentials in a neuron (motor or association neuron). (**a**) Three excitatory synapses (A, C, and D) and a single inhibitory synapse (B) are assumed. The summation of depolarizing and hyperpolarizing effects (antagonistic effects)—in terms of accumulating intracellular currents—is given in the trigger zone of the neuron (axon hillock). The line thickness of the indicated currents illustrates the local strength of current densities. If the resulting membrane voltage u at the axon hillock reaches the threshold level, an action potential is generated here. (**b**) The presynaptic action potentials at the synapses A to D. (**c**) The excitatory and inhibitory postsynaptic potentials at the synapses A to D. For the inhibitory postsynaptic potential, the target voltage is assumed to be $< -70\,\mathrm{mV}$ (Footnote 72). The time correspondence in between (**b**) and (**c**) is indicated by *dashed lines* (at the synapse A), signifying a slight delay of the postsynaptic potentials relative to the presynaptic action potentials. (**d**) The voltage u at the trigger zone as the result of a weighted integration of all postsynaptic potentials (Footnote 97). An action potential is generated here because the level of u surpasses the threshold level

- The larger the *synaptic area*
- The shorter the *distance* from the synapse to the trigger zone

the larger are the induced currents from this synapse toward the trigger zone. The strength of the induced currents proportionally determines the *weight of this synapse*. From this point of view, the weights of excitatory and inhibitory synapses are comparable with each other. While *excitatory synapses* (reside on dendrites) are farther from the trigger zone but result in stronger postsynaptic potentials, *inhibitory synapses* (reside on the cell body) are closer but with weaker postsynaptic potentials (Sect. 2.1.4.2). A single *neuronal synapse* is insufficient, by far, to drive u at the trigger zone to the threshold so that the neuronal synapse has a relatively *small weight* and a synchronous activation of numerous synapses is needed to evoke a neuronal response.[97] In general, the net effect of a single neuronal synapse depends on the synapse's type, location, size, and its proximity to the trigger zone. Some synapses are large and strong while others are small and weak. In analogy, the *neuromuscular synapse* can be described as having the *largest weight* because of its large size and its ability to postsynaptically generate an action potential without any additional input.

In fact, spatial and temporal summation is performed at the trigger zone to achieve the threshold for the generation of action potentials. As illustrated in Fig. 2.23 and Footnote 97, *spatial summation* refers to the summation of different postsynaptic potentials arriving from the different regions of the neuron. *Temporal summation* refers to the fact that *successive* action potentials in a *single axon terminal* yield successive graded postsynaptic potentials which interfere with each other. Provided that only a small delay occurs between successive action potentials (smaller than the time constant τ_A from Footnote 63), the propagating parts of the postsynaptic potentials constructively sum up at the trigger zone and thus increase the probability of neuron firing. A large τ_A favors the summation of successive postsynaptic potentials and thus facilitates the temporal summation. In analogy,

[97]Given the *spatial and temporal summation* of different postsynaptic potentials in Fig. 2.23 (summation of distant inputs), the voltage $u(t)$ at the trigger zone can be *approximated* by the linear *superposition principle*:

$$\begin{aligned} u(t) &= U_I + \left(\Delta u_A(t) \cdot e^{-x_A/\lambda} + \Delta u_B(t) \cdot e^{-x_B/\lambda} + \Delta u_C(t) \cdot e^{-x_C/\lambda} + \Delta u_D(t) \cdot e^{-x_D/\lambda}\right) \\ &\approx U_I + \Delta u_A(t) + \Delta u_B(t) + 2 \cdot \Delta u_C(t). \end{aligned}$$

Here U_I is the initial voltage at the trigger zone (for the resting state $U_I = U_R$), whereas Δu_A, Δu_B, Δu_C, and Δu_D correspond to graded changes of the postsynaptic potentials at the synapses A, B, C, and D, respectively; compare (2.14). The distance from the respective synapse to the trigger zone is denoted as x_A, x_B, x_C, and x_D; compare Footnote 29. The *length constant* λ (2.4) is in the range of 1 mm (Table 2.2), which is effectively large in comparison with the size of a neuron (about 10 μm), *limits losses* in the spread of the depolarization toward the trigger zone (i.e., $e^{-x/\lambda} \approx 1$ due to $x \ll \lambda$), and thus facilitates an effective spatial and temporal summation at the trigger zone. The above approximation neglects finite conduction velocity of the depolarization spread throughout the neuron and capacitive currents through the membrane. Both effects from above would manifest in a time delay of $u(t)$ related to $\Delta u_A(t) \ldots \Delta u_D(t)$.

synchronous action potentials in multiple axon terminals (connected synaptically to a single neuron) also yield a temporal summation of the respective postsynaptic potentials (Fig. 2.23). Given the *different adaptation speeds* of receptors from above, it can be derived that for slowly adapting receptors the temporal summation occurs for a longer time than for rapidly adapting ones if an input stimulus of the same duration is assumed in each case.

It is interesting that some *neurons* even *fire spontaneously* without any external stimuli. For instance, voltage-gated nonselective ion channels for cations are activated in these neurons by hyperpolarization of the membrane, whereas voltage-gated Ca^{2+} channels are activated by depolarization (Kandel et al. 2000). Such cells may exhibit self-generated *rhythmic firing* (Footnote 41), similar to the pacemaker cells (specialized cardiac muscle cells) in the sinoatrial node of the heart (Sect. 2.4.2).

Amplification of the stimulus in the receptor cells should be shortly addressed. For instance, a *single molecule* of a chemical stimulus (substance) may activate a receptor protein (residing outside the cell membrane) in terms of a *low energy* complementary interaction; e.g., a scent molecule may activate an odorant receptor (compare Fig. 2.5). This interaction triggers a chain reaction for the synthesis of numerous molecules which then act as *secondary transmitters* inside the cell. Consequently, the secondary transmitters begin to *gate ion channels* in the membrane (from intracellular site) or to vary the intracellular *amount of* Ca^{2+} *ions*. As a whole, an active response of the receptor cell is provoked in terms of a *high energy* response; e.g., a substantial change in $u(t)$ across the membrane is induced with a subsequent generation of an action potential. Usually amplification enzymes are involved, which activate specific stages of the chain reaction.

The *plasticity* on the synaptic level, as was introduced in Sect. 2.1.4.2, leads to important *functional* and (even) *anatomical transformations* in the network of interconnected neurons. In response to *appropriate stimuli*, the arising functional changes are usually *short term* and affect the efficiency of existing synaptic connections. In contrast, anatomical changes are *long term* where existing connections of neurons may be pruned or new connections established within the scope of learning and experience. In the words of Kandel et al. (2000), "It is this potential for plasticity of the relatively stereotyped units of the nervous system that endows each of us with our *individuality*."

2.3 Muscle

Muscle cells are excitable cells which are made up of contractile muscular tissue and serve as a source of electric and mechanic biosignals. There are three types of muscles:

- *Skeletal muscles* for *voluntary* body movements.
- *Smooth muscles* for *involuntary* control of inner organs and blood vessels.
- *Cardiac muscles* for *involuntary* pumping of blood.

Although there are structural and functional differences between these three types of muscles (as shown below), in each case the *muscle cells* involved *shorten* when they are excited. The contraction happens through the sliding of lengthy subunits within the muscle cell over each other.

2.3.1 Structure

As illustrated in Fig. 2.24, *skeletal muscles* are usually attached to bone on each end by connective tissue tendons. The muscle contraction typically rotates bones around their joints and thus produces body movements. The skeletal muscle contracts only when its *individual muscle cells are stimulated* by a motor neuron; compare Fig. 2.17b.

From a structural point of view, *skeletal muscle* is subdivided into parallel columns (resembling "strings" in stringy meat), known as *fascicles*, with a diameter of about 100–1, 000 μm each (Fig. 2.24). Each of the fascicles is surrounded by elastic fibers, blood vessels, and nerves to provide elasticity, nutrients, and control for the muscle cells, respectively (compare Footnotes 128 and 131). A single fascicle is a bundle of elongated *muscle cells*, known as *muscle fibers* or myofibers, with a diameter of about 10–100 μm and length from 1 mm up to 20 cm. Unlike most other cells in the body (Fig. 2.2a), skeletal muscle cells contain multiple nuclei.

The *muscle cells* are subdivided into even smaller subunits, known as *myofibrils*, with a diameter of about 1 μm (i.e., a few hundred myofibrils per muscle cell). The myofibrils extend in parallel rows in the muscle cell and exist in a *repeating pattern* in the longitudinal direction. The corresponding successive subunits of an individual myofibril are known as *sarcomeres*; compare Figs. 2.25a and 2.28. The sarcomere has a cylindrical shape and axial length of about 2 μm.

A myofibril consists of smaller subunits, called *myofilaments*. As shown in Fig. 2.25a, myofilaments contain relatively *thick filaments* (known as A bands) with a diameter of about 10 nm; more than 1,000 thick filaments can be found in the cross section of a single myofibril or within a single sarcomere (Silbernagl and Despopoulos 2007). The thick filaments are surrounded by partially *overlapping thin filaments* (nonoverlapping region known as I band), each 5 nm in diameter. As illustrated in Fig. 2.26, thin filaments form a hexagonal arrangement. The thick filaments are mainly composed of the protein *myosin* while the thin filaments of the protein *actin*. The succession of A and I bands forms a repeating pattern along the myofibril.[98] Numerous *cross bridges* (known as *myosin heads*) stick out from the

[98]The distinctive feature of *skeletal muscle cells* is their *striated appearance* in the longitudinal direction when viewed microscopically. The stripes are produced by alternating dark and light bands which are anatomically seen as successive dark A bands and light I bands, respectively (Fig. 2.25a). A similar striated pattern can also be observed in cardiac muscle cells, as illustrated in Fig. 2.28.

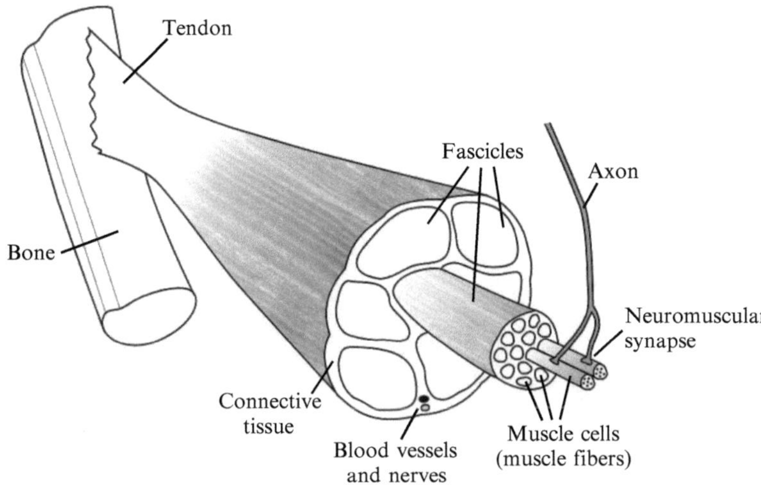

Fig. 2.24 Basic anatomy of the skeletal muscle attached to bone via tough tendons; compare Fig. 2.17b

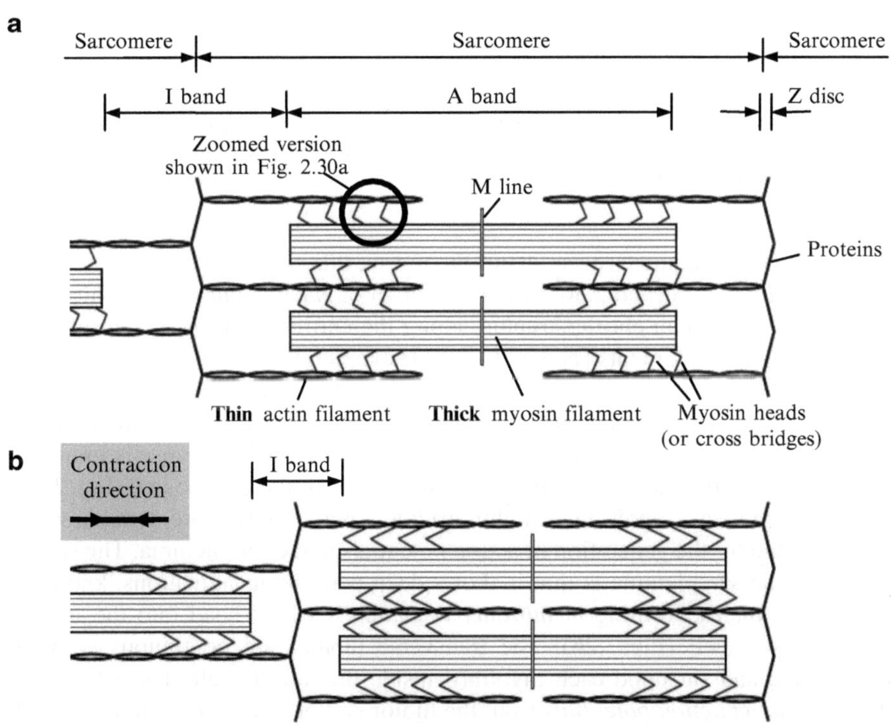

Fig. 2.25 Repeating units of thick and thin filaments in a myofibril section of a muscle cell, comprising sarcomeres. (**a**) Relaxed myofibril (in relaxed muscle cell) with a relatively small overlap in between thick and thin filaments. (**b**) Contracted myofibril (in contracted muscle cell) with a large overlap of the filaments due to advanced filament sliding

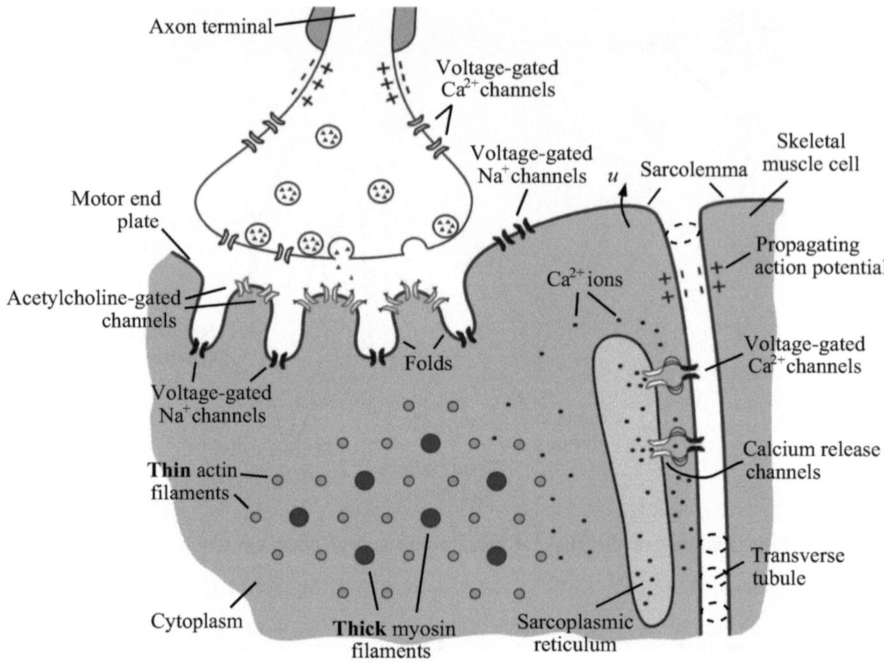

Fig. 2.26 Neuromuscular synapse which provides electromechanical coupling between the electrical excitation of the axon terminal and mechanical contraction of the skeletal muscle cell (compare Fig. 2.19 and 2.24). In the demonstrated case, contraction occurs in the normal direction to the image plane (compare Fig. 2.25)

sides of the thick filament (every 40 nm) to the six thin filaments in the overlapping region; see Figs. 2.25a and 2.30. In addition, a set of *elastic filaments* runs along the sarcomeres to form an elastic structure along the entire length of the muscle cell; for instance, these filaments account for an elastic recoil of muscles when stretched passively.

The membrane surrounding the muscle cell is called the *sarcolemma*. A specialized region of the sarcolemma at the neuromuscular synapse is known as the *motor end plate*, i.e., the postsynaptic membrane of the muscle cell (Fig. 2.26). The motor end plate represents a highly *excitable region* of the sarcolemma, which is responsible for the generation of action potentials in the sarcolemma. The special feature of the sarcolemma is that it shows deep periodic invaginations, known as *transverse tubules*, forming *membraneous channels* which head into the interior of the muscle cell (Fig. 2.26). The transverse tubules are continuous with the sarcolemma and surround each myofibril inside the muscle cell. The tubules are able to *conduct action potentials* from the motor end plate radially into the muscle cell, which actually triggers a contraction of the relevant muscle cell.

Another important feature is that *transverse tubules* are very *closely located* to a special endoplasmic reticulum of the muscle cell, known as the *sarcoplasmic reticulum*. The sarcoplasmic reticulum *stores* Ca^{2+} *ions* and consists of closed *sacs and tubes surrounding each myofibril*.

Each skeletal muscle cell is provided with a *neuromuscular synapse*, a single termination of a motor neuron on a muscle cell, usually near its midpoint (Sect. 2.1.4.2). The activation of the synapse results in the *full contraction* of this individual muscle cell. The innervating axon typically branches out at its end, as illustrated in Fig. 2.24. Therefore, a single motor neuron usually innervates numerous muscle cells, even as many as the number of collateral branches. A single motor neuron and all the muscle cells that it innervates are known as a *motor unit*.

In fact, a single motor neuron can innervate (and stimulate to contract) up to a few thousand skeletal muscle cells. The corresponding *innervation ratio* is low in *small motor units*; e.g., one neuron innervates only about 25 muscle cells in face muscles or in muscles that position the eyes. This small innervation ratio favors fine control of contractions of a particular muscle; e.g., about 2,000 motor units are in the outer eye muscles to allow for a smooth and fine control of the eye. In contrast, *large motor units* yield more powerful but less tunable contractions, in which more than 1,000 muscle cells can be innervated via a single motor neuron, as given, for instance, in the calf muscle. A single muscle is usually governed by numerous motor units (>100).

Smooth muscle consists of *smooth muscle cells*, another type of muscle cell, which lack cylindrical sarcomeres and are nonstriated (Footnote 98). As illustrated in Fig. 2.27, smooth muscle cells have a fusiform shape tapered on both ends with a length of about 0.1 mm. Smooth muscle cells may build circular arrangements (e.g., in the walls of arterial blood vessels, Sect. 2.5.1) and longitudinal arrangements (in stomach, intestine, and ureters). Within the cell, the discussed structure of

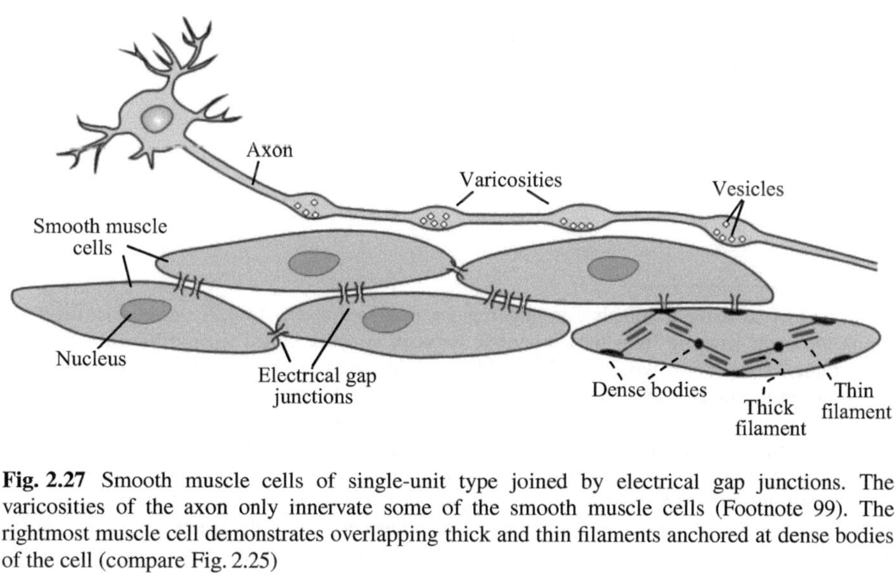

Fig. 2.27 Smooth muscle cells of single-unit type joined by electrical gap junctions. The varicosities of the axon only innervate some of the smooth muscle cells (Footnote 99). The rightmost muscle cell demonstrates overlapping thick and thin filaments anchored at dense bodies of the cell (compare Fig. 2.25)

the overlapping filaments exists (similar to Fig. 2.25a). Unlike skeletal muscle cells, the *myosin heads are densely arranged* next to each other. As shown in Fig. 2.27, the thin filaments are quite long and attach to special protein structures within the cell or to special regions of the cell membrane, known as *dense bodies*. The *overlapping filaments* are rather *loosely arranged* and predominantly oriented obliquely with respect to the longitudinal axis of the cell (Fig. 2.27). In addition, the membrane regions act as dense bodies to mechanically couple adjacent smooth muscle cells with each other.

In fact, there are *two types of smooth muscle cells*:

- Single-unit cells (e.g., found in uterus, digestive tract, and bladder)
- Multiunit cells (e.g., in arterioles, iris, ciliary body, and spermatic duct)

Adjacent *single-unit cells* are joined by *electrical gap junctions* (i.e., electrical synapses, see Footnote 67) for the propagation of membrane depolarization and accordingly of smooth muscle activation (Fig. 2.27). Some of the single-unit cells are innervated[99] by the autonomic nervous system so that a direct and individual innervation of each cell is not necessary (unlike skeletal muscle cells). In contrast, *multiunit cells* typically have no electrical gap junctions. Each multiunit cell *receives excitation* from the autonomic nervous system via "synapses in passing" (according to Footnote 99), similar to the individual excitation of skeletal muscle cells.

Cardiac muscle is built from *cardiac muscle cells*. Like skeletal muscle cells, cardiac muscle cells are striated, contain sarcomeres, and are tubular in shape with a length of up to 0.1 mm and diameter of about 30–50 μm. Figure 2.28 illustrates two cross sections of cardiac muscle cells with indicated morphological structure; compare with Fig. 2.25a. Numerous mitochondria indicate a high energy consumption of the cardiac muscle cells. However, while the skeletal muscle cells are structurally and functionally separated, the cardiac muscle cells—as illustrated in Fig. 2.29—are short, *branched*, and intimately *interconnected* with each other at the ends of each cell. Numerous mechanical and *electrical gap junctions* (Footnote 67) provide interconnections in between neighboring cardiac muscle cells.

[99] In contrast to axon terminals forming regular synapses (Sect. 2.1.4.2), *numerous swellings* along the axon (coming from the autonomic nervous system), called *varicosities*, are involved in forming "*synapses in passing*" (Fox 2011). As illustrated in Fig. 2.27, the swellings contain neurotransmitter molecules which can be released along the length of the axon toward target cells located at some distance, i.e., toward smooth muscle cells. The released acetylcholine molecules bind to muscarinic acetylcholine receptors (compare Footnote 71) which are expressed in the *entire membrane* of the smooth muscle cell; in contrast, nicotinic acetylcholine receptors of the skeletal muscle cell are only expressed in the motor end plate (Fig. 2.26). The neurotransmitter binding indirectly closes K^+ channels in the membrane and thus depolarizes the membrane of the smooth muscle cell. Typically, a *single varicosity* stimulates *numerous smooth muscle cells* because the resulting synaptic transmission is rather diffuse (Fig. 2.27).

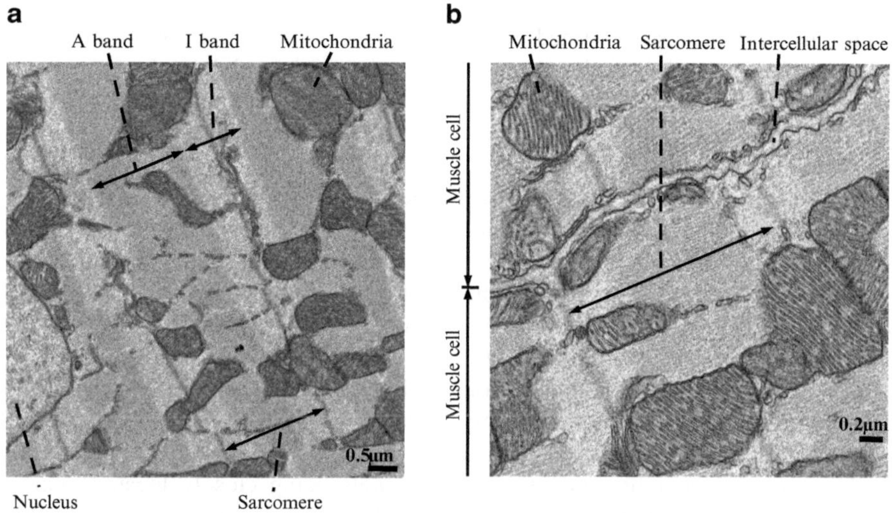

Fig. 2.28 Microscope images of cardiac muscle cells from mouse. The photographs (**a**) and (**b**) were taken by transmission electron microscopy (Wang and Sougrat 2011)

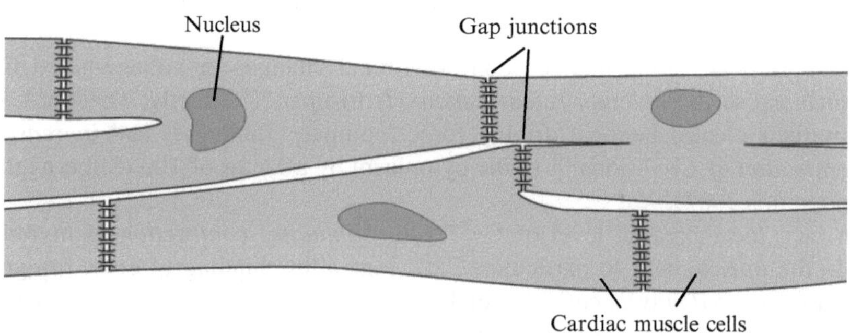

Fig. 2.29 Interconnected tubular cardiac muscle cells over mechanical and electrical gap junctions

2.3.2 *Function*

Skeletal muscle and its constituent *skeletal muscle cells* contract through the *sliding filament mechanism* with a maximum length reduction of about 40%. When stimulated, the individual myofibrils shorten as a result of *shortening of the successive sarcomeres*; compare Fig. 2.25. The width of overlapping regions in between thick and thin filaments increases and the I bands tend to become shorter, whereas the respective lengths of the thick and thin filaments remain the same. In fact, *cross bridges* in between the filaments cause mechanical *power strokes* and *sliding* of the thin filaments over and between the thick filaments.

To begin with, each and every *skeletal muscle cell is innervated* by an individual neuromuscular synapse governed by the *somatic nervous system.* A *single presynaptic action potential* yields an *all-or-none contraction* of the *relevant skeletal muscle cell*, as graded contractions are not possible (unlike smooth muscle cells, see below). In the course of the synaptic stimulation, excitatory postsynaptic potentials are generated in the motor end plate (Sect. 2.1.4.2). As shown in Figs. 2.19 and 2.26, voltage-gated Na^+ channels are opened in the junctional folds or next to the motor end plate because of the induced equalizing currents. *Action potentials* arise in the sarcolemma, propagating along the sarcolemma through their active *regeneration*; similar to the propagation of action potentials along an unmyelinated axon (Sect. 2.1.4.1). In consequence, the action potentials *propagate* into the transverse tubules, i.e., *into the interior* of the muscle cell.

In response to this membrane depolarization, *voltage-gated* Ca^{2+} *channels* in the transverse tubules open. Interestingly, there is a *mechanical–molecular coupling* between these voltage-gated Ca^{2+} channels and special channels in the membrane of the sarcoplasmic reticulum, with the channels being closely located against one another (Fig. 2.26). The latter *membrane channels,* known as *calcium release channels,* serve for rapid Ca^{2+} *release from the sarcoplasmic reticulum* (where Ca^{2+} is stored) into the cytoplasm. These release channels are 10 times larger than the voltage-gated Ca^{2+} channels, permitting the *rapid inflow of* Ca^{2+} ions into the cytoplasm.

Thus, depolarization-induced conformational changes in voltage-gated Ca^{2+} channels cause the *calcium release channels* to *open*[100] directly. The Ca^{2+} ions follow their electrochemical driving force (compare Table 2.1) and *increase the concentration of* Ca^{2+} ions[101] in the cytoplasm by a factor of 100 (Silbernagl and Despopoulos 2007).

In fact, the increased level of Ca^{2+} *ions stimulates contraction* of myofibrils within the muscle cell. In particular, Ca^{2+} ions allow binding of cross bridges to actin. In a *relaxed muscle cell*, a special protein (tropomyosin, Fig. 2.30) adhered to actin chains *blocks binding* of myosin heads to actin, namely, blocks specific binding sites on the actin; this blocking prevents contraction of the muscle cell. In the case

[100]In the case of the *skeletal muscle cell*, opening of the *voltage-gated* Ca^{2+} *channels* serves only as a *mechanical signal* to open the more efficient calcium release channels for rapid diffusion of Ca^{2+} into the cytoplasm. In the case of the *cardiac muscle cell*, the diffusion of Ca^{2+} ions through the voltage-gated Ca^{2+} channels serves as a *chemical signal* to open the calcium release channels (see Footnote 109). In contrast, the *smooth muscle cell* employs the diffusion of Ca^{2+} ions through the voltage-gated Ca^{2+} channels as the *primary source of* Ca^{2+} *ions* in the cytoplasm (see Footnote 106).

[101]To *intensify the inflow of* Ca^{2+} ions into the cytoplasm, there are Ca^{2+}*-activated* Ca^{2+} *channels* in the membrane of the sarcoplasmic reticulum. A rise in the Ca^{2+} concentration in the cytoplasm opens these channels and fortifies the inflow of Ca^{2+} ions even more; compare Footnote 109.

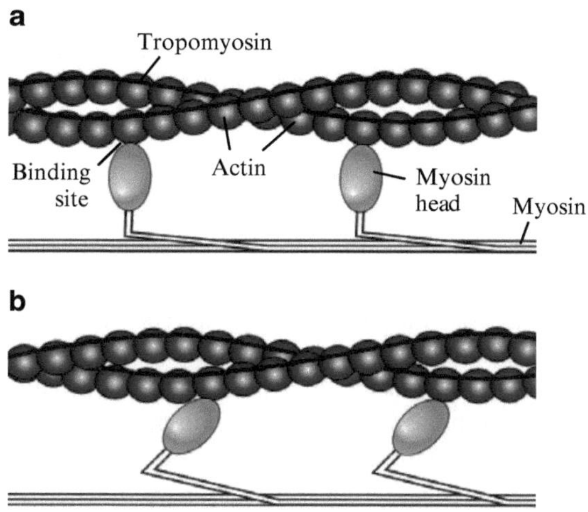

Fig. 2.30 Sliding filaments as a basis for mechanical power strokes and contraction of the muscle cell. (**a**) Cocked myosin heads bind to actin establishing a bridge before sliding; compare with Fig. 2.25a, (**b**) Myosin heads become flexed which causes the thin actin filaments to slide to the right; compare with Fig. 2.25b

where sufficient Ca^{2+} ions are available in the cytoplasm, some of this *Ca^{2+} binds* to another protein (troponin) associated with the tropomyosin. This binding of Ca^{2+} causes troponin's conformational change that moves the blocking tropomyosin out of the way and thus *clears the binding sites* on the actin for myosin heads. Once the binding sites are clear, arrested cross bridges can be established and the *sliding* of thick and thin filaments past each other can begin (or continue), as long as the *cytoplasmic Ca^{2+} is available.*

The sliding *cross-bridge cycle*, i.e., *cyclical interactions* between the myosin heads and binding sites on the actin, *begins* when

- The myosin heads are not attached to actin and have a *flexed shape*, which corresponds to *muscles cells at rest*. Before the heads can bind to actin,
- The heads bind ATP (source of *chemical energy*, Footnote 19), become phosphorylated under the influence of Ca^{2+} (by ATP hydrolysis), change their spatial conformation (heads become *cocked*), and thus have the *potential mechanical energy* for the *shortening of myofilaments*. Provided that the binding sites on actin are clear (i.e., cytoplasmic Ca^{2+} is sufficiently available, see above),
- The *heads bind* to actin, as illustrated in Fig. 2.30a (and Fig. 2.25a).
- Then the heads become dephosphorylated (bound phosphor ion from the previous hydrolysis is released). Consequently, the heads undergo a conformational change and again become *flexed*. It causes the cross bridges to produce a *mechanical power stroke* and the overlap between thick and thin filaments increases (Fig. 2.25b). As shown in Fig. 2.30b, the filaments slide along each other by about 4–12 nm per single power stroke (Silbernagl and Despopoulos 2007).

- The following release of adenosine diphosphate from the heads (Footnote 19) and binding of new ATP *release the heads*[102] from actin, which completes the power stroke. Herewith *another cross-bridge cycle* is initiated provided that Ca^{2+} and ATP are present in the cytoplasm in sufficient amounts.

Many cross-bridge cycles must be repeated to shorten the muscle cell, whereas mechanical power strokes from different cross bridges are asynchronous. As an approximation, the *shortening of the myofilaments continues* as long as cytoplasmic Ca^{2+} is available.

When the muscle cell is *no longer synaptically stimulated*, the voltage-gated Ca^{2+} channels in the transverse tubules close along with the calcium release channels. The *intracellular* Ca^{2+} *ions* are actively *transported out of the cytoplasm* back into the sarcoplasmic reticulum.[103] The contraction of the muscle cell stops and the whole muscle experiences *elastic recoil*, allowing the *muscle to relax.*

A single action potential only stimulates particular muscle cells within a single motor unit. A single muscle cell quickly contracts and then relaxes, with this mechanical response known as a *muscle twitch.* The twitch lasts for about 10–100 ms and is actually much longer than the duration of a single action potential (of about 2 ms, Fig. 2.14a). As shown in Fig. 2.31, the *twitch is delayed* by about 10 ms with respect to the corresponding action potential. This is because the action impulse has to propagate along the full length of the muscle cell with a finite conduction velocity of about 2 m/s (comparable with the conduction velocity in unmyelinated axons, Table 2.2) and, on the other hand, the formation of mechanical cross bridges also requires a finite amount of time.

With increasing *frequency of successive action potentials*, the relaxation time between successive twitches will get shorter and the individual twitches summate to produce a *sustained and prolonged muscle contraction* of increased strength, known as *tetanus*. The contraction force increases by up to a factor of 4 compared with a single muscle twitch (Silbernagl and Despopoulos 2007). On the cellular level, the amount of intracellular Ca^{2+} ions increases if another action potential occurs before all Ca^{2+}—released by the previous action potential—has been removed from the cytoplasm. The Ca^{2+} *release* becomes much stronger than its continuous reuptake into the sarcoplasmic reticulum; Ca^{2+} remains in the cytoplasm for the *cross-bridge cycle to continue.* Additional binding sites on actin are activated and more cross bridges are formed, resulting in a *greater output of*

---

[102] The stiff state of muscles after death, termed *rigor mortis*, results from the fact that cross bridges cannot detach from actin. This is because the cell lacks ATP and adenosine diphosphate remains bound to the myosin heads.

[103] The Ca^{2+} ions are transported against their diffusion gradient from the cytoplasm *into the sarcoplasmic reticulum* and also *out of the cell* (Footnote 40). In analogy with section "Active Transport," *active* Ca^{2+} *pumps* are involved in this transport, whereas the pumps (built out of proteins) reside in the membrane of the sarcoplasmic reticulum and are *usually powered by ATP* (Footnote 19).

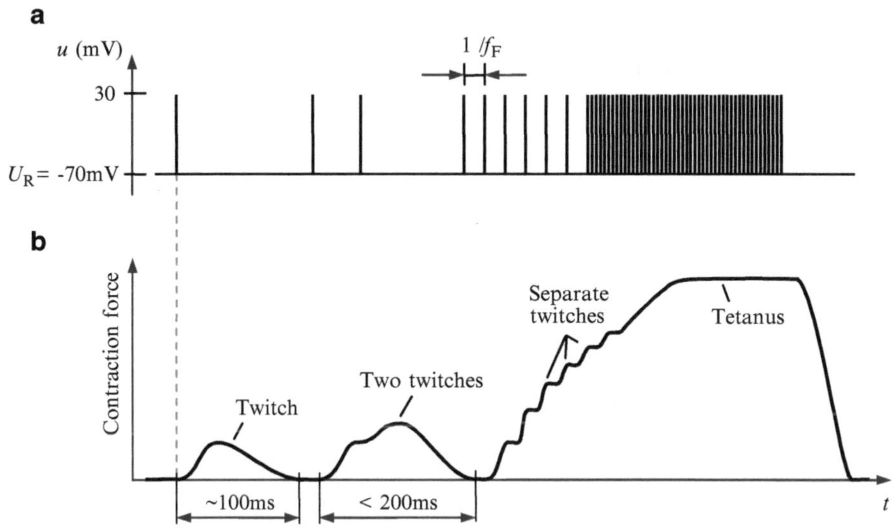

Fig. 2.31 Contraction of a muscle cell in response to a train of action potentials in the axon terminal (Fig. 2.26). (**a**) Innervating action potentials have a varying instantaneous frequency, i.e., a varying firing rate f_F. (**b**) Each action potential triggers a delayed contraction of the muscle cell, known as muscle twitch. The successive twitches interfere with each other and produce a prolonged and sustained muscle contraction, known as muscle tetanus

force.[104] Given successive action potentials, the muscle cell shortens even more *towards* its *maximum contraction*, where all possible cross bridges are continuously formed. Figure 2.31 illustrates a single twitch, the interferences of two time-shifted twitches and of multiple twitches at an increasing firing rate of action potentials. At a stimulation frequency of up to 10 Hz, separate muscle twitches can still be observed, whereas at frequencies *above 50 Hz*[105] a *smooth sustained contraction* results.

---

[104]In fact, the amount of active *muscle force* is related not only to the *frequency of action potentials* but also to the *length of muscle cell*. As the *sarcomeres* are stretched, the overlapping region of the thick and thin filaments decreases which results in a decreasing contraction force; compare Fig. 2.25. On the contrary, a strong reduction in the sarcomeres length yields progressive overlap of the thin filaments with each other that occludes binding sites for myosin heads and thus also decreases the contraction force. There is an *optimal length* of the sarcomere of about 2–2.2 μm (Silbernagl and Despopoulos 2007) at which maximum contractile force is generated. It is interesting to note that *skeletal muscle cells* usually work around their optimal length, whereas *cardiac muscle cells below their optimal length* [the sarcomere length of about 1.5 μm prior to heart filling (Fox 2011)]. In consequence, when cardiac muscle is prestretched—before its contraction due to an increased ventricular filling—the muscle develops a greater contraction force, determining the mechanism of *Frank–Starling law* (Footnote 225).

[105]A sustained muscle contraction at 50 Hz (or 60 Hz) has severe consequences in terms of *electrical accidents* with (alternating current) *power lines*. As a *disadvantage*, the arising *tetanus* impedes a voluntary release of the live metallic parts of the power lines that were touched.

The *skeletal muscle*—as a bunch of muscle cells, Fig. 2.24—is able to produce voluntary *graded contractions*, i.e., is able to generate a varying contraction force. This is attained by variations in

- The *number of motor units* being activated
- The *frequency of action potentials*

That is, a stronger muscle contraction requires a greater number of *simultaneously activated motor units* of this muscle and thus a greater number of simultaneously activated muscle cells. As already mentioned, graded contractions of individual muscle cells are not possible for a single stimulus (or a single action potential). On the other hand, a stronger muscle contraction can be achieved by a time *sum of successive muscle twitches* with increasing stimulation frequency (tetanus).

In fact, different motor units have to be *asynchronously* activated to attain a *smooth contraction* or a certain *muscle tone* of the skeletal muscle (complete tetanus). Some motor units begin to twitch while those previously activated begin to relax so that *pulsatile twitches* of individual muscle cells *become smoothed* over time (Fig. 2.31).

Smooth muscle cells also contract by means of the sliding filament mechanism.[106] Because of a relatively loose arrangement of the contractile apparatus within the cell (Fig. 2.27) and densely arranged myosin heads in the thick filament, the smooth muscle can even *contract when greatly stretched*. For instance, smooth muscle cells of the uterus are stretched up to eight times their original length by the end of pregnancy and are still able to contract (Fox 2011).

A smooth muscle cell of *single-unit* type—unlike a skeletal muscle cell—can produce a *graded depolarization* of its membrane and a *graded contraction*. In fact, the muscle tone is controlled by the intracellular Ca^{2+} level. The greater the depolarization level of the membrane, the more Ca^{2+} ions will enter the cell; consequently, it accelerates the depolarization even more and thus opens more voltage-gated Ca^{2+} channels. With the increasing Ca^{2+} level in the cell, more cross bridges will bind to actin and strengthen the contraction of the smooth muscle cell (Footnote 106). In addition, the contraction of smooth muscles is relatively slow and sustained, which allows *sustained contractions* under *resting conditions*. The smooth muscle is always in a state of a certain contraction yielding a certain *muscle tone* (Footnote 130).

---

[106]Unlike skeletal muscle cells, *smooth muscle cells* contract primarily in response to extracellular Ca^{2+} ions diffusing into the cell through the voltage-gated Ca^{2+} channels in the cell membrane. The sarcoplasmic reticulum is less developed there. The opening of these voltage-gated Ca^{2+} channels is graded by the depolarization level of the membrane so that the *depolarization level controls the contraction strength* of the smooth muscle cell.
In addition, cytoplasmic Ca^{2+} ions combine with *different proteins* (calmodulin instead of troponin) and activate a specific *enzyme* (myosin light-chain kinase). The enzyme catalyzes the *phosphorylation of myosin chains* within cross bridges that activates binding of the cross bridges to actin; thereby, a contraction of the smooth muscle cell is generated.

Any influence which leads to an increased *concentration of intracellular Ca^{2+} ions* increases the *contraction strength*. Besides the electrical *depolarization* (even without an action potential), diverse chemical *transmitters*, *hormones*, and even mechanical *stretching*[107] of the smooth muscle cell regulate the intracellular Ca^{2+} level and thus control the contraction strength of this smooth muscle cell (Footnote 130). Typically the mechanical response of the smooth muscle cell is an *integrative response* to

- The different types of stimuli with *electrical*, *chemical*, and *mechanical* origin
- The *excitatory* and *inhibitory* stimuli, e.g., due to a simultaneous stimulation via sympathetic and parasympathetic axons (Footnote 99)

In *single-unit* muscle cells, the membrane depolarization can be conducted from cell to cell over the electrical gap junctions (Fig. 2.27) so that all interconnected smooth muscle cells *contract* in a coordinated fashion and the whole smooth muscle contracts *as a single unit* (similar to the cardiac muscle, see below). Some smooth muscle cells display *spontaneous and autonomous membrane depolarization*, i.e., an intrinsic dynamic electrical activity in their membranes. The level of $u(t)$ rhythmically changes by about 10–20 mV with a frequency of 3–15 per minute (Silbernagl and Despopoulos 2007). If $u(t)$ exceeds a certain threshold level, *regular and periodic action potentials* are generated. Such smooth muscle cells act as *pacemaker cells* (Footnote 41) stimulating others in the smooth muscle via electrical gap junctions. Even at low frequencies of periodic stimulation the relatively inert contractions of the smooth muscle cells easily merge into *tetanus*. For instance, a periodic excitation of smooth muscles in the wall of arterial vessels yields a rhythmic pattern in the peripheral blood perfusion (Sect. 3.2.3).

Some *single-unit muscle cells* are *innervated* by axons, as shown in Fig. 2.27. However, this external stimulation only modifies the aforementioned spontaneous behavior of these cells, leading to an acceleration or inhibition of the depolarization level. In contrast, each *multiunit muscle cell*[108] must receive an innervation from outside, triggering its mechanical contraction. The synaptic transmission involved (after Footnote 99) is slower and more diffuse than in the regular neuromuscular

[107]*Stretching of a smooth muscle* cell opens stretch-gated Ca^{2+} channels in its membrane, which favors Ca^{2+} inflow and *depolarization* of the membrane. The depolarization promotes a *graded contraction* of the cell in response to its *graded stretch*. For instance, the *ureter* contracts in response to its increased volume, without any external stimulation involved. In analogy, the above mechanism autonomously maintains a certain constriction of *arterial vessels* under resting conditions, which upholds a constant blood flow independent of the actual blood pressure (Footnote 130).

[108]It is interesting to note that *multiunit cells* in the uterus *mutate into single-unit cells* by the end of pregnancy (Silverthorn 2009). Electrical gap junctions arise in between smooth muscle cells and thus individual contractions of the involved cells become synchronized so that the uterus as a whole can contract more efficiently.

synapses (Sect. 2.1.4.2). Single-unit muscle cells (if at all) and multiunit muscle cells are under control of the *autonomic nervous system.*

Cardiac muscle cells also contract by means of the sliding filament mechanism.[109] However, the cardiac muscle can only *contract as a whole muscle*, i.e., full contraction each time, and the number of cardiac muscle cells to contract can not be varied because the cells are electrically joined (Fig. 2.29). The action potentials spread primarily in the axial direction of the muscle cell, from one cell to another over the electrical synapses. That is, the cardiac muscle yields an all-or-none response without any graded contractions; nevertheless, the *contraction strength* of the cardiac muscle cells can be varied by *hormones*, *prestretching* the muscle cells, and the *Ca^{2+} level* in the cytoplasm determining the number of activated binding sites on the actin (compare Footnotes 104, 107, and 225).

In contrast to skeletal muscle cells which require an external stimulation (over neuromuscular synapses), *some cardiac muscle cells* (less contractile) can spontaneously produce action potentials (Footnote 41). These action potentials lead to spontaneous contractions of (electrically interconnected) cardiac muscle cells downstream. The action potentials arise in the so-called *pacemaker cells* (Footnote 111) whose activity is also influenced by the *autonomic nervous system* (Sect. 3.1.1). In general, the origin of the spontaneous excitation may be at any point in the cardiac muscle, leading to *contraction of the entire cardiac muscle.* As a special feature, the corresponding action potentials are much longer in duration compared with those in skeletal muscle cells (about 300 ms vs. <5 ms).[110] In consequence, the *refractory period* of the cardiac muscle cell ends almost immediately after its mechanical contraction; thus *no tetanus* and *no self-excitation* (of electrically interconnected cardiac muscle cells, e.g., connected in a cycle) are normally possible here.

---

[109] Unlike skeletal muscle cells, *cardiac muscle cells* do not have a mechanical–molecular coupling between the voltage-gated Ca^{2+} channels in the transverse tubules and calcium release channels in the sarcoplasmic reticulum. Instead, *Ca^{2+} ions* that enter the cell through the voltage-gated Ca^{2+} channels *stimulate opening of the calcium release channels*, referred to as Ca^{2+}-activated Ca^{2+} release (Footnote 101). Here *Ca^{2+} ion acts as a second messenger* to the calcium release channel which releases a great amount of Ca^{2+} into the cytoplasm (Footnote 36). The intracellular Ca^{2+} binds to troponin and *stimulates muscle contraction.* Consequently, the excitation–contraction coupling in the cardiac muscle cells is slower than in skeletal muscle cells.

[110] When a nonpacemaker *cardiac muscle cell* is stimulated by an excited pacemaker cell, the cardiac muscle cell becomes depolarized to its threshold. Thereby, the *up-shoot phase* of the induced action potential is due to the rapid Na^+ inflow (through voltage-gated Na^+ channels), then the potential declines a bit and *maintains* a certain *depolarization level* at about -15 mV for up to 300 ms (plateau phase) due to a relatively slow Ca^{2+} inflow (through voltage-gated Ca^{2+} channels); see Footnotes 109 and 111. *Repolarization* is achieved through a rapid K^+ outflow (through voltage-gated K^+ channels). For instance, a single action potential in the ventricle extends typically from the Q wave until the end of the T wave (Fig. 2.37).

2.4 Heart

The heart is the *motor for the circulatory system*, pumping blood through systemic and pulmonary circulation loops (Sect. 2.5). The heart receives blood from pulmonary and systemic *veins* and pumps it into systemic and pulmonary *arteries*, respectively. The heart is a muscular organ whose rhythmic contractions force blood to circulate within both circulation loops. As illustrated in Fig. 1.15, the heart represents a *major source of numerous biosignals*.

2.4.1 Structure

The heart, which is about the size of a fist, is a *double pump* with *four chambers*:

- *Right* and *left atria*
- *Right* and *left ventricles*

As shown in Fig. 2.32, the right and left atria receive blood through veins from the body and the lungs, respectively. The right and left ventricles pump blood into the arterial system, respectively, to the lungs and the body; compare Fig. 2.39.

The right and left sides of the heart are kept separate and comprise two pumps working in unison (Fig. 2.32). The atria and ventricles are interconnected by *one-way valves*, the so-called *atrioventricular valves*. These valves ensure a *unidirectional flow of blood* from atria to ventricles on both sides of the heart. On the right side, the atrioventricular valve is called the *tricuspid valve*, whereas on the left side the *mitral valve* resides.

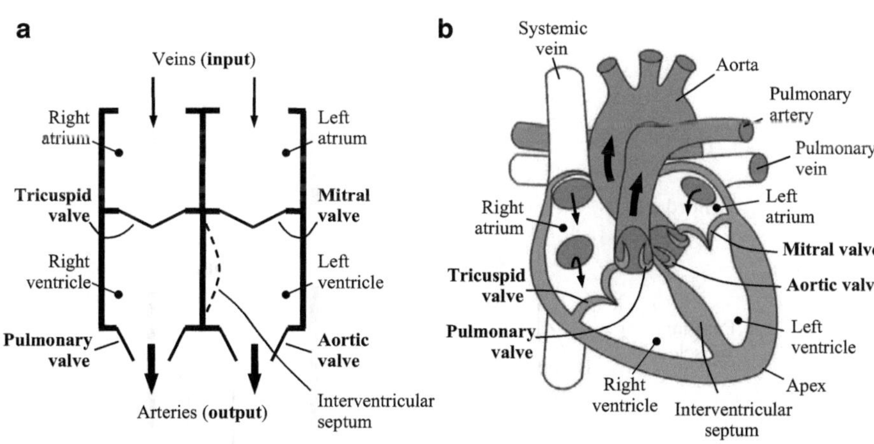

Fig. 2.32 The heart during blood ejection phase. (**a**) Schematic representation of the heart as four separate chambers with interconnecting valves. The leftward displacement of the interventricular septum is demonstrated during inspiration (section "Normal Respiration" in Sect. 3.2.1.1). (**b**) A section through the heart, showing its basic structure. Image data partly taken from (Wikipedia 2010)

The blood flow out of the ventricles, i.e., out of the heart, is also governed by *one-way valves*, the so-called *semilunar valves.* On the right side, the semilunar valve is called the *pulmonary valve* and is located at the point at which the pulmonary artery leaves the right ventricle toward the lungs (Fig. 2.32b). On the left side, the *aortic valve* is at the entrance into the aorta which conducts oxygenated blood to the upper and lower body.

The heart is composed of *cardiac muscles*, known as *myocardia.* The heart contains *two distinct myocardia* (two heart muscles), one myocardium forming both atria and another forming both ventricles. Both myocardia are structurally and functionally separated and are also *electrically isolated* from each other by the so-called fibrous skeleton. The walls of the left ventricle are much thicker than those of the right ventricle (about 10 mm vs. 3 mm) because the blood pressure in the left ventricle is much higher than in the right ventricle.

The *valves* are formed from a dense *connective nonmuscular tissue.* All valves open and close in response to the *pressure difference* between atria and ventricles (atrioventricular valves, Fig. 2.33b) or between ventricles and connected arteries (semilunar valves, Fig. 2.34b). The tricuspid valve has three flaps while the mitral valve has only two flaps; the mitral valve is also referred to as the bicuspid valve. In order to *prevent eversion of leaflets* of atrioventricular valves, the leaflets—as shown in Fig. 2.33a—are connected to *papillary muscles* within ventricles by strong tendinous cords. During contraction of ventricles (with closed atrioventricular valves), papillary muscles contract simultaneously and keep the flaps tightly closed. As illustrated in Fig. 2.34a, both semilunar valves have three cusps each while a single cusp is shaped as a *pocket.* Semilunar valves appear to be stiffer and close more rapidly than atrioventricular valves.

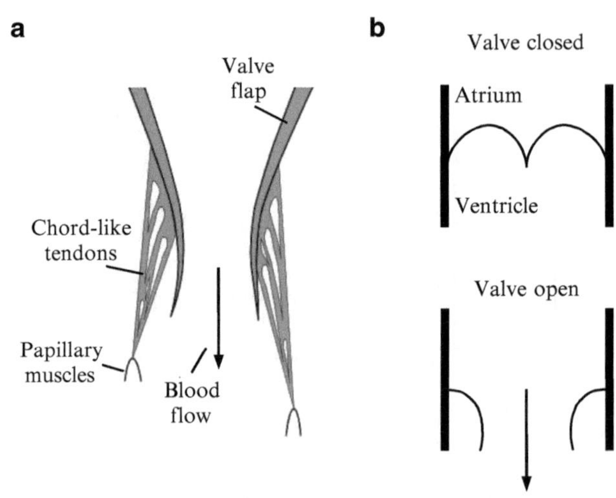

Fig. 2.33 (**a**) Atrioventricular valves whose flaps are supported by tendinous cords. (**b**) Schematic closure and opening of atrioventricular valves; compare Fig. 2.34b

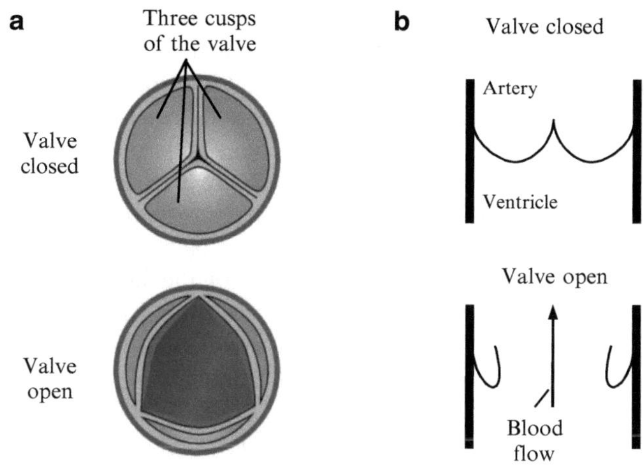

Fig. 2.34 (**a**) Semilunar valves build out of three pocketlike cusps. (**b**) Schematic closure and opening of semilunar valves; compare Fig. 2.33b

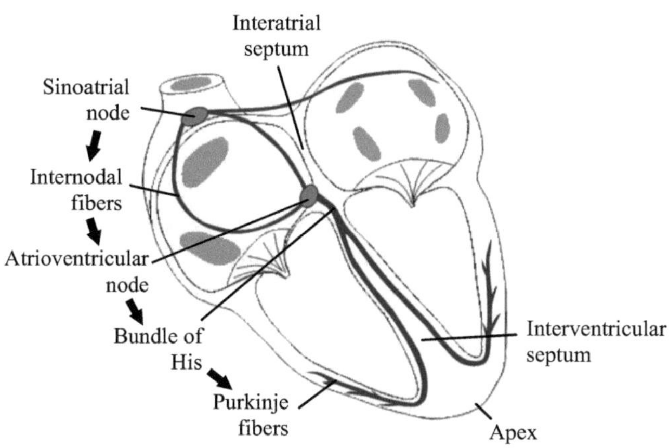

Fig. 2.35 Electrical excitation of the heart through a conductive system consisting of specialized cardiac muscle cells. *Arrows* approximately indicate the propagation direction of excitation in the course of the cardiac cycle. Image data partly taken from Malmivuo and Plonsey (1995)

The right atrium and right ventricle are separated from the left atrium and ventricle by a muscular wall, known as the septum. The lower part of the septum separating the ventricles is referred to as the *interventricular septum* (Fig. 2.32b), the upper part as the interatrial septum (Fig. 2.35).

A sophisticated *conductive system* resides in the heart, governing the propagation of electrical excitation along myocardia. The conductive system coordinates the pumping action of each chamber so that the heart can efficiently pump blood through both circulation loops. As illustrated in Fig. 2.35, the conductive pathway begins with the *sinoatrial node*, an accumulation of specialized cardiac muscle cells, located in the right atrium near the venous opening. The action potentials are generated in the sinoatrial node, which spread to and throughout adjacent cardiac muscle cells of the *right and left atria. Electrical gap junctions* between cardiac muscle cells facilitate this spread, whereas special electrically conducting tissues—specialized cardiac muscle cells, known as internodal fibers (Fig. 2.35)—support this spread as well. Since both myocardia are electrically isolated from each other, special conducting tissues electrically *interconnect atria and ventricles*, i.e., transmit action potentials from atria to ventricles. Namely, action potentials pass the *atrioventricular node* located in the region of the interatrial septum, which leads to the only conducting path from atria to ventricles. Then, the conducting pathway continues with the *Bundle of His*, a conducting tissue which *penetrates* the electrically *isolating fibrous skeleton* and descends along the interventricular septum toward the apex of the heart (Fig. 2.35). The Bundle of His separates into two branches which are continuous with the *Purkinje fibers* within ventricular walls. Within *ventricles*, action potentials spread again over electrical gap junctions between cardiac muscle cells.

There are three regions in the heart that contain *specialized cardiac muscle cells* able to spontaneously generate action potentials and to act as *pacemakers* of the heart (Fig. 2.35):

- *Sinoatrial node* acts as a *spontaneous primary pacemaker*[111] with the heart rate of about 70 per minute (in the normal heart).
- *Atrioventricular node* can act as a *secondary pacemaker* (potential pacemaker) with a lower heart rate of about 50 per minute.
- *Bundle of His* and *Purkinje fibers* can act as *tertiary pacemakers* (potential pacemakers) with an even lower heart rate of about 30 per minute.

[111]*Specialized cardiac muscle cells* in the *sinoatrial node* do not maintain a constant resting membrane potential but a rhythmic *pacemaker potential* (Footnote 41). A *slow spontaneous membrane depolarization* occurs in these cells toward their threshold level in order to generate an action potential (diastolic depolarization). Namely, *voltage-gated channels for both* Na^+ *and* K^+ ions are opened in response to a (previous) membrane hyperpolarization. The inflow of Na^+ ions predominates the outflow of K^+ ions, which depolarizes the membrane (compare sections "Regulatory Mechanisms" and "Synaptic Propagation"). As soon as the *threshold* is reached, *voltage-gated* Ca^{2+} *channels* in the membrane open, which *accelerates the depolarization* in progress and forms the upward phase of an arising action potential in the pacemaker cells (compare with the inflow of Na^+ ions forming the upward phase in excitable neurons, Fig. 2.14a, b). In addition, the inflow of Ca^{2+} leads to a partial contraction of these cardiac muscle cells (Footnote 109). Then *voltage-gated* K^+ *channels* are opened in response to the depolarization, generating an outflow of K^+ and thus a subsequent *repolarization* of the membrane. At a certain level of the hyperpolarization, the voltage-gated channels for Na^+ and K^+ open again and a new period of the rhythmic pacemaker potential begins.

It is important to note that the rate of the diastolic depolarization in the *sinoatrial node* can be slowed by *parasympathetic axons*, facilitating an outflow of K^+ ions out of pacemaker cells, as discussed in Footnote 71. On the other hand, *sympathetic axons* can accelerate the heart rate by increasing an inflow of Ca^{2+} ions (from the extracellular side) into pacemaker cells.

2.4.2 Function

Electrical excitation of the heart and its mechanical pumping of blood are closely interrelated. Obviously, electrical events precede and initiate mechanical events. Thus the timing and spatial distribution of action potentials in the heart are crucial for the genesis of the pumping action. In fact, an interplay of *pacemaker* cells, conductive cardiac muscle cells of the *conductive system*, and contractile *cardiac muscle cells* determines the heartbeat, as will be shown below.

As illustrated in Fig. 2.36a, the *right side* of the heart pumps blood through the *pulmonary circulation*. That is, venous *deoxygenated blood from the body* enters the right atrium; i.e., blood whose oxygen amount has been partially depleted by bodily tissues and whose carbon dioxide amount has increased (Footnote 112). Then this blood enters the right ventricle which presses it out of the heart into pulmonary arteries carrying blood *to the lungs*. In the lungs, pulmonary arteries branch out into pulmonary *capillaries* in which blood becomes enriched with oxygen and depleted in carbon dioxide via gas diffusion across thin capillary walls.

The *oxygenated blood* returns via pulmonary veins to the *left side* of the heart, entering the *systemic circulation* (Fig. 2.36b). That is, blood *from the lungs* enters the left atrium and then the left ventricle, whereas the ventricle pumps it into a systemic artery, i.e., the aorta. Arterial branches transport oxygen-rich blood *to the body* (upper and lower body). In bodily tissues, arterial vessels branch further into systemic *capillaries* in which blood becomes depleted in oxygen and enriched with

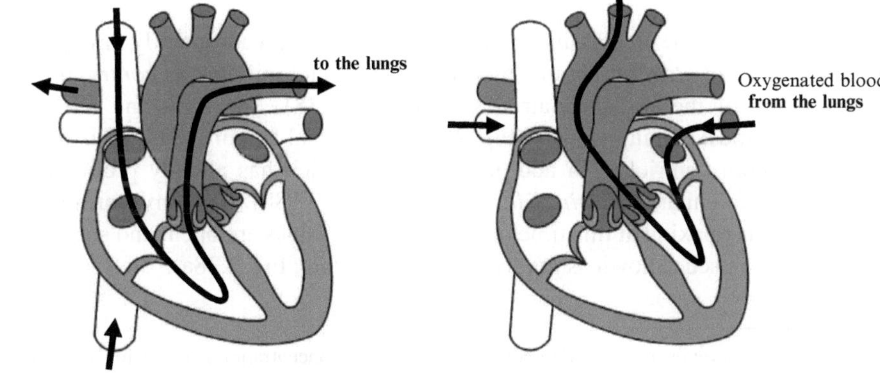

Fig. 2.36 Double pump of the heart. (**a**) The right side of the heart pumps blood through the pulmonary circulation. (**b**) The left side pumps blood through the systemic circulation. Closed atrioventricular valves and open semilunar valves indicate blood ejection phase during systole. Image data partly taken from (Wikipedia 2010)

carbon dioxide.[112] The deoxygenated blood is emptied into large veins and returns to the right atrium (Fig. 2.36a).

In short, *blood circulates* in a *closed loop*, comprising

- *Pulmonary circulation* that consists of *right atrium*, tricuspid valve, right ventricle, pulmonary valve, pulmonary arteries, pulmonary capillaries, pulmonary veins
- *Systemic circulation* that consists of left atrium, mitral valve, left ventricle, aortic valve, systemic arteries, systemic capillaries, systemic veins, and lastly *right atrium*

It should be stressed that the rate of *blood flow* (2.30) is identical in the *right and left sides* of the heart, i.e., identical in the *pulmonary and systemic circulations* (Footnote 124). Likewise, the amount of transported blood per time unit is identical in both circulations; compare the series connection of these circulations in Fig. 2.39. In fact, Frank–Starling law mainly accounts for the equalization of both blood flows (Footnote 225). A single drop of blood passes through the entire circulation loop in about 1 min (Fox 2011), which can be easily derived from the total blood volume (of about 5 l) and the blood flow [of about 5 l/min, (2.30)].

Since the electrical excitation of the heart is tightly interrelated with its mechanical action, both *electrical and mechanical behaviors* should be simultaneously considered from a functional point of view. Figure 2.37 illustrates the electrical behavior of the heart, whereas Fig. 2.38 illustrates the corresponding mechanical behavior.

Heartbeats—the repeating pattern of contraction and relaxation of heart muscles—are normally initiated and controlled only by the primary pacemaker, the *sinoatrial node* (Fig. 2.35). To start a single heartbeat,

- The sinoatrial node creates an *action potential* which spreads quickly through the *myocardium of atria* [at about 1 m/s (Fox 2011)] and initiates its *contraction*[113] (beginning of the so-called atrial systole, Fig. 2.38). The right and left atria contract almost simultaneously (Fig. 2.37a–c). A final amount of blood is forced to enter both ventricles, in addition to blood which has already prefilled the ventricles (compare the *time instant E* in Fig. 2.38b). The myocardium of ventricles is relaxing at this time, atrioventricular valves are open, and semilunar valves are closed (known as *late ventricular diastole*, Fig. 2.38a).

---

[112]Because of *cellular respiration* (Footnote 19), the *oxygen* concentration is lower in tissues than in the *capillary blood*, whereas the reverse is true for *carbon dioxide*. The latter concentration gradients facilitate loss of oxygen and a simultaneous enrichment with carbon dioxide in blood passing along capillaries in tissues.

[113]It is interesting to note that the *contraction of atria* is not vital for the heart function because this contraction contributes only about 20% to the end-diastolic ventricular blood volume (Fox 2011); compare a minimal increase of the ventricular blood volume at the *time instant E* in Fig. 2.38b. However, during *exercise* missing atrial contractions would compromise ventricular filling and thus compromise a needed increase in cardiac output.

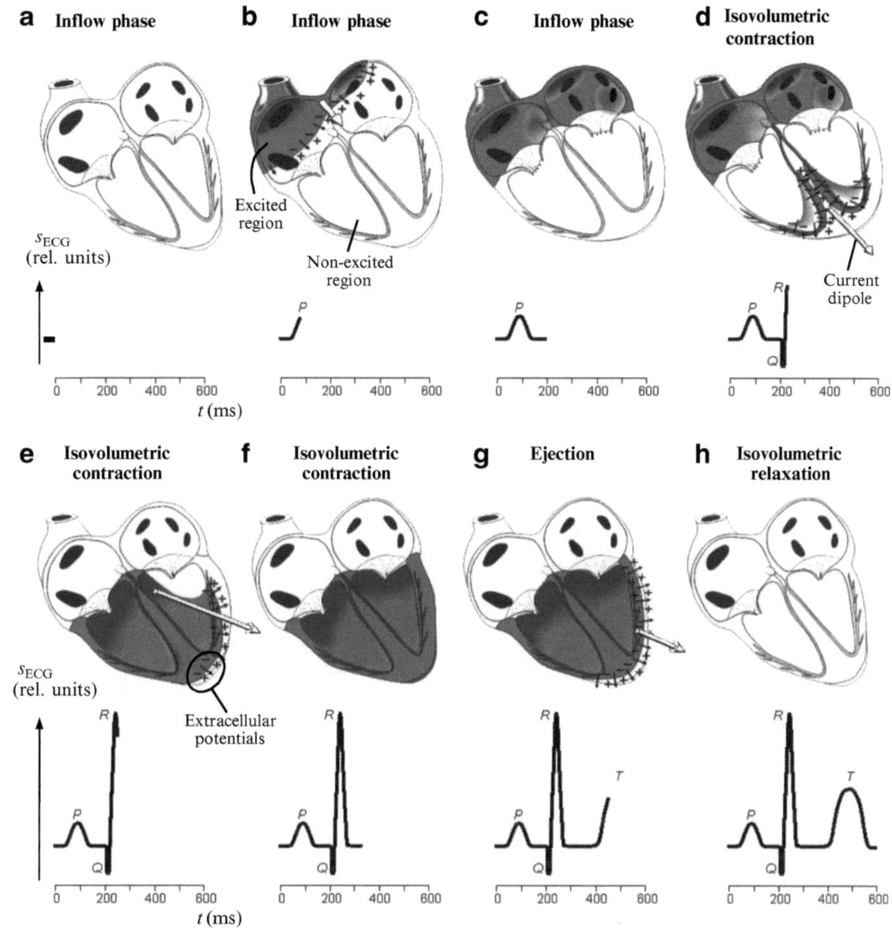

Fig. 2.37 Electrical excitation of the heart during a single cardiac cycle. *Top figures* in (**a**)–(**h**) show propagation of the excitation from atria to ventricles. *Gray color* indicates excited regions of myocardia. *Bottom figures* in (**a**)–(**h**) illustrate the corresponding formation of an electric biosignal electrocardiogram s_{ECG} (lead I Einthoven); compare with the induced mechanical pumping of blood from Fig. 2.38. Typical waves and peaks of s_{ECG} are marked with *letters*. Image data partly taken from Malmivuo and Plonsey (1995)

- The propagating action potential reaches the *atrioventricular node* where the propagation undergoes a slight *time delay*[114] to ensure that atria have already ejected their blood into ventricles before the myocardium of ventricles starts to contract. Likewise, a relatively fast electrical propagation of excitation has to be coordinated with a relatively slow mechanical pumping action.

[114]The atrioventricular node delays action potentials by about 0.1–0.2 s with a conduction velocity through the node of only about 0.05 m/s.

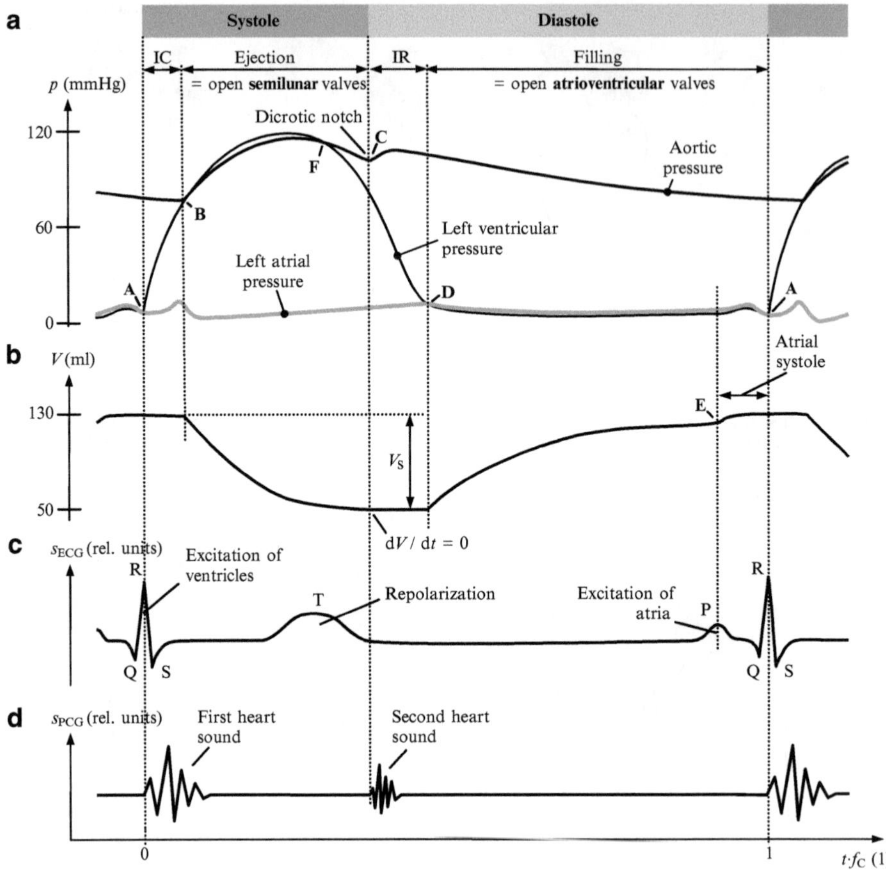

Fig. 2.38 Mechanical pumping action of the heart in response to the electrical excitation of the heart (Fig. 2.37). (**a**) Synchronous waveforms of the blood pressure p in the aorta, the left ventricle, and the left atrium during the cardiac cycle with the duration $1/f_C$ (compare with Fig. 2.48). Typical phases of the cardiac cycle are denoted including the isovolumetric contraction phase (IC) and the isovolumetric relaxation phase (IR). (**b**) The corresponding volume V in the left ventricle with V_S as the left ventricular stroke volume. (**c**) Electric biosignal electrocardiogram s_{ECG} with typical waves and peaks. (**d**) Acoustic biosignal phonocardiogram s_{PCG} with two consecutive heart sounds

- Then, the action potential propagates quickly down the *Bundle of His* toward ventricular *Purkinje fibers* (at about 1–4 m/s), which stimulates both *ventricles to contract* (end of atrial systole and *beginning of ventricular systole*); compare with Fig. 2.37d–f. Ventricular pressures rise above atrial pressures while atrioventricular valves close (at the *time instant A* in Fig. 2.38a). As soon as ventricular pressures exceed pressures in the pulmonary artery (at about 10 mmHg) and the aorta (at about 80 mmHg), semilunar valves open (at the *time instant B* in Fig. 2.38a).

- Blood is separately ejected into the arteries (Fig. 2.48c) while the right and left ventricular pressures rise[115] even more up to about 25 and 120 mmHg, respectively (corresponds to Fig. 2.37g). About two thirds of the blood the ventricles contain, known as the stroke volume, leave the heart (Fig. 2.38b). As the myocardium of *ventricles begins to relax*, ventricular pressures start to decrease and fall below arterial pressures just outside semilunar valves (beginning with the *time instant F* in Fig. 2.38a). Semilunar valves close a bit later at the *time instant C* because the *blood flow* through these valves has to cease for the valves to close (i.e., $\mathrm{d}V/\mathrm{d}t = 0$ at the time instant C in Fig. 2.38b). In fact, semilunar valves close in the course of a temporary backflow of blood into ventricles catching and filling the pocket-like cusps (*end of ventricular systole* and *beginning of ventricular diastole*); compare Fig. 2.48c. This backflow yields a prominent drop in the arterial pressure waveform, known as dicrotic notch; compare Fig. 2.38a and Footnote 158. Ventricular pressures continue to fall (corresponds to Fig. 2.37h).
- Later, when ventricular pressures have decreased below atrial pressures (at nearly atmospheric pressure, i.e., 0 mmHg), atrioventricular valves open and *ventricular filling* starts again (at the *time instant D* in Fig. 2.38a). Ventricular pressures increase slightly during the filling once ventricles are completely relaxed. Now, both *myocardia of atria and ventricles are relaxed* until a new action potential from the sinoatrial node initiates another heartbeat.

As illustrated in Fig. 2.38a, *ventricular systole* (usually referred to as *systole*) is known as the time period during which the myocardium of both *ventricles contracts and ejects* blood into arterial vessels. During systole, atrioventricular valves are closed; i.e., blood enters atria[116] but not ventricles. *Ventricular diastole* (or simply *diastole*) is known as the time period during which the myocardium of both *ventricles relaxes*. In fact, diastole is the time interval from aortic valve closure to mitral valve closure. During most of the diastolic period, blood enters ventricles through atrioventricular valves while atria receive blood from veins[117] (Fig. 2.38a, b).

---

[115] The *ventricular pressure* increases not only because of an active ventricular contraction but also because of a decreasing *ventricular volume* during the initial part of the *ejection phase* (Fig. 2.38a). Equation (3.4) illustrates this fundamental *relationship* (from Laplace law), i.e., the transmural pressure p_T increases with a decreasing ventricular radius r and an increasing wall thickness h for a given tensile stress σ in the ventricular walls. The level of σ is determined by the contraction level of ventricles.

[116] Filling of atria is strongly supported by the *ejection phase* of ventricles, during which atrioventricular valves are drawn downward toward the apex of the heart, inducing *suction of blood into atria* out of veins.

[117] The *venous return* is determined by

- The remaining *blood pressure* after blood passes capillaries, related to the atrial pressure;
- The *suction* of blood during the ejection phase (Footnote 116);
- Thex *sympathetically* driven contraction of smooth muscles in the venous walls, which leads to constriction and emptying of veins; and

Typically, *systole* is subdivided into

- *Isovolumetric contraction phase* (known also as pre-ejection phase)
- *Ejection phase*

while *diastole* is subdivided into

- *Isovolumetric relaxation phase*
- *Filling phase*

As shown in Fig. 2.38a, the *isovolumetric contraction phase*[118] begins when ventricles start to contract and *atrioventricular valves close*; this phase ends when *semilunar valves open*. The *ejection phase* follows during which blood is ejected and the ventricular volume decreases (Fig. 2.38b); this phase ends with the *closure of semilunar valves*. The *isovolumetric relaxation phase* follows in which all valves are closed—the ventricular volume does not change—until *atrioventricular valves open*. Afterward, a rapid *filling* of ventricles occurs, initiating the *filling phase*. The final period of the filling phase consists of an atrial contraction, delivering the final amount of blood into ventricles immediately before the *next isovolumetric contraction phase* (Footnote 113).

As illustrated in Fig. 2.37, *depolarization* spreads actively from atria to ventricles. Also the subsequent *repolarization* boundary, i.e., the boundary between already resting and still excited regions, can be approximately considered as moving passively from atria to ventricles, starting the relaxation phase of cardiac muscle cells. The spatial spread of the depolarization within a single myocardium is indicated by arrows in Fig. 2.37, pointing from a negative extracellular charge (an excited region, compare Fig. 2.18a) to a positive extracellular charge (a nonexcited region). It is important to stress that a single *myocardium* is a *single functioning unit*. That is, an action potential that originates in any cardiac muscle cell of the myocardium leads to excitation and contraction of all the other cardiac muscle cells of this myocardium (Sect. 2.3.2).

It is important to note that *excitation and contraction of ventricles* go *from the inner side* of ventricles to their outer side. In parallel, the excitation begins at the apex of the heart and goes in the direction of atria (compare Fig. 2.37d with Fig. 2.37e). This is because blood has to be pumped *toward semilunar valves* away

- The *reduced intrathoracic pressure* during *inspiration* leading to suction of blood into intrathoracic veins (section "Normal Respiration" in Sect. 3.2.1.1). In parallel, the *intraabdominal pressure is increased* during inspiration because of lowering of the diaphragm. This lowering compresses intraabdominal veins and thus supports filling of intrathoracic veins with blood from intraabdominal veins. Furthermore,
- *Venous valves* facilitate a unidirectional flow (return) of venous blood in combination with closely aligned (and radially) *pulsating arterial vessels* and *contracting skeletal muscles* close to veins (Sect. 2.5.1). Arterial pulsations and muscle contractions provide a massaging action to veins, temporarily opening venous valves and thus aiding the return of venous blood.

[118]The duration of the isovolumetric contraction phase depends on the stretching of ventricular muscles prior to contraction (preload), central arterial pressure, and cardiac contractility (Footnote 234).

from the apex to be injected into arterial vessels (the pulmonary artery and the aorta). In analogy, the *excitation front in atria* propagates *toward atrioventricular valves* in order to pump blood into ventricles away from veins filling atria (Fig. 2.37b).

The electrocardiogram in Figs. 2.37 and 2.38, an *electric biosignal*, has rhythmic waves and peaks which are related to *electrical excitation of myocardia*. However, the electrocardiogram is not directly proportional to the voltage u across the membrane of a cardiac muscle cell (compare Fig. 2.14a and Fig. 2.26); for details see Sect. 4. In short, the *spread of the depolarization* (and the repolarization) determines the waveform of the electrocardiogram, whereas the spatial *direction of the spread* determines the sign of waves and peaks.

The first upward deflection of the electrocardiogram corresponds to the spread of atrial depolarization *toward the apex* and is known as the *P wave* (Fig. 2.37b). When about half of the *atria* is already depolarized, the P wave reaches its maximum. On the other hand, when both *atria are fully depolarized*, the electrocardiogram is at its *zero baseline* because all extracellular regions of atria show the same charge polarity (negative polarity).

Later, portions of the *ventricular* myocardium are excited further down in the interventricular septum but not at the anatomical passage from atria to ventricles. For a short period, the depolarization propagates *toward atria* (toward the base of the heart) which yields a subsequent downward deflection of the electrocardiogram, the so-called *Q wave* (Fig. 2.38c). The depolarization then propagates *toward the apex* yielding a sharp upward deflection, the so-called *R wave* (Fig. 2.37d, e). Afterward the direction of the depolarization's spread in the ventricular walls *reverses again* for a short period that generates a downward deflection immediately after the R wave, the so-called *S wave* (Fig. 2.38c). Now that the *entire mass of ventricles is depolarized*, the electrocardiogram levels off again at its *zero baseline* (Fig. 2.37f).

The *repolarization of ventricles* starts from the outer side of ventricles, whereas the repolarization boundary moves *inward toward atria* (Fig. 2.37g). Although the outer side of ventricles is the last to depolarize, it *recovers first* because action potentials in the outer side are relatively short. In consequence, an *inward spread of the repolarization* generates an upward deflection of the electrocardiogram, the so-called *T wave*. That is, this deflection has the same sign as if it was generated by an *outward spread of the depolarization*. It should be noted that *repolarization of atria* occurs during the prominent QRS complex so that a relatively small contribution of this repolarization cannot be recognized in the course of the electrocardiogram. In summary[119] (compare Fig. 2.38c),

- *Atrial excitation* is reflected by the *P wave*
- *Ventricular excitation* by the *QRS complex*
- *Ventricular repolarization* by the *T wave*

[119]The *time interval* from the onset of the Q wave until the end of the T wave reflects depolarization and repolarization of ventricles and its duration depends on the instantaneous *heart rate*. In contrast, the time interval from the onset of the P wave until the onset of the Q wave does not depend on the heart rate.

It should be stressed that a *pressure difference* between the sides of a *valve* determines its opening or closure. When ventricles relax, for instance, the pressure in atria increases (due to venous return) relative to decreasing pressure in ventricles. As soon as the pressure in an atrium exceeds that in a ventricle downstream, the interconnecting *atrioventricular valve* opens and onsets filling of this ventricle with blood from the atrium (at the time instant D in Fig. 2.38a). In contrast, when the pressure in this ventricle exceeds that in the atrium upstream, the atrioventricular valve closes and prevents backward flow of blood (at the time instant A). In analogy, *semilunar valves* open when the ventricular pressure is higher than the arterial pressure outside the valves (at the time instant B), allowing blood to flow out of the heart. In an approximation, both atrioventricular valves and, on the other hand, both semilunar valves open and close in synchrony (compare section "Normal Respiration" in Sect. 3.2.1.2).

The phonocardiogram in Fig. 2.38d, an *acoustic biosignal*, illustrates the first and second heart sounds (Sect. 5), which arise in synchrony with the mechanical pumping action of the heart. The *first heart sound* arises during the closure of atrioventricular valves immediately after the R wave. This sound is induced by an abrupt change in the tension of the valve's flaps, vibrations of heart muscles, and deceleration of blood close to valves. The *second heart sound* indicates the closure of semilunar valves and begins at the end of the T wave.

Pacemaker potentials of the *secondary and tertiary pacemakers* are typically overridden by the pacemaker potential of the *primary pacemaker* so that the sinoatrial node governs the timing of the heartbeat. The primary pacemaker dominates because its *firing rate* is higher than that of any other potential pacemakers (Sect. 2.4.1). Likewise, an action potential *rapidly* originating in the sinoatrial node propagates and reaches areas of potential pacemaker cells. This action potential stimulates potential pacemakers before these pacemaker cells can stimulate themselves up to their threshold through their own pacemaker potentials (spontaneous potentials but *less rapidly* changing). The slow spontaneous depolarization of the secondary and tertiary pacemakers is suppressed (or reset) as long as action potentials from the primary pacemaker can reach areas of these potential pacemakers. However, in case the primary *pacemaker fails* or the conducting pathway to potential pacemakers is interrupted, a potential pacemaker with the next highest firing rate can serve as a source of action potentials while still maintaining vital heart beats (at a lower heart rate).[120] In addition, it is important to note that sympathetic and parasympathetic branches of the *autonomic nervous system* (Sects. 3.1.1 and 3.2.2.1) regulate

- The rate of spontaneous depolarization of the primary pacemaker cells, i.e., *instantaneous heart rate* (Footnote 111).
- The *propagation speed* of action potentials in the heart, especially in the atrioventricular node (by regulating a local Ca^{2+} inflow and K^{+} outflow).

[120]If a pacemaker other than the sinoatrial node governs heart contractions, occurring heart beats are known as *ectopic beats* (Footnote196).

- The *contraction force* of both myocardia (by regulating Ca^{2+} inflow into cardiac muscle cells, i.e., an increased Ca^{2+} inflow increases the contraction force).

The length/force relationship of the cardiac muscle cell—as briefly described in Footnote 104—determines the *pressure/volume relationship* of chambers in the heart, especially that of ventricles. That is, an increased diastolic filling yields a larger contraction force and a larger stroke volume during the subsequent contraction of ventricles; compare Footnote 225. In addition, the *contractility of the heart* can be increased through an increased *Ca^{2+} level* in the cytoplasm and is also governed by the *heart rate.*[121]

2.5 Circulatory System

The circulatory system[122] is a blood *distribution network* transferring gases, nutrients, and other liquid substances to and from the cells. The functions of the circulatory system include

- *Transportation* of diverse substances (respiratory, nutritive, and excretory)
- *Protection* against abnormalities (clotting and immune mechanisms)
- Hormonal and body temperature *regulation*

The circulatory system represents a vital physiological system[123] and is a rich source of biosignals used for diagnosis and therapy. It should be stressed that only those structures and functions will be considered in detail, which are relevant for *biosignal generation* and *propagation* (compare Fig. 1.3a), in particular for pulse wave genesis.

---

[121] *Several local mechanisms* determine the relationship between the heart rate and contractility of the heart:

- Given a relatively low heart rate, i.e., a low number of action potentials per time unit, the total inflow of Ca^{2+} ions into cardiac muscle cells is relatively weak (Footnote 109); there is quite a lot of time for the transport of Ca^{2+} out of the cytoplasm (Footnote 103). With *increasing heart rate*, the concentration of Ca^{2+} ions in the cytoplasm *increases* as well as the *contractility of the heart.*
- In contrast, a *high heart rate* reduces the end-diastolic volume in ventricles and thus *reduces the contractility of the heart* (Footnote 225).

[122] In fact, the circulatory system consists not only of the *cardiovascular system* for blood transport but also of the *lymphatic system*. The lymphatic system transports interstitial fluid toward the heart and empties it into the subclavian veins. From a functional point of view, the lymphatic system removes interstitial fluid (i.e., maintains fluid balance in the body), transports fats from the small intestine to the blood, and provides immunological defense.

[123] The importance of the topic can be illustrated in the words of the Canadian physician William Osler (1849–1919) in his book *The principles and practice of medicine* (1892): "... a man is only as old as his arteries. ..." Likewise, cardiovascular diseases comprise the leading cause of mortality in western countries.

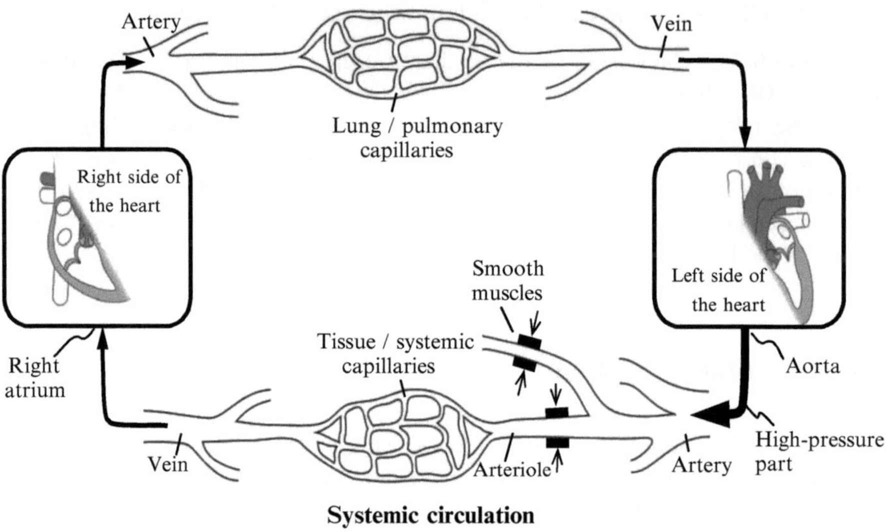

Fig. 2.39 The closed loop comprising systemic and pulmonary circulation. The thickness of the *arrows* indicates the level of the blood pressure

2.5.1 Functional Structure

The blood circulatory system, as shown schematically in Fig. 2.39 and anatomically in Fig. 2.40a, comprises

- *Systemic circulation* delivering oxygenated blood to the upper and lower body, and transporting deoxygenated blood back to the heart. The circulation of blood through the heart muscles, termed *coronary circulation*, is also a part of the systemic circulation.
- The other part of the blood circulatory system comprises *pulmonary circulation* delivering deoxygenated blood to the lungs and oxygenated blood back to the heart.

The *systemic circulation* deserves an extended description, for it acts as a rich *source of biosignals* used in diagnosis and therapy. In contrast, the pulmonary circulation is rather regionally restricted and its descriptive biosignals, though unique,[124] are hardly accessible by unobtrusive methods.

[124]The *pulmonary circulation* differs from the systemic circulation in the following ways. In short, vascular lengths in the pulmonary circulation are shorter and more symmetric with respect to the heart; arterial wall thickness is less showing a higher distensibility and lower pulse wave velocity (compare Sect. 2.5.2), the mean arterial pressure and vascular resistance are only about 15% of the

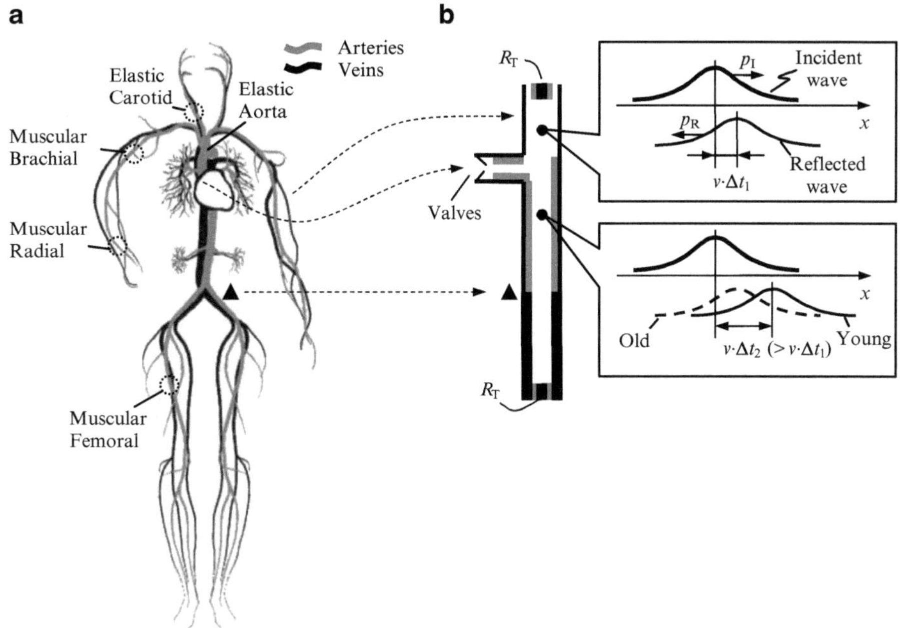

Fig. 2.40 (**a**) Circulatory system of human with indicated main arteries for biosignal establishment. Image data partly taken from (Wikipedia 2010). (**b**) The corresponding reflection model of asymmetric T-tube (Footnote 164) with schematically indicated arterial stiffness and peripheral resistance R_T; qualitatively represented by line thickness, *black color* for young person, *black* and *gray colors* for elderly persons. The upper and lower part of the body corresponds to the upper shorter and lower longer limb of the T-tube, respectively. The aortic valves serve as input while both incompletely occluded ends (with the total peripheral resistance R_T) and the passage from the elastic aorta to the muscular arteries (denoted as *filled triangle*) serve as reflecting sites; compare Fig. 2.48. The incident pressure p_I waves (at the valves) and the reflected pressure p_R waves (with R_T as reflecting site) are sketched for the upper and lower limb. The time delay Δt_2 is larger than $\Delta t_1 (< \Delta t_2)$ because the lower limb is longer than the upper limb, provided that the pulse wave propagation velocity v is constant. Aging is indicative of the gradual diminishing of the passage from the aorta to the muscular arteries and thus the near disappearance of the (distinct) lower reflection site (at *filled triangle*). In addition, $\Delta t_2 (\approx \Delta t_1)$ is reduced with aging because of the increased v (Footnote 164)

systemic values with a lower peripheral reflection coefficient in the lungs (Nichols and O'Rourke 2005). Likewise, the right ventricle of the heart has to generate much less pressure than the left ventricle (15 mmHg versus 100 mmHg in average) in order to create identical blood flows in the pulmonary and systemic circulations. This is because the vascular resistance of the pulmonary circulation [or the total peripheral resistance of the pulmonary circulation, compare (2.20)] is relatively low.

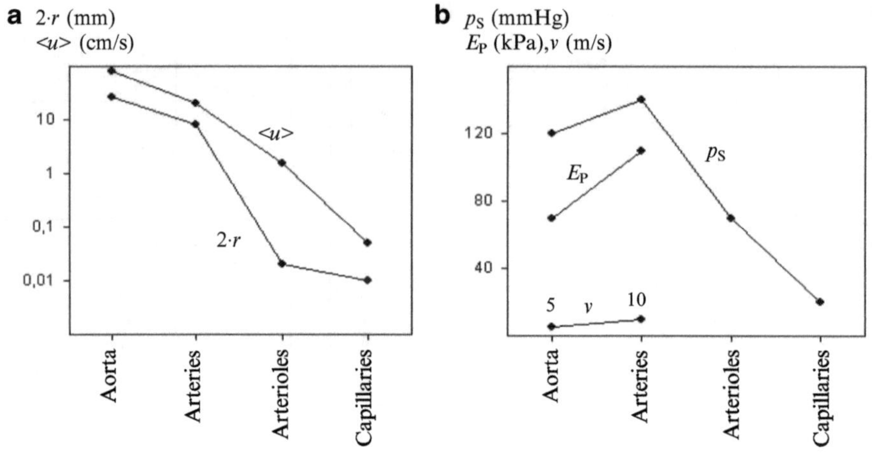

Fig. 2.41 Approximate properties of the arterial circulatory system. (**a**) Vessels diameter $2 \cdot r$ and mean blood flow velocity $\langle u \rangle$ on logarithmic scale. (**b**) Systolic blood pressure p_S, pressure–strain modulus E_P (for definition see Footnote 133), and propagation velocity v of pulse waves on linear scale. Data extracted from (Silbernagl and Despopoulos 2007; Nichols and O'Rourke 2005; McMillan 2006)

The circulatory system has a *high-pressure part*, see Fig. 2.39, which yields strong cardiac *pulsations* and a mean pressure of about 100 mmHg.[125] It includes the *arterial network* of vessels carrying the blood away from the heart, i.e., from the left heart side, thus serving the needs of *cushioning pulsations* and *blood supply*. The arterial network begins with the aorta, passes arteries, arterioles, and ends up by capillaries; compare also Fig. 2.41. The total blood flow is distributed unequally to different organs according to their unequal resistances to blood flow (namely, resistances of organ's arterioles), governed by their actual needs.

Correspondingly, the *low-pressure part* of the circulatory system yields nearly absent cardiac pulsations, with a mean pressure of about 15 mmHg. The system includes the *venous network* of vessels carrying the blood toward the heart, the right side of the heart, and vessels of pulmonary circulation (Fig. 2.39). About 80% of blood remains in the low-pressure system which serves as a major site of *blood storage* performing capacitance role in the circulation.[126] Consequently, the *blood volume* is the primary determinant of the *venous pressure* (e.g., in the right atrium).

[125]In terms of a comparison, the level of 100 mmHg corresponds to about 1/7 of the atmospheric pressure, i.e., 1 Bar = 10^5 Pa = 750 mmHg.

[126]For instance, if *blood volume* is increased (due to blood transfusion) more than 99% of the added blood volume will be found in the *low-pressure part* of the circulatory system (Silbernagl and Despopoulos 2007). On the other hand, a reduced blood volume (due to blood loss) almost exclusively impacts the low-pressure part only.

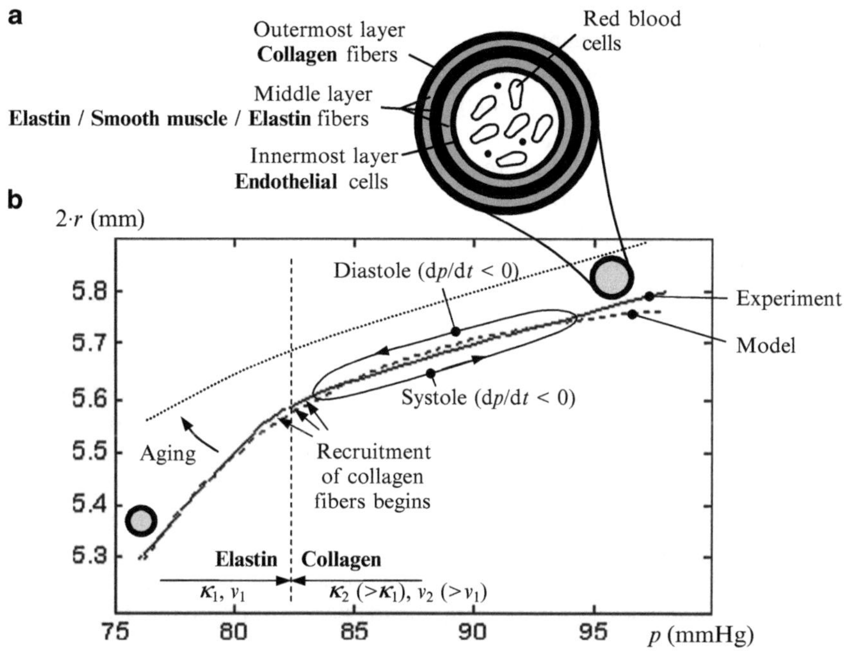

Fig. 2.42 (**a**) Qualitative structure of arterial vessel and its wall. (**b**) Relationship of the inner vessel diameter $2 \cdot r$ and blood pressure p of the human carotid artery with the experimental data (*solid line*) taken from (Shau et al. 1999), modeled data (*dashed line*) from (3.2), and expected changes because of aging of this elastic artery (*dotted line*). Relations of the module κ of volume elasticity (2.23) and the velocity v of propagating blood pressure waves are given in vivo for the regions dominated by elastin or collagen. In addition, the hysteretic behavior of r over the cardiac cycle is schematically indicated

The structure and properties of the involved *blood vessels* play a crucial role in the physiology of the circulatory system. The *arterial* vessels, carrying blood away from the heart, and *venous* vessels, carrying blood toward the heart, exhibit sophisticated properties to facilitate an efficient blood circulation, as described below.

Considering the *structure* of the *arterial vessels*, the composition of the *arterial walls* should be noted. As illustrated in Fig. 2.42a, the walls are composed of three layers: tunica intima (latin *tunica* coat and *intima* inner), tunica media (latin *media* middle), and tunica externa (latin *externa* outside). The *innermost layer*, tunica intima, is = comprised of a smooth lining of endothelial cells[127] and connective tissue that contains elastic fibers. The *middle layer*, tunica media, is the thickest

[127]It should be noted that the *endothelium* plays an important role in the functional regulation of *arterial stiffness* and thus of the blood pressure. The endothelial cells release vasoconstrictive mediators (as endothelin) as well as vasodilating mediators (as nitric oxide).

layer in arteries and is responsible for the *change in lumen* size of the vessel; i.e., it is the main determinant of the vessel's mechanical properties. The middle layer contains *smooth muscles* in circumferential orientation, *elastic* fibers and *collagen* fibers.[128] This layer is mainly stimulated by *sympathetic nerve fibers* (Footnote 188) so that smooth muscles[129] can contract to *vasoconstrict* or can relax to *vasodilate*[130] the vessel, i.e., decrease or increase the vessel's radius, respectively. As a result of vasoconstriction (due to an increased sympathetic stimulation), the blood flow is decreased and blood pressure is increased, whereas the reverse is true during vasodilation (at a reduced sympathetic stimulation); see also section "Blood Pressure and Flow" and Footnote 175. The *outermost layer*, tunica externa, consists of thick collagen fibers and some elastic fibers that *stabilize* and anchor the blood vessel to surrounding tissues.

The *structure* of the *venous walls* is similar to that of the arteries and is also composed of three layers. The *innermost* layer consists of endothelial cells, the *middle* one a smooth muscle layer and elastin, and the *outermost layer* is made of connective tissue. The *sympathetic activation* of the *smooth muscle* may help to increase the volume of blood filling the heart from the blood reservoir, the veins (Footnote 117); however, veins have less muscle than do comparably sized arteries.

In contrast to arteries, medium-sized veins possess *venous valves* that prevent blood from flowing back (away from the heart) because of gravity. The prevailing *parallel arrangement of veins and arteries*[131] in the body, as indicated in Fig. 2.40a, facilitates indirect pumping of venous blood by radially pulsating arteries through periodic activation of venous valves. The return of venous blood to the heart is also assisted by *skeletal muscles* close to veins, *respiratory activity* (Sect. 3.2.1), and (sympathetic) activation of *smooth muscles* in the venous walls; see Footnote 117.

The *venous walls* are 50% thinner than arterial walls, e.g., about 0.5 mm thick for large veins and 1 mm thick for large arteries (Silverthorn 2009); this allows veins to bulge more readily when a large volume of blood enters. The outer *radius* of veins

---

[128]In numerical terms, collagen with a Young's modulus E in the range of 3–100 MPa according to (Milnor 1989; Nichols and O'Rourke 2005) has a much higher stiffness than elastin with E in the range of 0.1–1 MPa.

[129]For instance, *smooth muscles* are missing in the vascular wall of cancerous vessels, explaining why they do not respond to hormones regulating normal vessel diameter in a circadian profile (Moser et al. 2008); compare Sect. 3.2.3.

[130]Arterial *vasodilation* and *vasoconstriction* induced by drugs is usually attained by their effects on smooth muscle, causing *relaxation* or *strengthening*, respectively. However, the nonlinear increase in stiffness with increasing vessel diameter (of dilated central elastic artery, compare Footnote 142) should be considered as a counter effect (Fig. 2.42).
It is important to note that under *resting conditions* almost all arterioles are constricted to about half of their maximal diameter because of *resting sympathetic activity* (Kandel et al. 2000) and *passive stretching* of their smooth muscles; compare Footnote 107 and Sect. 3.2.2.1. This allows for targeted *local control* of vessel diameter through dilation and constriction.

[131]In fact, *arteries* and *veins* proliferate *in parallel* with *nerves* (Sect. 2.2.1). Vascular and neural networks are closely aligned to each other, serving the mutual needs of local information processing and local supply of nutrients (Carmeliet and Tessier-Lavigne 2005).

is also larger than that of arteries to facilitate volume needs, e.g., 8 mm for large veins and 4 mm for large arteries (Silbernagl and Despopoulos 2007). Figure 2.41a demonstrates the change of the outer radius along the arterial tree on a logarithmic scale. The *ratio* of *wall thickness* to *radius* along the arterial tree is about 0.1 (Nichols and O'Rourke 2005). During *diastole*, the arterial wall of *large arteries thickens* and the *radius decreases* by about +10% and −5%, respectively, while the reverse is true for *systole* (Nichols and O'Rourke 2005).

Buffering blood pulsations from the heart demands a prominent *elasticity of arteries*. This elasticity is mainly influenced by the distribution of the elastic and collagen fibers (compare Footnote 128) within the arterial wall.

- In the *proximal arteries*, particularly in the *thorax region*[132] (Nichols and O'Rourke 2005), elastin is the *dominant component*. For instance, in the thoracic ascending aorta *elastin* makes up 41% of the dry tissue while *collagen* makes up another 20% (Milnor 1989). The remainder consists of smooth muscle, water (accounts for 70% of an artery's weight), and nonfibrous composite.
- In the *distal arteries*, the relative composition is reversed; e.g., elastin makes up 20% and collagen another 51% in the carotid artery or 15% and 35% in the abdominal aorta (Milnor 1989). Besides the small amount of the elastic tissue, an important mechanoelastic feature of the distal arteries is their relatively thick *smooth muscle* layer which actively helps to regulate the value of blood pressure.

In short, *distal* arteries are stiffer than *proximal* arteries. For example, the mean pressure–strain modulus[133] of aorta, carotid, and femoral artery (Fig. 2.40a) is about 70, 90, and 110 kPa, respectively (Milnor 1989; Nichols and O'Rourke 2005); compare Fig. 2.41b. Generally, elastin predominates in the *large and medium arteries* to facilitate *elastic storage* in these vessels while the smooth muscle predominates in the *small arteries* (vessels) to facilitate *control of vessel stiffness* (or resistance).

Thus the *arteries* located *centrally* (e.g., aorta, carotids, and iliac, Fig. 2.40a) are predominantly *elastic arteries* and experience mainly *passive changes* in their wall properties (compare Fig. 2.42b and Sect. 2.5.2.1). In contrast, *peripheral arteries* (e.g., femoral, brachial, and radial, Fig. 2.40a) are predominantly *muscular arteries*, expand less under pressure, and experience mainly *active changes* in wall stiffness with smooth muscle activation.

---

[132]The *thorax region*—in the context of proximal arteries from above—is roughly delineated by the diaphragm on the lower side and by arterial branches leaving the arch of the aorta on the upper side.

[133]The *pressure–strain modulus*, also known as Peterson modulus, is defined as $\Delta p \cdot r / \Delta r$ with p as pressure and r as radius. The Peterson modulus is a simple expression of actual experimental measurements; compare Fig. 2.41b. The modulus often provides a way of comparing the results of investigators who may have chosen differing assumptions about the influence of artery wall thickness and other factors in calculating the classical *stress–strain modulus*, i.e., the Young's modulus (Milnor 1989).

It should be noted that *arterial* vessels are *tapered* moving from proximal to distal regions. The reduction of their cross-sectional area can be approximated by a function with an exponential dependence on the distance from the upstream site (Nichols and O'Rourke 2005); this is particularly important for pulse wave reflections and arterial vascular resistance (Sect. 2.5.2).

Furthermore, *arteries* in the body are naturally under a condition of *longitudinal tension*, which stabilizes and anchors them in the tissue by preventing their longitudinal movements during passage of the cardiac pulse. When dissected they shorten by more than 20% and cause the arterial wall to thicken (Pedley 1980; Nichols and O'Rourke 2005).

The *elasticity* of arteries is strongly determined by their *anisotropic* and *viscoelastic*[134] wall material. In particular, viscoelastic properties and the viscosity of the blood within the vessel determine the *frequency-dependent* and *hysteretic* behavior of wall elasticity and thus are important as the pressure pulse passes the artery. As shown in (Pedley 1980; Gribbin et al. 1976; Nichols and O'Rourke 2005), the *dynamic Young's modulus*[135] increases with frequency and is effectively constant for frequencies above 2 Hz, the modulus being greater by a factor of 1.1 (aorta) up to 1.7 (carotid artery) than the static value (for 0 Hz) of the static Young's modulus. Interestingly, the presence of the dynamic Young's modulus yields a dependence of the artery stiffness on the instantaneous value of the heart rate. In other words, with *increasing heart rate*, the time allowed for a vessel to distend is reduced, which actually results in an *increased Young's modulus.*[136]

Compared to arteries, thin-walled collapsible *veins* are much more *elastic*, *facilitating their blood storage function*. For instance, venous volume compliance $\Delta V/\Delta p$ ($=\mathrm{V}/\kappa$, compare (2.23) in section "Pulse Propagation"), with vessel volume denoted as V, averages about 160 ml/mmHg for 80 kg body weight while that of the arteries is only about 4 ml/mmHg, with large differences among different body parts (Bronzino 1995). The *compliance* is highly *nonlinear* in both cases because of the heterogeneous structure of the vessel's walls; the compliance decreases at higher p and V (Sect. 2.5.2.1). In addition, at lower p the compliance

---

[134] *Viscoelastic* material demonstrates both viscous and elastic behavior under applied stress; i.e., it requires a finite time to reach the state of deformation appropriate to the stress and a similar time to regain their unstressed shape. In particular, viscoelastic materials exhibit hysteresis in the stress–strain curve (compare Fig. 2.42b); i.e., they show *stress relaxation* and *creeping* over time. In other words, a stepwise increase in strain causes a gradual decrease in stress while a stepwise increase in stress causes a gradual increase in strain because of creeping (Milnor 1989; Pedley 1980).

[135] Thomas Young (1773–1829) was an English physician and physicist after which the *Young's modulus* was named, which describes the ratio of the mechanical stress over the strain in a solid material.

[136] There is another *indirect mechanism* which relates *heart rate* and *arterial stiffness*. That is, increased sympathetic activity (Footnote 188) increases heart rate and, on the other hand, vasoconstricts arterial vessels. This vasoconstriction yields an increased stiffness (Footnote 130).

of veins is about 20 times greater than that of arteries, whereas at higher p the compliances of veins and arteries become similar.

Lastly, the effects of *aging* on arterial vessels should be mentioned, for they strongly affect biosignals of humans derived from the systemic circulation. These effects may easily lead to a misinterpretation if the age of the person is disregarded; compare section "Reflected Pulse Propagation" and Sect. 3.1.3. As people age, their arteries *dilate and stiffen* progressively, causing the arterial wall to become thicker and larger in diameter; e.g., an increase in diameter of about 40% from 20 to 80 years is reported in elastic arteries (Nichols and O'Rourke 2005); compare Fig. 2.42. In addition, arteries become less flexible with increased reflected wave amplitude and increased pulse wave velocity (see Sect. 2.5.2).

Interestingly, the large *predominantly elastic arteries*, centrally located in the body, experience significant passive stiffening of their wall over time, i.e., they *stiffen with age*, with a greater degree of stiffening in the *lower body* (Fig. 2.40). The stiffening is a result of progressive histological degeneration of elastin fibers and atherosclerosis[137]; in addition, the pulsatile strain contributes to the breakage of elastin fibers. In contrast to the elastic arteries, the *predominantly muscular arteries*, located peripherally, show much *less change over time* because of their active muscular control. Because of these uneven effects of aging on elastic and muscular arteries, the *aortic stiffness* even exceeds that of peripheral arteries *after age 60* (O'Rourke 2009; Nichols and O'Rourke 2005), which distinctively affects the pulse transmission along arteries; see section "Reflected Pulse Propagation."

In order to assess *stiffening* over time in *quantitative* terms, the characteristic impedance [for definition see (2.28) in section "Blood Pressure and Flow"], pulse wave velocity [see (2.22)], and the Young's modulus [see (2.24) and Footnote 133] of the aortic wall should be considered. For the aorta, the characteristic impedance and pulse wave velocity at least double from 20 to 80 years (Nichols and O'Rourke 2005) while the Young's modulus also more than doubles between the ages 20 and 60 (Milnor 1989).

To *demonstrate* the effects of *aging*, Fig. 2.40b shows stiffening of the arteries in the lower body with increasing age. Its causal impact on the pulse propagation will be discussed in section "Reflected Pulse Propagation." Furthermore, Fig. 2.42b illustrates the relationship of the inner vessel diameter and blood pressure within the vessel of the human carotid artery, i.e., of a predominantly elastic artery. The magnitude of stiffness is given by the slope, as discussed in Sect. 2.5.2.1, and its demonstrated flattening with age is indicative of increased arterial stiffness.

---

[137]*Atherosclerosis* refers to local deposition of fatty plaque inside the blood vessels and to inflammation in the vessels wall, which causes the vessels to narrow and to harden. Atherosclerosis constricts and stiffens vessels while aging, i.e., senile *arteriosclerosis*, dilates and stiffens in a diffuse way; compare Sect. 2.5.1.

2.5.2 Phenomena

As an introduction into the structure and basic function of the circulatory system was given in Sect. 2.5.1, the phenomena behind will now be discussed. The focus of the discussion will be on physiologic behavior which determines derived (pulsatile) biosignals.

The arterial system actually serves two functions:

- Low resistance *conduit* of blood
- *Buffering* of its pulsations

as already introduced in Sect. 2.5.1. The pulsations are mainly driven by time-dependent pressure gradients due to periodic *heartbeats*, yielding *pulsatile blood flow* q and oscillations of the *blood pressure* p. The inertial forces due to elasticity of arteries convert (damp) the *pulsatile* q leaving the heart into the smooth *steady* q[138] in the periphery, as is needed for blood supply.

2.5.2.1 Arterial Behavior

The elastic *large arteries* predominantly serve as a *cushioning reservoir* or "Windkessel"[139] that stores blood during systole and expels it to the periphery during *diastole*, thus performing a smoothing function. The aforementioned prevalence of *elastin* in the *proximal arteries* (Sect. 2.5.1) can be explained by their need to cope with the sudden high p produced during *systole*; that is, the *pressure pulse is absorbed* by the elastic wall. As shown in Fig. 2.43, the proximal arteries must have thick, elastic walls containing abundant elastic material, allowing stretching, so that the vessel lumen may accommodate the blood volume change. The elastic recoil of the elastic arteries during diastole is responsible for maintaining a *continuous flow of blood* to smaller vessels even while the heart is relaxing; that is, stored energy in the elastic wall is returned to the blood flow (Fig. 2.43). On the other hand, proximal arteries also have a thick, outer coat of *collagen* fibers whose tensile strength and high stiffness *prevent over-distension* of the artery wall at higher p.

---

[138]Within the physiological flow conditions, two velocities are actually of interest: the *velocity* u of *blood flow* and the *velocity* v of propagating *blood pressure waves* along the vessels. Usually, the value of v is about one order higher than $\langle u \rangle$, the average of u, e.g., $v \approx 5\,\mathrm{m/s}$ and $\langle u \rangle \approx 0.2\,\mathrm{m/s}$ along the aorta (Silbernagl and Despopoulos 2007); compare Fig. 2.41. It should be noted that the absolute values of u are not reasonable here for comparison aims, since u shows a nearly parabolic profile over the arterial diameter $2 \cdot r$ with the peak values of about 1 m/s in the middle of the aorta and zero values at the inner wall (Nichols and O'Rourke 2005); see Sect. 2.5.2.2 and later Fig. 2.45.

[139]The *Windkessel* is a German word for "air chamber" and refers to the air-filled dome of a historical fire engine which acts as a cushion or buffer, converting intermittent pumping of water into continuous flow. The term stems from the German translation of Stephen Hales (1677–1761) works, an English physiologist.

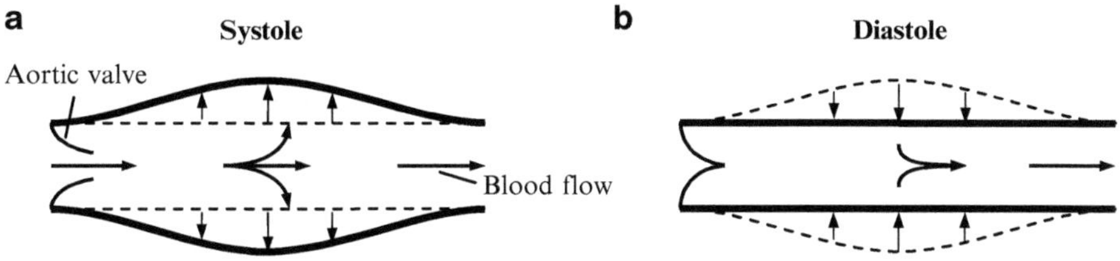

Fig. 2.43 Large elastic arteries act as a cushioning reservoir or "Windkessel" (from Footnote 139). (**a**) The pressure pulse is absorbed by the elastic arterial wall during systole, after the opening of the aortic valve. (**b**) The stored energy in the elastic wall is used to drive blood flow during diastole, after the closure of the aortic valve

While the large arteries perform the aforementioned passive smoothing function, *small arteries* perform active control through their *smooth muscle* activation (mediated by sympathetic fibers or even locally induced, Footnote 130). In particular, arterioles alter their fluid resistance R [see later (2.18)] and thus the total peripheral resistance R_T (2.20) by changing their *radius* r (see Sect. 2.5.2.2). Thus, *arterioles* actively

- Control *mean arterial pressure*[140] $\langle p \rangle$
- Maintain *continuous* q
- Govern the *flow distribution* among different *organs and tissues* (Fig. 2.39)

Figure 2.42b demonstrates the nonlinear *relationship between* r *and* p of the carotid (central elastic) artery (Fig. 2.40a), displaying convexity in relation to the diameter axis. It can be observed that with increasing p, corresponding to increasing transmural pressure,[141] the magnitude of the slope decreases nonlinearly, indicating that the artery walls become stiffer. For lower p values, (low) stiffness is maintained by elastin fibers, whereas for higher p values the (stiffer) collagen fibers begin to be recruited (compare Footnote 128) and the artery stiffness increases.

The latter behavior applies for both *proximal* and *distal* arteries; arteries become *more resistant* to stretch, that is less distensible, at higher p and r[142] (Milnor

[140]In clinical praxis, the *mean arterial pressure* is defined as $p_D + 1/3 \cdot (p_S - p_D)$ with p_D and p_S as the *diastolic blood pressure and systolic blood pressure*, respectively (section "Pulse Waveforms of Pressure and Flow"). This expression accounts for the fact that approximately 1/3 of the cardiac cycle is normally spent in systole, constituting a good approximation for peripheral arteries in general. Strictly speaking, this multiplier is in the range from 0.33 up to 0.41 (Zheng et al. 2011).

[141]The *transmural pressure* p_T is defined as the pressure difference between the inside and outside of the blood vessel, as demonstrated in Fig. 2.44b.

[142]Interestingly, artificial *dilation by drugs* has been shown to decrease stiffness of the peripheral *muscular arteries*, whereas any dilated central *elastic artery* gets stiffer because of increased r, according to Fig. 2.42 and Footnote 130 (Nichols and O'Rourke 2005). These *paradoxical* effects on the muscular arteries can be explained on the basis of smooth muscle arrangement in the muscular arterial wall, i.e., the drug-induced relaxation of the muscles transfers stresses from stiff collagenous fibers to more extensible elastin fibers. However, common drugs for vasodilation

1989; Pedley 1980). For instance, the Young's modulus of the aorta increases exponentially with increasing intravascular p. It is important to note that with increasing p, the value of v will increase as well since v is dependent on the stiffness of the artery wall (see section "Pulse Propagation").

Interestingly, changes of p within the cardiac cycle induce a *hysteretic behavior* of r (Sugawara et al. 2000), as illustrated by Fig. 2.42b. The changes of r follow the cardiac changes of p with some delay, thus yielding a delayed increase in r with increasing p during systole ($\mathrm{d}p/\mathrm{d}t > 0$) and a delayed decrease in r during diastole ($\mathrm{d}p/\mathrm{d}t < 0$). The width of the hysteresis of the r–p curve (up to 5% of p) was observed to increase with increasing dominance of the viscoelastic properties (Footnote 134) of the smooth muscle layer (Shau et al. 1999).

2.5.2.2 Steady Flow

The blood *flow q* is actually determined by the *pressure p gradient* along the artery. Thus q is not directly related to p but to the spatial derivative $\mathrm{d}p/\mathrm{d}x$ along the vessel's axial direction x. As demonstrated in Fig. 2.44a, in a homogenous section of a vessel of length l and inner vessel radius r, the *steady flow*, i.e., q not varying over time, depends on the pressure gradient $p_1 - p_2$ over the vessel section, as described by the Poiseuille equation[143]:

$$q = \langle u \rangle \cdot A = \langle u \rangle \cdot \pi \cdot r^2 = \frac{\pi}{8\mu} \cdot \frac{r^4 \cdot (p_1 - p_2)}{l}. \tag{2.17}$$

Here $\langle u \rangle$ is the average blood flow velocity over the cross-sectional area A ($= \pi \cdot r^2$) of the vessel and μ is the dynamic viscosity of the liquid. Equation (2.17) assumes a parabolic profile[144] of u, as demonstrated in Fig. 2.44b, with its maximum at the axis of the vessel ($y = 0$) and zero values at the inner vessel wall ($y = r$), since the lamina in contact with the wall is at rest.

Equation (2.17) can be simplified by

$$q = \frac{p_1 - p_2}{R} \quad \text{or} \quad Q = \frac{P_1 - P_2}{R}, \tag{2.18}$$

with R as the *fluid resistance* (or vascular, longitudinal resistance). For the sake of completeness, the real amplitudes P_1, P_2, and Q of the (general) time functions

cause a decrease in R_T and thus a decrease of mean arterial pressure, the latter pressure passively decreasing the r of central elastic arteries and consequently their stiffness.

[143]Jean Louis Marie Poiseuille (1799–1869) was a French physician and physiologist who first described the relation between the flow, radius, and pressure during laminar flow of fluids in circular tubes.

[144]The *parabolic profile* results from the assumptions that u is constant along x (compare Fig. 2.44a) and that the force opposing the flow over unit area is proportional to μ and the velocity gradient $\mathrm{d}u/\mathrm{d}y$ of the blood (compare Fig. 2.44b).

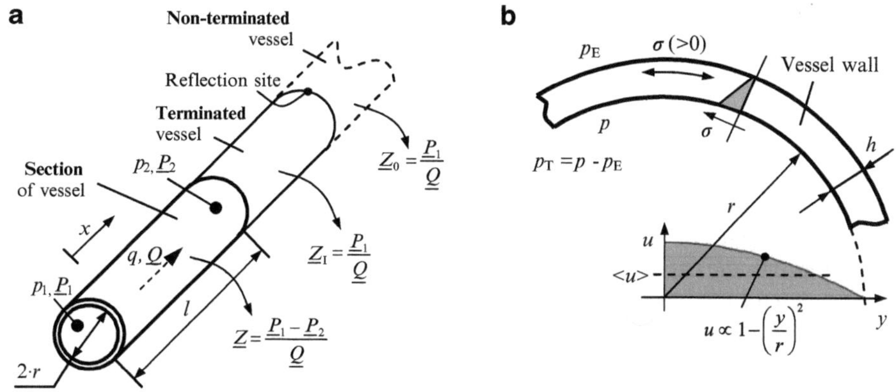

Fig. 2.44 (**a**) Schematic representation showing how concepts of the vascular complex longitudinal impedance $\underline{Z}$, input impedance $\underline{Z}_I$ (considering terminated vessels and reflections), and characteristic impedance $\underline{Z}_0$ (open vessels without reflections) for pulsatile blood pressure p and blood flow q are defined. (**b**) Concepts of the circumferential stress σ and the transmural pressure p_T in the arterial wall, as well as of the parabolic profile of the blood flow velocity u in the inner lumen of the vessel [compare Fig. 2.45 and (3.16)]

$p_1(t)$, $p_2(t)$, and $q(t)$, respectively,[145] were introduced; compare later (2.26), (2.28), and (2.29) for the pulsatile flow. Thus the resistance R can be given as

$$R = \frac{8\mu \cdot l}{\pi \cdot r^4}, \tag{2.19}$$

representing the opposition to steady-state flow along the artery (compare Fig. 2.44a). According to (2.19), the value of R is largely *dependent on* r^4, even to the fourth power. Even small changes in arterial r may induce strong changes in q, e.g., to reduce q by 50% only a 16% reduction of r is necessary. Thus *small vessels* found in the *periphery* have a disproportionately *high R*.

Equation (2.18) for a single segment of vessel can be adapted to the total *systemic circulation*, as shown in Fig. 2.39. The value of p_1 can be approximated as *mean arterial pressure* in the aorta $\langle p^{\mathrm{Ao}} \rangle$, which corresponds to the *input* of the systemic circulation (Fig. 2.39). In analogy, p_2 can be approximated as *mean venous pressure* in the right atrium $\langle p^{\mathrm{At}} \rangle$, which corresponds to the *output* of the

[145]Generally, the involved time functions can be defined as $p_1(t) = P_1 \cdot g_1(t)$, $p_2(t) = P_2 \cdot g_2(t)$, and $q(t) = Q \cdot g_3(t)$ with P_1, P_2, and Q as respective amplitudes and $g(t)$ as an arbitrary function having the physical unit of 1. In the case of *steady flow*, $g(t) = 1$ and therefore $p_1(t) = P_1$, $p_2(t) = P_2$, and $q(t) = Q$ in (2.18). In the case of *pulsatile flow*, the *sinusoidal components* of $p_1(t)$, for instance, can be expressed as $p_1(t) = P_1 \cdot g_1(t) = P_1 \cdot \cos(\omega t + \varphi)$ $= \mathrm{Re}[\underline{p}_1] = \mathrm{Re}[\underline{P}_1 \cdot e^{j\omega t}] = P_1 \cdot \mathrm{Re}[e^{j(\omega t + \varphi)}]$ with $\underline{p}_1$ as the complex waveform, $\underline{P}_1$ as the *complex amplitude*, ω as the angular frequency ($\omega = 2\pi \cdot f$ with f as frequency), and φ as the phase angle.

systemic circulation. Lastly, q represents systemic *cardiac output* $\langle q \rangle$, and R the *total peripheral resistance* R_T. It yields

$$R_T = \frac{\langle p^{Ao} \rangle - \langle p^{At} \rangle}{\langle q \rangle} \approx \frac{\langle p^{Ao} \rangle}{\langle q \rangle}, \tag{2.20}$$

with the reasonable assumption that $\langle p^{Ao} \rangle$ is much higher than $\langle p^{At} \rangle$, i.e., about 100 mmHg versus only a few mmHg. Thus $\langle p^{Ao} \rangle$ *is the driving pressure* that drives blood through the capillary beds in the systemic circulation; the driving pressure in the pulmonary circulation amounts to only about 10 mmHg (Footnote 124). From a practical point of view, (2.20) constitutes a simple approach to estimate R_T with measured values for p and q (Sect. 3.1.3).

Since R_T accounts for the total *systemic vascular resistance* which is a result of *arteries and veins* connected *in series*, compare Figs. 2.39 and 2.41, the contribution of the different vessels to R_T is of interest. In particular, the respective contributions are relevant in terms of active q and p control by the *smooth muscle activation* in the respective vessels' wall (Sect. 2.5.1). That is, the complete vascular bed is made up of arteries, *arterioles*, capillaries, and veins, whereas the respective contributions to R_T are 10%, *60%*, and 30% contributed by the capillaries and veins (Nichols and O'Rourke 2005).

Thus the contribution of *arterioles*, found in the *peripheral vessels*, to *active* R_T *and p control* is the greatest (Milnor 1989) because of their *narrow lumina*[146] (2.19). Accordingly, p drops significantly along the arterioles (Fig. 2.41b). Likewise, the rate of q toward a particular organ or tissue (*local blood perfusion*) is controlled by R of the *arterioles*, i.e., by activation of *smooth muscles* in arterioles[147] (Fig. 2.39).

For instance, an induced *vasoconstriction* of arterioles, e.g., drug related (Footnote 130), causing a decrease in r by as much as 80% may yield a manifold (>20) *increase in the corresponding R* (2.19). In consequence, the vasoconstriction contributes significantly to an increase in R_T, *impaired outflow* of blood out of arteries, and constitutes an *effective method to increase p*, particularly, the level of $\langle p \rangle$ (2.20) *upstream the arterioles*. In contrast, the levels of p and q are reduced in the capillaries, i.e., *downstream the arterioles*. The reverse is true for the *vasodilation*.

---

[146]Although *capillaries* are narrower than *arterioles* (Fig. 2.41a), the capillary resistance to blood flow is lower than the arteriolar resistance. This is because the total cross-sectional area of numerous capillaries is much greater than that of arterioles.

[147]For instance, *vasodilation in a strongly perfused organ* (e.g., in exercising large muscles) may significantly decrease R_T and, in turn, decrease $\langle p \rangle$ (2.20). However, in order to maintain the driving force for blood flow (given by $\langle p \rangle$) through other organs at a constant level, the *cardiac output is typically increased* while other organs and tissues experience an *enhanced vasoconstriction* (e.g., in the viscera) as compensatory effects.

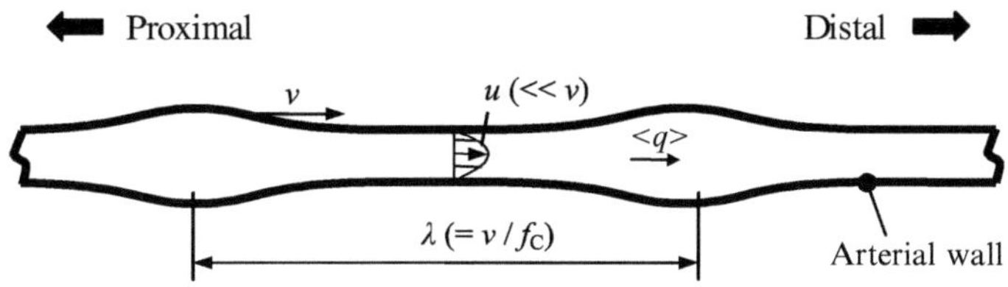

Fig. 2.45 Schematic pulse wave propagation with the heart rate f_C along the artery with velocity v, without considering reflections and the actual length of the arterial tree with respect to the propagation wavelength λ. The parabolic profile of the blood flow velocity u is indicated, compare Fig. 2.44

2.5.2.3 Pulsatile Flow

Only the *steady flow* component is considered by (2.20); i.e., the gradient of $\langle p \rangle$ determines $\langle q \rangle$ at zero oscillating frequency $f = 0$. However, the arterial tree exhibits a highly *pulsatile flow* with the *heart rate* f_C, for the *pressure*, *flow*, and *arterial diameter pulses* (or waves) are generated by the periodic blood surge in the aorta with $f = f_C$, as illustrated in Figs. 2.39 and 2.45.

Pulse Propagation

Pulse waves travel along the arteries with v (a *time–spatial* characteristic), oscillate with f in time domain (*time* characteristic), and oscillate with wavelength λ along a propagation path (*space* characteristic) according to

$$v = \lambda \cdot f. \tag{2.21}$$

Here it is important to be aware of the values of λ applicable to the human arterial tree. That is, a typical cardiac cycle shows $f = f_C \approx 1$ Hz with v in the range of 5 m/s (compare Footnote 138). Equation (2.21) yields a λ value of about 5 m which is *significantly larger* than average *body dimensions* of less than 2 m. Actually the pulsatile wave reaches the peripheral vessels before the cardiac pulsation period is over and, in most cases, even before the blood ejection out of the heart is completed. Such relatively large values of λ have a significant impact on the pulse *wave reflections*; see section "Reflected Pulse Propagation." It should be noted that Fig. 2.45 does not convey the proper relationship between λ and the longitudinal dimension of an artery.

In particular, the *velocity* v of p and q waves is determined by the physical *properties of the arterial wall* and the contained blood within. That is

$$v = \sqrt{\frac{\kappa}{\rho}}, \tag{2.22}$$

where κ is the module of *volume elasticity* and ρ is the density of the propagating medium. The stiffer or the less compressible the medium is, the higher the resultant value of v. The value of κ is given by the ratio of compressive stress $\mathrm{d}p$ to volumetric strain, i.e., to the relative change $\mathrm{d}V/V$ in volume V:

$$\kappa = \frac{\mathrm{d}p}{\mathrm{d}V/V} \approx \frac{\mathrm{d}p}{\mathrm{d}A/A} \approx \frac{\mathrm{d}p}{2 \cdot \mathrm{d}r/r}. \tag{2.23}$$

If l is constant, $\mathrm{d}V/V$ can be approximated by the relative change in $\mathrm{d}A/A$, which is a reasonable assumption because the artery shows a very high Young's modulus in the longitudinal direction and, on the other hand, the artery is longitudinally constrained and prestressed in vivo (Pedley 1980; Nichols and O'Rourke 2005). Furthermore, if the arterial cross section is assumed to be circular, i.e., $A = \pi \cdot r^2$, then κ can be expressed in terms of r (2.23). Usually the concept of *arterial compliance* is used, defined as a change in volume $\mathrm{d}V$ for a given change in pressure $\mathrm{d}p$, or in terms of (2.23), *defined as* V/κ.

An alternative way to *approximate velocity* v considers E as the *Young's modulus* of the wall in the *circumferential direction*, and h as the wall thickness in terms of Moens–Korteweg equation,[148]

$$v = \sqrt{\frac{E}{\rho} \cdot \frac{h}{2r}}, \tag{2.24}$$

with the underlying assumptions that there is a thin wall, i.e., $h/2r \ll 1$, normally less than 0.1 (Nichols and O'Rourke 2005), incompressible inviscid blood, and $l \gg 2r$ (Milnor 1989; Pedley 1980); compare Fig. 2.44b. In (2.24), ρ is the blood density which can be assumed to be nearly constant (Pedley 1980). Thus v is considered a good surrogate for arterial distensibility given by E.

Actually the *values of* v are in the range of 2–12 m/s (Milnor 1989; Nichols and O'Rourke 2005); compare Fig. 2.41 and Footnote 138. Since E from (2.24) or κ from (2.22) increases from the proximal sites, e.g., heart, to the distal sites (Sect. 2.5.1), the value of v increases as well; e.g., in the aorta v is 3–5 m/s while in the radial artery 5–12 m/s (Silbernagl and Despopoulos 2007).

During *pulse propagation*, if only the fundamental harmonic (sinusoidal wave) of the pulsatile wave is considered, then $f = f_\mathrm{C}$ in (2.21). However, the real pulses contain *basic* and *higher harmonics*, which—in terms of the Fourier analysis[149]—sum[150] to yield a *nonsinusoidal* shape of the resulting pulse. Figure 2.46 offers

[148] The equation was independently derived by Adriaan Isebree Moens (1847–1891), a Dutch physiologist, and Diederik Johannes Korteweg (1848–1941), a Dutch mathematician, whereas its fundamental principle was already given by T. Young (see Footnote 135).

[149] Joseph Fourier (1768–1830) was a French mathematician after whom the Fourier analysis and series was named. The analysis represents *general* functions by sums of harmonic sinusoidal functions. The Fourier series is famous for decomposing a *periodic* function into a sum of simple oscillating functions, namely sines and cosines; compare Footnote 150.

[150] According to the Fourier series (Footnote 149), the pressure waveform $p(t)$ with the (fundamental) cardiac period $1/f_\mathrm{C}$; compare Fig. 2.46, can be given as a sum of *real sinusoidal functions* p^k at multiple oscillating frequencies $k \cdot f_\mathrm{C}$ of the fundamental frequency f_C, so that

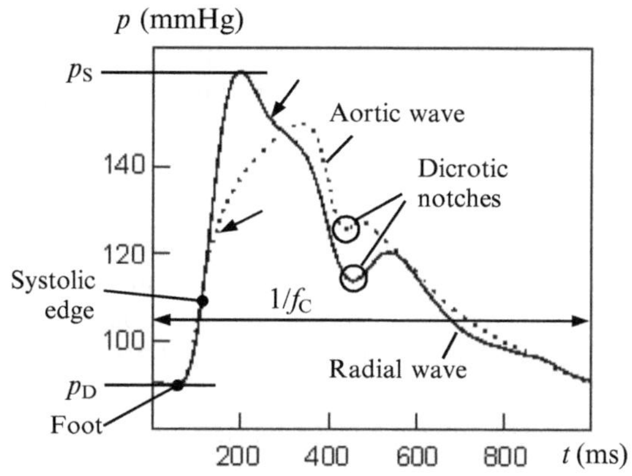

Fig. 2.46 Cardiac blood pressure p pulses, with heart rate f_C, in the aorta and radial artery simultaneously recorded and aligned over time. The approximate start of the reflected waves (inflection points) is indicated by *arrows* (compare Fig. 2.48b). Data taken from (Chen et al. 1997)

evidence for the nonsinusoidal shape if a pressure pulse in the aorta or radial artery is considered. To give a quantitative example, for typical pressure pulses similar to those in Fig. 2.46, the fundamental (basic, first) harmonic at f_C exhibits the highest magnitude and contributes nearly 90% to the total pulse energy while the second harmonic at $2 \cdot f_C$ has the second highest magnitude and contributes only about 8%.

It should be stressed that v *varies with blood pressure*. In particular, v increases with $\langle p \rangle$ and p_D because the artery becomes stiffer with increasing pressure, as discussed in Sect. 2.5.2.1 and shown in Fig. 2.42. The main influence on v is p_D,

---

$$p(t) = \langle p(t) \rangle + \sum_{k=1,2,\ldots}^{\infty} p^k = \langle p(t) \rangle + \sum_{k=1,2,\ldots}^{\infty} \left(P_C^k \cdot \cos(k \cdot 2\pi f_C \cdot t) + P_S^k \cdot \sin(k \cdot 2\pi f_C \cdot t) \right),$$

where $\langle p(t) \rangle$ is the mean value of $p(t)$ over the cardiac period and ${P_C}^k$ and ${P_S}^k$ are the respective amplitudes of the additive sinusoidal functions at multiple oscillating frequencies $k \cdot f_C$ (compare Footnote 145). The amplitudes are calculated as

$$P_C^k = 2f_C \int_0^{1/f_C} p(t) \cdot \cos(k \cdot 2\pi f_C \cdot t)\,dt \quad \text{and} \quad P_S^k = 2f_C \int_0^{1/f_C} p(t) \cdot \sin(k \cdot 2\pi f_C \cdot t)\,dt.$$

A more *compact* and attractive realization of the Fourier series can be given by *complex functions*, so that

$$p(t) = \langle p(t) \rangle + \sum_{k=1,2,\ldots}^{\infty} |\underline{P}^k| \cdot \cos\left(k \cdot 2\pi f_C \cdot t + \gamma^k\right), \quad \gamma^k = \arg(\underline{P}^k),$$

with $\underline{P}^k$ as the *complex amplitude* at the frequency $k \cdot f_C$ and its magnitude $|\underline{P}^k|$ and phase $\arg(\underline{P}^k)$ obtained by

$$\underline{P}^k = 2f_C \int_0^{1/f_C} p(t) \cdot e^{-jk \cdot 2\pi f_C \cdot t}\,dt.$$

since p_D is closer in value to $\langle p \rangle$ than p_S[151] (Footnote 140). For instance, v increases *exponentially* along the aorta with an increase in arterial pressure, since E of the aorta increases exponentially with increasing intravascular pressure; see (2.24). In addition, *v increases with frequency* in the range *up to about 2 Hz* and then levels off, because E of the arterial wall increases with frequency in this frequency range, as described in Sect. 2.5.1.

As with any propagation phenomena, in a *reflectionless* case (section "Reflected Pulse Propagation"), the waves of p and q are *exponentially attenuated* along their pathways because of the viscous component of the arterial wall and the viscosity of the blood (Nichols and O'Rourke 2005). The viscous component can be mainly attributed to the smooth muscles; compare Footnote 134. The attenuation in a viscoelastic tube can be expressed mathematically as

$$p = p_1 \cdot e^{-\alpha \cdot x}, \tag{2.25}$$

where α is the attenuation coefficient, x is the distance downstream from the origin, and p_1 is the pressure at the origin $x = 0$ (Fig. 2.44a); the term $1/\alpha$ can also be interpreted as the propagation distance at which p has already decayed to 37% ($=1/e$) of the original p_1. For instance, the product $\alpha \cdot \lambda$ (for $x = \lambda$), i.e., the attenuation per wavelength, is in the range between 0.8 and 1.5 for the carotid artery and shows no significant variation in the frequency range 40–200 Hz; for the thoracic aorta $\alpha \cdot \lambda$ tends to be lower (Nichols and O'Rourke 2005).

Interestingly, with *higher* v, the wave *damping is less significant* because of *viscous effects*; this has a significant impact on the reflections' behavior with aging (section "Reflected Pulse Propagation"). In addition, *low-frequency* components of the pulse wave are attenuated less than *high-frequency* components under the *assumption of constant* v, e.g., the transmission coefficient of the pressure wave is 0.95 and 0.8 for 1 Hz and 10 Hz, respectively, over the 10 cm segment of the thoracic aorta (Nichols and O'Rourke 2005). That is, α for the low-frequency component is smaller than for the high-frequency component, which is in line with the previously mentioned constant value of $\alpha \cdot \lambda = \alpha \cdot v/f$ under the assumption of plateaued v above 2 Hz. This frequency dependence of α is expected due to the viscous component of the wall and the viscosity of the blood.

Lastly a few remarks should be made concerning *empirical assessment of v*. Although v can be obtained from the propagation of the blood flow wave, blood pressure wave, or that of the vessel's diameter wave, it is technically much easier to measure pressure and diameter in a noninvasive and unobtrusive way; compare Sect. 3.1.3. In particular, the traveling speed of a certain *fiducial point*—within a unique pressure or diameter waveform, see section "Pulse Waveforms of Pressure

[151] Actually there are numerous approaches to estimate both p_D and p_S out of v (as reviewed in Sect. 3.1.3.1), while p_D is more closely related to v, as shown, for instance, in (Kaniusas et al. 2006).

and Flow"—is assessed through the analysis of two pressure or diameter waveforms recorded at two *different locations* along the arterial tree; see section "Estimation from Pulse Running Time" in Sect. 3.1.3.1 for more details. Usually foot-to-foot velocity or rising (systolic) edge velocity is calculated or that of the dicrotic notch (if present) is obtained; compare Fig. 2.46. The aforementioned fiducial points are not arbitrary selected but are chosen in such a way as to *avoid influence of reflected waves*, which significantly change the shape of the traveling waveform during wave propagation; see section "Reflected Pulse Propagation." For instance, the rising systolic edge from above does not contain any reflected wave, for the incident wave traveling with a finite speed v has not yet been reflected or the reflected wave has not yet reached the site of recording.

Blood Pressure and Flow

Steady flow is induced by the *pressure gradient* which yielded an introduction to the resistance R; see (2.18). In a similar way, the so-called complex *longitudinal impedance* $\underline{Z}$ has to be introduced to describe the *pulsatile flow*. The assumption for this aim is that the pulses are reduced into the *sum of specific sinusoidal waves*, according to the Fourier analysis in the frequency domain; see below and Footnote 150. The *complex amplitudes* of the respective sinusoidal waves are denoted as $\underline{P}^k$ for the pressure and $\underline{Q}^k$ for the flow (Footnote 145), with the index k indicating their dependency on the frequency $k \cdot f_C$; compare example from section "Pulse Propagation." Thus $\underline{Z}$ is defined as the ratio of the pressure gradient $\underline{P}_1^k - \underline{P}_2^k$ to $\underline{Q}^k$ for a given harmonic at $\mathrm{k} \cdot f_C$ along a *segment of an artery*, compare Fig. 2.44a, whereas $\underline{Z}(= \underline{Z}^k)$ is then also a frequency-dependent quantity. In analogy with (2.18), we get

$$\underline{Q}^k = \frac{\underline{P}_1^k - \underline{P}_2^k}{\underline{Z}^k} = \frac{\underline{P}_1^k - \underline{P}_2^k}{Z^k \cdot e^{j\varphi^k}}. \tag{2.26}$$

The impedance $\underline{Z}$ is given in terms of its magnitude Z and phase φ. The value of Z is calculated by taking the modulus of $\underline{P}_1 - \underline{P}_2$ divided by the modulus of $\underline{Q}$ for a given frequency. In a complimentary way, φ is given to be the delay between waves of $\mathrm{Re}[(\underline{P}_1 - \underline{P}_2) \cdot \mathrm{e}^{\mathrm{j}\omega t}]$ and $\mathrm{Re}[\underline{Q} \cdot \mathrm{e}^{\mathrm{j}\omega t}]$ for given angular frequency ω $(= 2\pi \cdot f)$; in other words, φ is the delay between the sinusoidal wave of the pressure gradient $\mathrm{d}p/\mathrm{d}x$ and the sinusoidal wave of the flow q for a given frequency; compare Fig. 2.44a. It should be noted that $\underline{Z}^0 = R$, for (2.18) is applicable for the mean values of pulsatile p and q, i.e., for $k = 0$.

Equation (2.26) considers the ratio of $\mathrm{d}p/\mathrm{d}x$ and q in a formal way by introducing the impedance $\underline{Z}$, assuming sinusoidal waves composing pulses. Actually, (2.26) can be rewritten by considering the real arterial properties. That is, if the *pressure gradient* $\mathrm{d}p/\mathrm{d}x$ is considered as a *harmonic* function over time of type $P_G \cdot \cos(\omega t)$ with the gradient amplitude P_G being constant, then q can be given as (Nichols and

O'Rourke 2005)

$$q = \frac{\pi}{\mu} \cdot \frac{r^4 \cdot M(\alpha)}{\alpha^2} \cdot P_G \cdot \sin(\omega t + \varphi) \quad \text{with} \quad \alpha = r \cdot \sqrt{\frac{\omega\rho}{\mu}}. \tag{2.27}$$

Here α describes the kinematic behavior of the liquid. The function M with $0 < M < 1$ and phase φ with $0 < \varphi < \pi/2$ represent expressions containing Bessel functions and α as parameters. In comparison with steady flow (2.17), the amplitude of the sinusoidal flow no longer varies linearly with the pressure gradient and there is a flow delay of $\pi/2 - \varphi$. It can be assumed that $\varphi > 0$ because of the finite *flow momentum* and *blood viscosity*; thus the waveform of $\mathrm{d}p/\mathrm{d}x$ leads the waveform of q (see below).

However, when α is less than 0.5, i.e., in *small arteries* ($r \to 0$) or at *low frequencies* ($\omega \to 0$), M/α^2 approximates $1/8$ and φ tends to $\pi/2$; actually, (2.17) and (2.27) become identical and q shows steady flow behavior with no phase delay between q and $\mathrm{d}p/\mathrm{d}x$. In contrast, in *large arteries* or at *higher frequencies*, when α is large, M/α^2 and φ decrease; furthermore, the deflection of q decreases for given P_G, and the delay of q versus $\mathrm{d}p/\mathrm{d}x$ approaches $\pi/2$.

It should be noted that p and $\mathrm{d}p/\mathrm{d}x$ are pulsatile, i.e., *nonsinusoidal*, as already mentioned; compare Figs. 2.46 and 2.47b. The contributing sinusoidal components p^k and $(\mathrm{d}p/\mathrm{d}x)^k$, which are time functions for (2.26) and (2.27), can be resolved by the Fourier analysis; compare Footnote 150. Then the resulting sinusoidal contributions of q^k are summed up to synthesize the nonsinusoidal q $(= \Sigma q^k)$. It is generally agreed that nonlinear terms[152] relating p (or $\mathrm{d}p/\mathrm{d}x$) and q are very small (Nichols and O'Rourke 2005), which justifies the above synthesis term by term.

Figure 2.47b, c illustrates the waveform of $\mathrm{d}p/\mathrm{d}x$ and the corresponding *delayed* q in a qualitatively accurate manner. The observable delay $\varphi' > 0$ demonstrates also the aforementioned behavior of $\varphi > 0$. However, it should be noted that $\varphi \neq \varphi'$ because $\varphi(= \varphi^k)$ is defined for a single harmonic at $k \cdot f_C$ [(2.26) and (2.27)] while φ' already considers the sum of harmonics yielding nonharmonic shapes of $\mathrm{d}p/\mathrm{d}x$ and q (Fig. 2.47b, c).

From a practical point of view, the pressure gradient $\mathrm{d}p/\mathrm{d}x$ in arteries, as considered in (2.26) and (2.27), is difficult to measure. Thus absolute values of pressure p and their relation with q are of high interest. Considering the complex amplitudes $\underline{P}_1$ and $\underline{Q}$ at the input of an artery in the *absence of wave reflections* (section "Reflected Pulse Propagation"), compare Fig. 2.44a and (2.26), and omitting the indices k for the sake of convenience, the *characteristic impedance* $\underline{Z}_0$ is defined as

[152] Strictly speaking, the *arterial* system is highly *nonlinear* concerning the relationship between p and q. The nonlinear terms result, for instance, from the nonlinear changes in r with p (Fig. 2.42) in that the wall gets stiffer with increasing strain (Sect. 2.5.2.1).

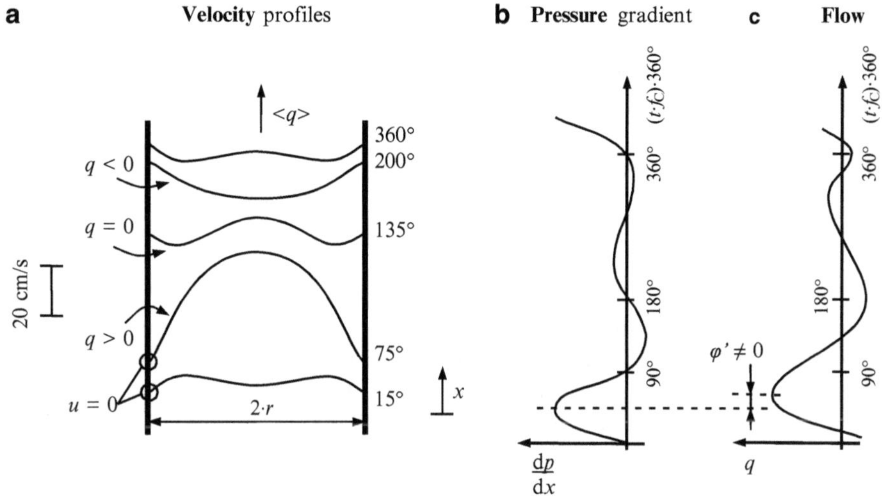

Fig. 2.47 (**a**) Schematic velocity u profiles of the pulsatile blood flow q over the arterial radius r with an indicated approximate amplitude of u for an aortic segment in resting state; compare with Figs. 2.41a and 2.44a. The phase within the cardiac cycle is indicated with a single cycle lasting 360°. (**b**) The corresponding spatial gradient of pressure p with f_C as heart rate. (**c**) Estimated q. Data are partly taken from (Nichols and O'Rourke 2005); compare with Fig. 2.50b, c

$$\underline{Z}_0 = \frac{\underline{P}_1}{\underline{Q}} = \underline{Z} \cdot \frac{v}{j\omega} \approx \frac{\rho \cdot v}{A} = Z_0, \tag{2.28}$$

where ρ is the density of blood. That is, the impedance $\underline{Z}_0$ describes the relationship between the pulsatile pressure and pulsatile flow if the time functions p_1 and q from Fig. 2.44a do not include any reflected waves returning back to the input. In other words, the arterial tree is either approximated to be very long, the vascular bed is maximally dilated, or frequencies of the signal components involved are relatively high, with all three *approximations* yielding *negligible wave reflection* effects; for details see section "Reflected Pulse Propagation." Likewise, the approximation in (2.28) is valid only for large arteries, at higher frequencies ω, or for large values of α in (2.27).

Equation (2.28) shows that the phase of $\underline{Z}_0$ is practically zero, i.e., $\underline{Z}_0 \approx Z_0$, which means that the phase shift between p_1 and q is zero under the above assumptions. In addition, (2.24) and (2.28) yield that Z_0 varies directly with v and E of the arterial wall and inversely with A.

In contrast to $\underline{Z}$, when considering only a *segment of artery*, the impedance $\underline{Z}_0$ accounts for the whole *vascular tree*, both $\underline{Z}$ and $\underline{Z}_0$ excluding wave reflections. According to (2.28), the relationship between $\underline{Z}_0$ and $\underline{Z}$ is similar to how absolute values are related to gradient values. The value of $\underline{Z}_0$ is principally determined by the *distensibility of major arteries* immediately distal to the site of measurement. For instance, a high distensibility of the proximal aorta, i.e., a low E of its wall, is responsible for low $|\underline{Z}_0|$ which is "seen" by the left ventricle of the heart. The low value of $|\underline{Z}_0|$ reduces the heart load during systole and thus the energy which the heart has to expend per beat.

In an analogous way, the *input impedance* $\underline{Z}_I$ can be defined as the ratio of $\underline{P}_1$ to $\underline{Q}$ at the input to any region of the circulation when *considering reflections* and all of the *vascular tree*; compare Fig. 2.44a. That is,

$$\underline{Z}_I = \frac{\underline{P}_1}{\underline{Q}}, \tag{2.29}$$

yielding similarities with (2.26) and (2.28). In contrast to $\underline{Z}_0$, the impedance $\underline{Z}_I$ considers formation of nodes and antinodes of p and q in terms of pulse wave reflections (Footnote 170); see section "Reflected Pulse Propagation."

As can be seen from the above definitions of the *impedances* $\underline{Z}$, $\underline{Z}_0$, and $\underline{Z}_I$, their modulus and phase generally *depend on frequency*. In actuality, their frequency behaviors are completely *different.*[153] Interestingly, the impedance $\underline{Z}_I$ is equal to R_T at $f = 0$, i.e., at the steady state, determining the mean blood pressure (2.20). On the other hand, the impedance $\underline{Z}_I$ is equal to $\underline{Z}_0$ at $f = \infty$, i.e., with strongly diminished (absent) reflections; see section "Reflected Pulse Propagation." If both extremes of $\underline{Z}_I$ over f are compared, then $|\underline{Z}_0|$ can be approximated at about 5% of R_T (Nichols and O'Rourke 2005).

It was shown that *dp/dx precedes q* in the time domain [Fig. 2.47b, c and (2.27)], which is actually revealed by a positive phase of $\underline{Z}$. In contrast, the absolute pressure *p lags behind q*[154]; compare Figs. 2.48b, c and 2.49 for the phase shift in between p and q, and lags behind the peak value of u. The latter phase relationship is actually given by (mainly) the negative phase of $\underline{Z}_I$ (2.29); see Footnote 153.

---

[153]The *modulus of* $\underline{Z}$ increases over f while its *phase* is positive and increases from zero at low f and then levels off, i.e., dp/dx gradient precedes q [compare (2.26)]. The *modulus of* $\underline{Z}_0$ is nearly constant over f while its *phase* is zero, i.e., $\underline{Z}_0 = Z_0$. Note from (2.28) that if $\underline{Z}$ increases with f, the value of Z_0 is not supposed to change with f. Lastly the *modulus of* $\underline{Z}_I$ falls from its high initial value at $f = 0$ (referred as the elastic Windkessel model at this frequency, Footnote 139) to low values at high f with superimposed fluctuations. The *fluctuations* diminishing in amplitude over f arise because of *reflections*, standing waves (compare Footnote 170), and frequency-dependent wave attenuations. The first minimum in the fluctuations (at about 4 Hz) corresponds to the node of p at the distance of $\lambda/4$ [$= v/(4 \cdot f) \approx v/16$ Hz, (2.21)] from the functionally discrete single reflection site. These fluctuations and the whole $|\underline{Z}_I|$ spectrum are shifted to the right for early reflections and increased v, e.g., at hypertension (Footnote 166). The *phase of* $\underline{Z}_I$ is zero at $f = 0$, becomes negative at higher f, and then fluctuates between negative and positive values with even higher f (Nichols and O'Rourke 2005). With increasing *age*, the modulus $|\underline{Z}_0|$ increases while the spectrum of $|Z_I|$ is shifted upward and to the right because of arterial stiffening (Sect. 2.5.1).

[154]During vasodilation (with reduced reflections, Footnote 171), for instance, the phase in between p and q is strongly reduced, i.e., the phase of $\underline{Z}_I$ is flattened (Footnote 153), and the contours of p and q become more similarity (Nichols and O'Rourke 2005).

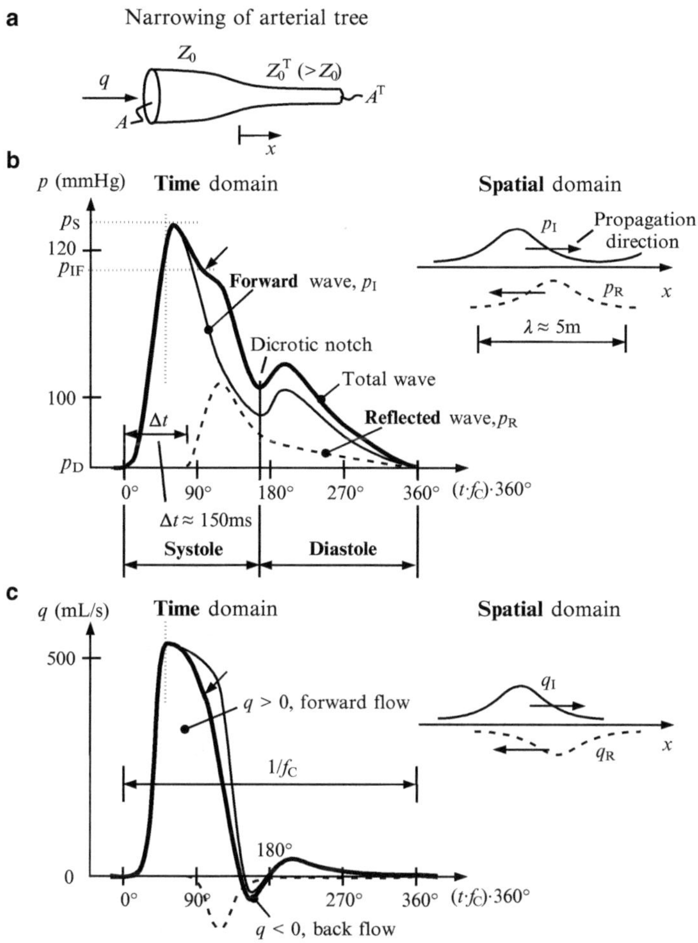

Fig. 2.48 (**a**) Narrowing of arterial tree which represents the site of pulse wave reflection of closed-end type. (**b**) Sketched positive reflection of blood pressure p waves with Δt as round-trip travel time from the aorta to the major reflecting site and back; compare Footnote 161 and Fig. 2.40. (**c**) Negative reflection of blood flow q waves. Approximate starts of the reflected waves (inflection points) are indicated by *arrows*. The absolute (approximate) values for p and q refer to the aorta

As already mentioned, pulsatile blood *flow* is actually determined by the *pressure gradient* along the artery, as given by (2.26); thus it is not directly related to p itself but to the spatial derivative $\mathrm{d}p/\mathrm{d}x$. In order to *illustrate* this in a simplified manner, Fig. 2.50a shows two identical but about 30 ms delayed waveforms of p, which can be assumed to represent blood pressure recordings at two arterial sites with a short distance in between; i.e., a distance of about 15 cm if v is assumed to be 5 m/s [(3.6) and Fig. 2.41b]. The calculated pressure gradient Δp—or the estimated q—in Fig. 2.50b oscillates about the mean with a reversed sign over the cardiac cycle

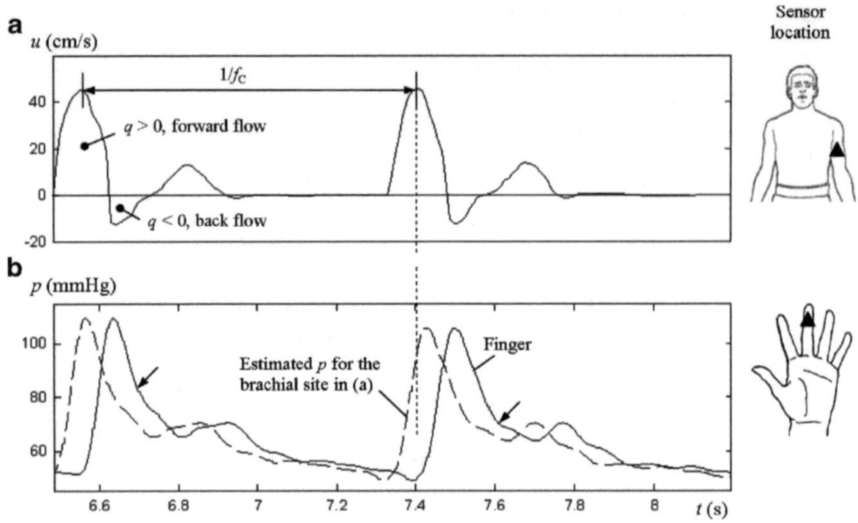

Fig. 2.49 (**a**) Blood flow velocity u in the brachial artery from the upper left hand, as recorded by the echocardiographic method (pulsed Doppler, for methodology see Sect. 3.1.3.2). (**b**) The course of the simultaneous blood pressure p from a finger on the left hand, as exemplified by barocardiogram (Sect. 3.1.3.1). The *dashed line* corresponds to a 70 ms advanced version of p, whicht roughly estimates p upstream in the brachial artery where u was measured. The delay results from the assumption of 7 m/s and 50 cm for the pulse wave propagation velocity and the distance to the brachial site, respectively (3.6). Approximate starts of the reflected waves (inflection points) are indicated by *arrows*; see also Fig. 2.54. Compare measured waveforms u and p with schematic behavior of q and p from Fig. 2.48

though the mean $\langle q \rangle$ is obviously nonzero. There is a significant similarity between, on the one hand, the *estimated* q from Fig. 2.50b and, on the other hand, *schematic* q from Fig. 2.47, Fig. 2.48, and even *real* q from Fig. 2.49, which qualitatively proves the derivation of q out of $\mathrm{d}p/\mathrm{d}x$. In addition, the waveform of $\mathrm{d}p/\mathrm{d}x$ precedes that of p, as expected from above. Obviously, this discussion does not consider effects like attenuation losses during wave propagation, wave reflections, and waveform changes along the arterial tree (sections "Pulse Waveforms of Pressure and Flow" and "Reflected Pulse Propagation").

To be more precise, the *mass* of the fluid and its *inertia* have to be considered in order to assess q and $\mathrm{d}p/\mathrm{d}x$ behavior. When a positive $\mathrm{d}p/\mathrm{d}x$ is applied to resting blood, compare (2.27), the blood will first appear to resist the movement ($q = 0$, $u = 0$); in analogy, at $\mathrm{d}p/\mathrm{d}x = 0$ the momentum of blood would keep it moving until the opposing *viscous forces* will bring it to rest. With increasing local flow velocity, the viscous drag also increases. As a result, there is an obvious aforementioned *phase shift in between* q *and* $\mathrm{d}p/\mathrm{d}x$ (Fig. 2.47b, c) with the gradient leading.

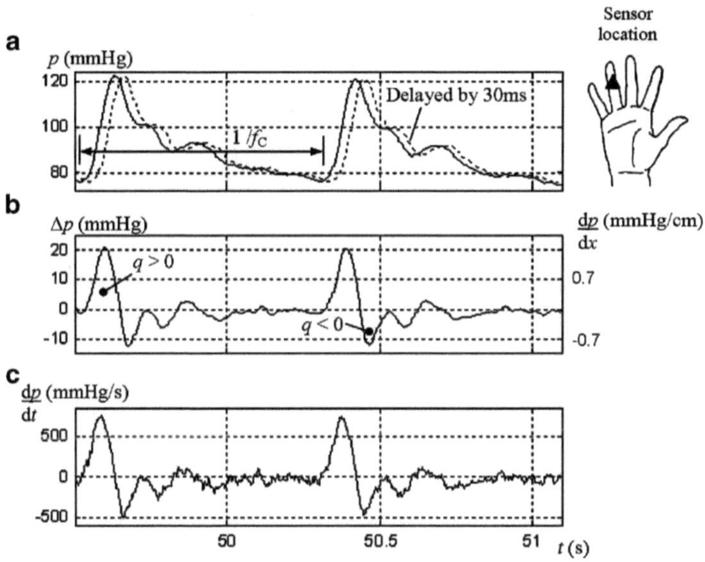

Fig. 2.50 (**a**) The pulsatile changes of the peripheral blood pressure p, recorded as barocardiogram from a finger on the right hand (for methodology see Sect. 3.1.3). The *dashed line* corresponds to a 30 ms delayed version of p. (**b**) The difference of the nondelayed and delayed version of p from (**a**), which approximates the spatial derivative of p (e.g., along coordinate x in Fig. 2.44) and thus the blood flow q (compare Fig. 2.47a). (**c**) The time derivative of the nondelayed version of p from (**a**)

Figure 2.47a illustrates the resulting shape and *behavior of u profiles* over the spatial gradient $\mathrm{d}p/\mathrm{d}x$, the latter shown in Fig. 2.47b. The *flow near the walls* follows the pressure gradient most closely and reverses easily when the gradient reverses because of relatively low velocity here. As shown in Fig. 2.47a, a lamina of zero velocity is at the wall that facilitates the flow reversal; the lamina close to the walls responds and moves first in response to the pressure gradient from Fig. 2.47b, the flow successively involving the lamina toward the axis of the vessel.

Figure 2.49 demonstrates the *real behavior* of u and p, where u was recorded in the brachial artery and p in the finger (Fig. 2.40a). By advancing p by the pulse transit time from the brachial site to the finger, an estimation of p in the brachial artery can be attained; see dashed line in Fig. 2.49b. The transit distance of about 50 cm and an estimated $v \approx 7\,\mathrm{m/s}$ yield roughly 70 ms as the effective *propagation time* (3.6). It should be noted that the course of *u approximates that of q*; the level of q is attained by integration of the nearly parabolic u profile (Figs. 2.44b and 2.47a) over the nearly constant cross-section A of the vessel during the cardiac cycle; compare (2.17). As mentioned in Sect. 2.5.1, the radius r changes by about $\pm 5\%$; thus $A = \pi \cdot r^2$ changes by only about $\pm 10\%$.

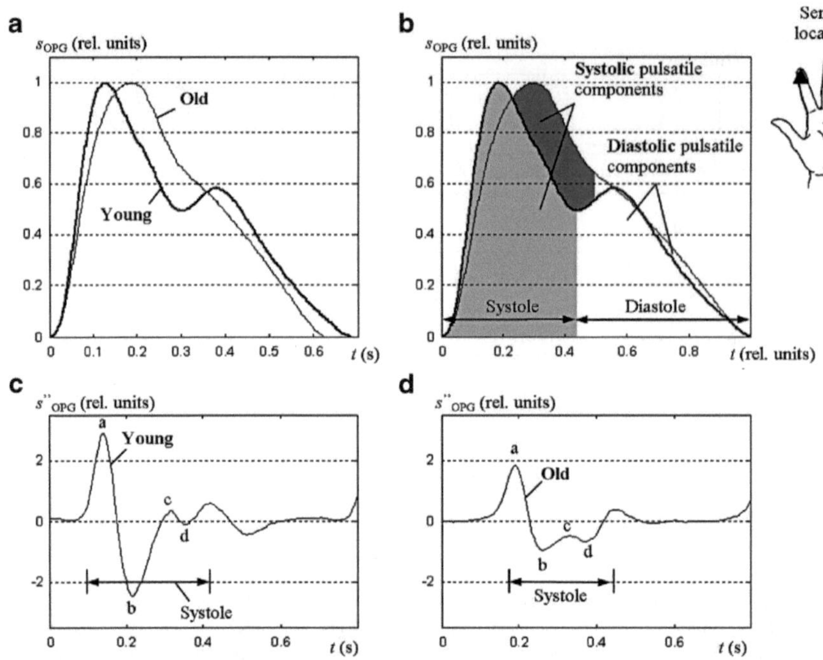

Fig. 2.51 Morphological changes in the pulse waveform of optoplethysmogram s_{OPG} from a finger (Sect. 3.1.3.3) by comparing a young person of age 19 and an elderly one of age 64. (**a**) Absolute time t axis. (**b**) Relative t axis. The efficiency of the pulsatile flow is demonstrated by the systolic and diastolic pulsatile components (related to local pulse wave); the less of a difference there is in between, the more efficient the arterial system will be in generating smooth blood flow; thus the waveform from a young person is more efficient. (**c, d**) The corresponding second time derivative s''_{OPG} of s_{OPG} with the discrete waves a–d (Sect. 3.1.3.3)

It is interesting to consider the *pulsatile components* of p and q from an energetic point of view. The pulsatile components, as a consequence of the heart's intermittent contraction, represent essentially *wasted work* lost in large arteries, with the work being proportional to $\int (p - \langle p \rangle) \cdot (q - \langle q \rangle) \cdot dt$. In contrast, the *useful work* $\langle p \rangle \cdot \langle q \rangle$ because of the steady components of p and q represents physiological energy used (lost) mainly in the arterioles for perfusion of the tissues, whereas p and q are almost steady. In normal circumstances, about 10% of the total work is pulsatile or wasted (Nichols and O'Rourke 2005). Thus the efficiency of the arterial system can be seen through comparison of *steady* and *pulsatile* components.

For instance, Fig. 2.51b demonstrates the decreasing efficiency in the *elderly* in comparison with *adolescents*. Here optoplethysmogram is considered as a biosignal, with its contour bearing a strong similarity to that of p; see Sect. 3.1.3.3. The *inefficiency* manifests as a large difference between $\langle p \rangle$ *during systole* and $\langle p \rangle$ *during diastole*. The $\langle p \rangle$ during systole is graphically approximated as area during systole,[155] which is a major determinant of (disadvantageous) myocardial oxygen demand and blood demand. On the other hand, $\langle p \rangle$ *during diastole*, as enclosed area during diastole, is a major determinant of (advantageous) ventricular perfusion in terms of *coronary blood flow* because the myocardium squeezes the coronary arteries during systole.[156] An *excessive pulsatility* in p and q may affect the *vascular bed of end organs* which have a low vascular resistance and a relatively high q, particularly that of brain and kidney (Sola et al. 2010). In addition, the excessive pulsatility in peripheral p is associated with *microvascular damage* and impaired function (Mitchell 2009); it is particularly apparent after age 60 when the pulsatile power is more weakly reflected at the interface between central and muscular arteries (Fig. 2.40a) and therefore is largely transmitted into peripheral microcirculation; see section "Reflected Pulse Propagation."

Lastly, a practical equation for the estimation of the *pulsatile* q from the left ventricular *stroke volume* V_S (Fig. 2.38b), i.e., the volume of blood ejected with each heartbeat, should be given,

$$q = V_S \cdot f_C. \tag{2.30}$$

The *pulsatile* q is usually referred to as *cardiac output*, i.e., the volume of blood pumped by the heart over a particular period of time. For the human arterial tree, the value of q can be easily estimated; usually f_C is in the range of 1 Hz in rest and V_S about 80 ml which yields q of about 5 l/min (Silbernagl and Despopoulos 2007).

Pulse Waveforms of Pressure and Flow

The *waveforms of p and q* over the cardiac cycle with the duration $1/f_C$ depend on

- *Cardiac properties* comprising ventricular properties, ventricular filling, and ejection.

---

[155] The integral of p during systole is termed *systolic time index* while that during diastole is termed *diastolic time index*. The ratio of latter indices is a proportional measure of the propensity for myocardial ischemia, i.e., of the restriction in coronary blood supply. In analogy, mean p_D is another measure for the efficiency of the coronary perfusion. To give an example, the ejection period in the elderly does not decrease appropriately with tachycardia (increasing f_C) so that the diastolic time index decreases, increasing the likelihood of myocardial ischemia (Nichols and O'Rourke 2005). In other words, a reasonable strategy for the optimization of the ventricular–vascular interaction should target the *minimization of the systolic time* index and the *maximization of the diastolic time* index.

[156] For instance, an augmented central p_S is associated with an increased opposition to systolic ejection (increased afterload) and thus with development of left ventricular *hypertrophy*. Conversely, a diminished central p_D compromises *myocardial oxygen supply*. Both effects may lead to a *vicious cycle of events* because the increased left ventricular mass will require an increased oxygen supply (Sola et al. 2010).

- *Vascular properties* which refer to physical properties of the arterial tree, as stiffness of vessels and degree of vasomotor tone affecting v and reflections (Sects. 2.5.1 and 2.5.2.1).

In fact, the influence of these cardiac and vascular properties is approximated in (2.20). It could be descriptively rewritten as

$$\langle p \rangle = \langle q \rangle \cdot R_T \approx V_S \cdot f_C \cdot R_T. \tag{2.31}$$

In an *approximation*, the most important parameters *affecting p* are V_S, f_C (cardiac properties), and R_T (vascular properties). That is, an increase in any of these parameters, if not compensated by a simultaneous decrease in another parameter, will yield an increased $\langle p \rangle$; compare hypertension from Footnote 166. Likewise, V_S is inversely proportional to R_T; however, an increased R_T lowers V_S for only a few beats because the heart adjusts and beats more strongly to eject more blood.[157]

However, the behavior of p and q is *pulsatile* with their particular waveforms deserving an extended description. As shown in Figs. 2.46, 2.48b, 2.49, and 2.52a the *contour of p* exhibits a strong rise at the beginning of systole which corresponds to the blood ejection, a *primary peak* during *systole*, an optional *secondary peak* during *systole* (e.g., in the elderly, see section "Reflected Pulse Propagation"), a small dicrotic notch,[158] a *diastolic wave*, an optional *secondary peak* during diastole (e.g., in adolescents), and a slow *exponential* decrease during diastole. The secondary peaks arise because of *reflections*, as laid down in section "Reflected Pulse Propagation." Actually, the *p level during systole* and the *duration of systole* determine the *myocardial oxygen demand*, while the p level during *diastole* and the duration of diastole determine the efficiency of the *coronary perfusion*; compare Footnote 155.

The highest *peak* and lowest *nadir* (trough) of the *pressure* wave are considered to represent the following traditional wave properties or characteristics (Figs. 2.46 and 2.48b):

- *Systolic blood pressure* p_S is defined as the peak arterial p. It is given as a sum of initial pressure value before heart contraction (i.e., diastolic value) and maximum pressure deflection during systole (i.e., pulse pressure). The level of p_S is determined by the amount of blood being forced into the aorta with each heart contraction. In addition, reflections may play a significant role if the secondary peak comprises the highest peak of the pressure wave. That is, the level of p_S is mainly *determined by* V_S and the *force of contraction*. The force is proportional to the resistance to blood flow determined by the *elasticity* [or stiffness, (2.23)]

---

[157] Given a lowered V_S, more blood remains in the ventricle and the end-diastolic volume is larger before the next contraction. Thus the ventricle is stretched to a greater degree and then it contracts more forcefully (Footnote 225) to maintain a constant $\langle q \rangle$.

[158] *Dicrotic notch* describes a transient of p for the time interval, when p in the aorta exceeds that in the left ventricle, backflow of arterial blood begins, and the aortic valve closes.

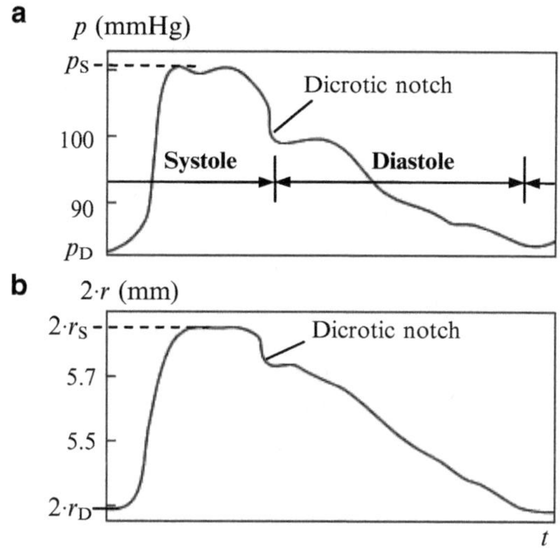

Fig. 2.52 Comparison of signal waveforms from the carotid artery over a single cardiac cycle. (**a**) Blood pressure p using data from (Shau et al. 1999). (**b**) The corresponding inner diameter $2 \cdot r$ with (maximum) systolic diameter $2 \cdot r_S$ and (minimal) diastolic diameter $2 \cdot r_D$

of the aorta adjacent to the heart. An increase in either will increase p_S; for instance, if the arterial wall becomes stiffer, e.g., in atherosclerosis (Footnote 137), the vessels are not able to distend with the pulsatile blood volume and so p_S increases for the same V_S.

- *Diastolic blood pressure* p_D is defined as the arterial p that exists at the end of the diastolic pressure decay. The *exponential decay* starts at the closure of the aortic valve and ends at its opening. The arterial *time constant* of the pressure decay is proportional to the *level of* R_T (2.20), aortic pressure p_S, and arterial compliance (2.23), as can be seen from the deBoer model (deBoer et al. 1987). Thus the final pressure p_D of this decay is proportional to p_S built up during systole (*start* of the exponential decay), proportional to the time constant (*rate* of the decay), and inversely proportional to the duration $1/f_C$ of the cardiac cycle (*duration* the decay continues for). The level of p_D is significantly affected by the degree of the *arteriolar tone*; if there is an increased *arteriolar vasoconstriction* (increased R_T), this will impede blood flowing out of the arterial system to the capillaries, and p_D will rise. In analogy, with lower f_C the resulting p_D decreases as there is greater time for blood to flow out of the arteries.
- *Systolic–diastolic blood pressure* or *pulse pressure* $p_S - p_D$ is *determined by* V_S and *aortic stiffness*; an increase in either increases $p_S - p_D$ and potentiates multiple cardiovascular diseases. Interestingly, increasing f_C tends to raise the pulse pressure because of dynamically increased stiffness of the arteries (Sect. 2.5.1), under the assumption that the level of V_S does not change.
- *Mean arterial blood pressure* $\langle p \rangle$ is proportional to the product of *mean blood flow* $\langle q \rangle$ [cardiac output, (2.30)] and R_T; compare (2.20). In other words, if q (measure for blood *inflow*) and R_T (*outflow*) change reciprocally and proportionately then $\langle p \rangle$ will not change. For the estimation of $\langle p \rangle$ from p_S

and p_D, see Footnote 140. It is important to note that $\langle p \rangle$ is considered to be the *driving pressure* for blood perfusion of vital organs.

It should be noted that the level of p_S and the pulse amplitude $p_S - p_D$ are directly and *positively related* to the *arterial stiffness* or arterial elastic modulus,[159] whereas p_D is *inversely related* to the stiffness (Nichols and O'Rourke 2005). That is, the stiffer the artery, the higher is the p_S, the lower the p_D, and the higher the pulse deflection $p_S - p_D$.

The *contour of q* shows a positive wave at the beginning of the ejection phase and a negative wave at the end of the ejection phase due to *forward flow* and *backward flow*, respectively. As illustrated in Fig. 2.48c, see also Figs. 2.47 and 2.49, the backward flow wave coincides with the dicrotic notch and is due to the *negative reflection* of the flow wave in the periphery, subtracting from the incident flow wave[160]; see section "Reflected Pulse Propagation" for the involved reflection phenomena of closed-end type. The negative reflection causes disadvantageous deceleration of systolic flow. In *adolescents*, the deceleration portion of q is convex to the right, as shown in Fig. 2.48c, by *middle age*, the deceleration becomes concave to the left, and in the *elderly* the deceleration is linear from peak q to zero.

In contrary to the pressure pulse, the oscillation of the *flow velocity diminishes* along the propagation pathway, compare Fig. 2.47a, because of both attenuation and negative reflection. Similarly, the *flow oscillation* also decreases because of the decreasing arterial cross section toward the periphery; see (2.17).

Thus the shapes of p and q waves are dramatically different; while p in the aorta continues to increase during the ejection, q decreases because of the previously described negative reflection (Fig. 2.48b, c). The resulting increase in the *ratio* of the pulsatile *p amplitude* to the pulsatile *q amplitude* along the arterial pathway is reflected by the corresponding increase of $\underline{Z}_I$ at low frequencies [(2.29) and Footnote 153], i.e., an increase of the low-frequency components $|\underline{P}_1|$ with respect to $|\underline{Q}|$ (Nichols and O'Rourke 2005). In other words, the change in the waveform of p due to that of the q wave depends on the changes of $\underline{Z}_I$ over frequency. Actually $\underline{Z}_I$ also depends on the distance from the *measurement site* to the *reflection site*, compare Fig. 2.44a, for $|\underline{Z}_I|$ is at minimum at $\lambda/4$ distance from the reflection site (Footnote 153). Generally, the amount the waveform changes while being transmitted is less in elderly persons with degeneration of elastic arterial structures and thus less reflections; see Sect. 2.5.1 and section "Reflected Pulse Propagation."

Since the shape of the waveform is strongly influenced by the reflections (section "Reflected Pulse Propagation"), a *classification scheme* of the *contours* of the ascending aortic p has been established (Nichols and O'Rourke 2005). This scheme is *based* on both the *prominence of the reflections* within the resulting contour and

[159]The tendency of increasing p_S and $p_S - p_D$ with *increasing elastic modulus* could be derived from (2.23), if Δp is approximated by $p_S - p_D$, or even from the definition of the pressure–strain modulus (Footnote 133).

[160]In severe *heart failure*, for instance, the negative wave reflection (Fig. 2.48c) may even cause premature termination of ventricular ejection and thus decreased V_S.

the *timing* between the secondary peaks and the systolic peak (with p_S). Here the prominence is quantitatively assessed by the *augmentation index* I_A, as defined later in section "Reflected Pulse Propagation." There are four types of waves to be distinguished:

- *Type A* in which case the inflection point (or shoulder), which indicates the onset of the reflected wave, *precedes* p_S and $I_A > 12\%$ (2.33), being typical in *elderly* persons between 40 and 65 years, e.g., aortic wave in Fig. 2.46.
- *Type B* similar to type A but $I_A < 12\%$, typical in *adults* between 30 and 45 years.
- *Type C* in which case the inflection point *follows* p_S and $I_A < 0$ (Footnote 169), typical in *adolescents* under 30 years, e.g., schematic wave in Fig. 2.48b.
- *Type D* in which case the reflected wave is completely *merged* into the incident wave, typical in the *elderly* over 65 years or in patients with hypertension (Footnote 166), e.g., pulse waveform in Fig. 2.51a considering old person.

Reflected Pulse Propagation

As discussed in section "Pulse Propagation," the p and q waves are exponentially attenuated along their pathway if reflections and changing properties of the arterial pathway are not considered. However, in reality the *pulse of* p $(= p_S - p_D)$ tends to *increase* as it travels from the heart to the periphery along the *arterial tree* (Fig. 2.41b) while p_D and $\langle p \rangle$ decrease. The *q pulse markedly diminishes* along the arterial tree. The pulse of p increases because of increased *arterial stiffness* (Fig. 2.40b, Fig. 2.42, and Sect. 2.5.1) yielding an amplified forward wave (traveling away from the heart) and, on the other hand, because of emerging reflections[161] yielding a *constructively interfering* reflected waves. Figure 2.53a demonstrates rising forward wave while propagating from the aorta to the periphery. The pulse of q diminishes because of increased arterial stiffness, reduced A [compare (2.28)], reduced $\langle u \rangle$ (Fig. 2.41a), and a *destructively interfering* reflected wave.

In order to address the above reflections, a narrowed arterial tree from Fig. 2.48a should be considered. The *reflection factor* Γ on this *discontinuity* with the impedance Z_0 from the input side and $Z_0{}^T$ from the output side can be defined as

---

[161] Waves of p and q get reflected wherever there is a *discontinuity* along the propagation path; compare (2.32), Fig. 2.48a. If the ratio $Z_0 = P_I/Q_I$ (2.28) of incident pressure waves $p = p_I$ to incident flow waves $q = q_I$ with the respective amplitudes P_I and Q_I (Footnote 145) has to change at the discontinuity with the impedance $Z_0{}^T > Z_0$ ($A > A^T$, Fig. 2.48a), reflected waves p_R and q_R with $Z_0 = -P_R/Q_R$ have to arise to fulfill $Z_0{}^T = (P_I + P_R)/(Q_I + Q_R)$, i.e., to fulfill the boundary condition. At the discontinuity with $Z_0{}^T > Z_0$, the resulting values are p_R, $P_R > 0$ (*positive reflection*) and q_R, $Q_R < 0$ (*negative reflection*). Actually, an *incomplete* occlusion of the arterial tree ($Z_0{}^T \neq \infty$) is responsible for the still pulsatile waveforms of p and q. In the case of a complete occlusion, an ascending p $(= p_I + p_R)$ would occur along the arterial tree because the wavelength λ ($v/f_C \approx 5\,\mathrm{m}$) (2.21) of the propagating pulse is much larger than the physical dimensions of the human body.

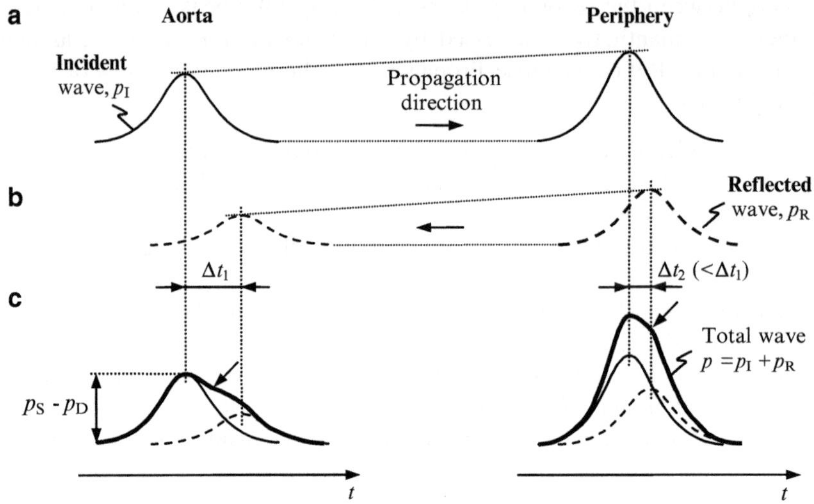

Fig. 2.53 Varying composition of (**c**) pulsatile blood pressure p waves out of (**a**) the incident wave p_I and (**b**) reflected wave p_R along the arterial propagation path, beginning at the aorta and ending in the periphery; compare Fig. 2.48b. The wave p_I typically increases along the pathway due to increasing arterial stiffness and despite attenuation of p_I over propagation distance (2.25). In the case of p_R, it is assumed that attenuation dominates its propagation from the periphery toward the aorta. The round-trip travel time Δt from the considered region to the major reflecting site and back is greater at the aorta than in the periphery, i.e., $\Delta t_1 > \Delta t_2$. Approximate starts of the reflected waves (inflection points) in p are indicated by *arrows*

$$\Gamma = \frac{P_R}{P_I} = -\frac{Q_R}{Q_I} = \frac{Z_0^T - Z_0}{Z_0^T + Z_0}. \tag{2.32}$$

Here P_I and Q_I are the respective amplitudes (=peak values) of the incident pulsatile pressure and flow waves p_I and q_I, whereas P_R and Q_R are the respective amplitudes of the reflected pressure waves p_R and q_R; compare Fig. 2.48b, c and Footnote 145. In the depicted case of Fig. 2.48a, $Z_0{}^T > Z_0$ holds because of a reduced A on the output side; consider (2.28). The value of Γ is always lower than 1 because of energy dissipation during the reflection event; in the arterial system $\Gamma < 0.8$ (Nichols and O'Rourke 2005). It should be noted that if reflections are considered at the entrance of the pulse wave into the peripheral vascular bed, then $Z_0{}^T$ in (2.32) can be substituted by R_T from (2.20).

From a practical point of view, *derivation*[162] of the *incident* and *reflected* p and q waves is usually of high interest, given the measured p and q waves (Sect. 3.1.3).

[162]In analogy to Footnote 161, *measured waves* of pressure p and flow q can be *dissected* into their incident components p_I, q_I and reflected components p_R, q_R with $p = p_I + p_R$ and $q = q_I + q_R$, as demonstrated in Fig. 2.54. Considering that the incident wave has sufficient time to travel to the periphery and back (as the reflected wave) in a single cardiac cycle, it can be deduced from

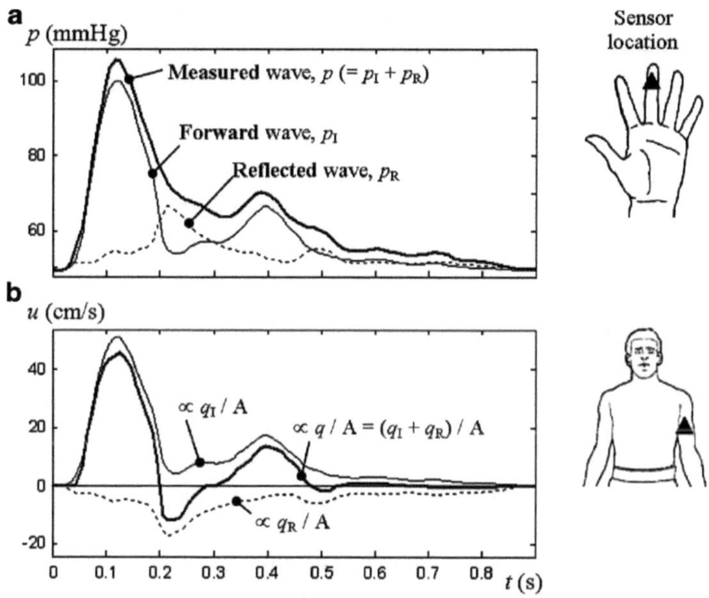

Fig. 2.54 Derivation of the incident and reflected blood pressure p and blood flow q waves, according to the procedure from Footnote 162. (**a**) The 70 ms advanced version of the blood pressure p from a finger on the left hand, which roughly estimates the blood pressure upstream in the brachial artery; see Fig. 2.49. The depicted p is identical to the second cardiac pulse from Fig. 2.49b. The incident forward wave p_I and the delayed, positively reflected wave p_R can be recognized; compare Fig. 2.48b. (**b**) The simultaneously recorded blood flow velocity u in the brachial artery from the upper left hand, with the depicted u identical to the second cardiac pulse from Fig. 2.49a. The incident forward wave q_I and the delayed, negatively reflected wave q_R can be recognized, compare Fig. 2.48c, scaled approximately with the cross-section area A of the brachial artery

Figure 2.54 demonstrates results from an appropriate calculation[163] of the incident and reflected waves, whereas an obvious correspondence of the derived waves is given with respect to the schematic representation from Fig. 2.48.

Generally, the absolute value of the amplitude of the *reflected* wave component in the *interfered wave* pattern is much smaller than that of the *incident wave* at *any arterial site* (Fig. 2.48b, c and 2.54), i.e., $|P_R| < |P_I|$ and $|Q_R| < |Q_I|$. This is

(2.32) and Footnote 161 that $p_I = (p + q \cdot Z_0)/2$, $p_R = (p - q \cdot Z_0)/2$ and, on the other hand, $q_I = p_I/Z_0$, $q_R = -p_R/Z_0$ (Wijngaard et al. 2009; Hughes and Parker 2009).

[163]In the model from Footnote 162 the term $q \cdot Z_0 = \langle u \rangle \cdot A \cdot Z_0$ was *approximated* by $u \cdot \rho \cdot v$; compare (2.28). Then the product $\rho \cdot v$ was estimated from the slope $\mathrm{d}p/\mathrm{d}u$ $(= \rho \cdot v)$ during early systole, as suggested in (Hughes and Parker 2009), and amounted to about 1 mmHg/(cm/s). Actually this estimation yields v of about 10 m/s for a ρ of blood of roughly 1.05 kg/m^3; furthermore, it confirms qualitatively the assumed level of v ($\approx$7 m/s) within the scope of Fig. 2.49b in section "Blood Pressure and Flow."

because $\Gamma < 1$ (2.32) and, on the other hand, the stronger attenuation of the reflected wave accumulated during its round-trip than that of the incident wave (2.25).

It should be noted that the level of Γ *decreases with increasing* f; e.g., a decrease of about 50% from 2 to 10 Hz is demonstrated in (Nichols and O'Rourke 2005). In addition, *high-frequency* components are more *strongly damped* during propagation than that of the low frequency, as described in section "Pulse Propagation." In consequence, *reflections* of *high-frequency components* are *nearly absent* while only that of low-frequency components prevail in the resulting pulse waveform. To illustrate this behavior, Fig. 2.48b depicts a forward wave superimposed by a (relatively) high-frequency oscillation at the dicrotic notch and a (relatively) smooth reflected wave without the superimposed high-frequency oscillation, the latter being lost during propagation.

So the increase in the *ratio* of the pulsatile p amplitude to the pulsatile q amplitude in the periphery is determined by the increase of $|\underline{Z}_{\mathrm{I}}|$ at low frequencies; see (2.29) and Footnote 153. In analogy, the *waveform of* p changes in relation to the *waveform of* q along the propagation pathway in accordance with $\underline{Z}_{\mathrm{I}}$ components at various frequencies.

In particular, *waveforms of* p change dramatically, as described later in more detail; compare Figs. 2.41b and 2.46 for *increased* p_{S} *during propagation*. The systolic increase gets steeper since E from (2.24) increases nonlinearly with increasing frequency; i.e., high-frequency components arrive earlier in comparison to low-frequency components (section "Pulse Propagation"), which develops an early systolic peak (Fig. 2.46). Furthermore, *positive reflections* of *closed-end* type arise at sites with *varying mechanical properties* of the arterial pathway; compare Fig. 2.40 and Footnote 161. The positive reflections add to the incident wave; observe Fig. 2.54a. In addition, note the *inflection points* in p, q, and u in Figs. 2.48, 2.49 and 2.53c, indicating the presence of reflected waves. In fact, *mechanical properties change* at the

- Onset of small peripheral vessels,[164] i.e., *arterial–arteriolar junctions* (compare contribution of arterioles to R_{T} in (2.20)
- Branching (e.g., iliac bifurcation), tapering
- Locally increased stiffness due to peripheral vasoconstriction

[164] According to (Nichols and O'Rourke 2005), the *reflection* and *re-reflection* (damped resonance) of the pulse occur between arterial–arteriolar junctions in the upper part of the body and those in the lower part of the body, not just between (closed) aortic valve and peripheral arterial–arteriolar junctions. The arterial tree can be represented by an asymmetric *T-tube* (Fig. 2.40b), whose shorter limb represents the *upper part of the body*, longer limb the *lower part of the body*, and the middle limb the connection to the aortic valve. Only this representation explains phase relationships of p and q fluctuations at different sites of the arterial tree. However, with *aging*, wave reflections occur almost simultaneously from the lower and upper part of the body, because of a relatively greater increase in v to the lower body; a stronger stiffening of proximal arteries in the lower body can be observed with age in comparison to the upper body; see Fig. 2.40b and Sect. 2.5.1. Thus the two effective reflecting sites appear as one; compare behavior of $\underline{Z}_{\mathrm{I}}$ from Footnote 153 and rationale behind an age-independent transfer function from Footnote 165.

The changes of the *waveform of* q toward the periphery are more *specific*, in addition to a generally *decreasing amplitude* of q. Basically, *negative reflections* of *closed-end* type arise, contributing to the diminishing of the q pulse; see Figs. 2.48c, 2.49a and 2.54b. The back flow wave, as described in section "Pulse Waveforms of Pressure and Flow," is usually more apparent below than above major branches of the abdominal aorta (Nichols and O'Rourke 2005). In small peripheral vessels there is no back flow at all due to their proximity to peripheral reflecting sites; i.e., a low round-trip travel time Δt from Fig. 2.48c prevents the *reversal of the total flow*. For instance, in the splanchnic branches of the abdominal aorta (e.g., renal artery), there is no back flow and the contour of q is similar to that of p; that was attributed to low R_{T} and hence to low Γ in vascular beds supplied by these arteries (Nichols and O'Rourke 2005); compare (2.32).

The *amplification of* p *waves* along their propagation path is *greatest* when v *is low*, i.e., when a relatively high level of damping occurs (section "Pulse Propagation"), and the reflection has the greatest positive effect on the peripheral p waves and the least positive effect on the central p waves (Nichols and O'Rourke 2005). That is, the timing of the reflection is such that the peripheral site is an *antinode* and the central site is a *node* at the distance $\lambda/4$ from the peripheral site in terms of standing waves; see Footnote 170.

If the opposite effects of the attenuation and amplification on the *pressure pulse* are considered, the *transmission coefficient*, for instance, from the aorta to the brachial artery (Fig. 2.40a) amounts to about 1.5, yielding a *50% amplification* in pulse amplitude or an increase in p_{S} of about 20 mmHg at rest (Fig. 2.41b and 2.53c). At the same time, the *levels of* p_{D} *and* $\langle p \rangle$ (Footnote 140) *fall slowly* starting at the aorta. Figure 2.46 demonstrates the amplification of the aortic pressure pulse by about 10% if recorded in the distal radial artery.

With *aging*, the *transmission coefficient* decreases to *about 1* at age 80 (Nichols and O'Rourke 2005). The coefficient decreases because of increased aortic stiffness (Sect. 2.5.1) and thus *increased* v (2.22), yielding reduced damping of reflected waves (section "Pulse Propagation"). In addition, the value of λ (2.21) increases, which progressively transforms the central site into an *antinode* so that the central and peripheral sites are almost equally affected by wave reflections. In fact, the degeneration levels of the upper and lower body limbs diverge with increasing age, even facilitating the introduction of a *generalized transfer function*[165] from the aorta to the upper limb.

The *waveform of* p includes numerous *secondary peaks* due to reflections, as mentioned in section "Pulse Waveforms of Pressure and Flow." Before their origin can be discussed, it is important to note that *incident* waves have *sufficient time* to travel to the periphery and return back as *reflected* waves in a single cardiac cycle, and usually even before the blood ejection period is over. Typically a cardiac

[165]Since the elasticity of the upper limbs does not degenerate over lifetime as strongly as that of the lower limbs (Fig. 2.40b and Sect. 2.5.1) and waveform changes in the upper limbs (brachial or radial pressure, Fig. 2.40) with age are mainly due to changes in ascending aortic pressure

wave covers a distance of about 5 m during its cycle duration of about $1/f_C = 1$ s for $v \approx 5$ m/s; see (2.21). Another important observation is that, in general, the particular *appearance and location* of the secondary peaks within the cardiac cycle vary along the propagation path. As illustrated in Fig. 2.53, this is because the *traveled distance* of the reflected wave increases with decreasing distance from the recording site to the aortic valve, i.e., to the origin of the incident wave. Obviously, the increased traveled distance increases both the *accumulated damping* of the reflected wave and its *travel time* Δt with respect to the onset of the incident wave; compare damped reflected wave and $\Delta t_1 > \Delta t_2$ in Fig. 2.53b. Figure 2.46 demonstrates varying waveforms from the aorta to the radial artery, which were recorded synchronously.

Thus superimposed *secondary peaks* may be present in the waveform during *systole* because of relatively *early reflections*; compare Fig. 2.48b with a relatively small round-trip *travel time* Δt. This is typical for the *elderly* for they show stiffened aortic arteries (Fig. 2.42), e.g., at least doubled $|\underline{Z}_0|$ from 20 to 80 years (Nichols and O'Rourke 2005), and/or elevated $\langle p \rangle$, e.g., with hypertension,[166] and thus *higher* v, e.g., at least doubled in value from 20 to 80 years [according to (2.22) and (2.28)], of the incident and reflected wave during the round-trip. An early *fusion* of the *incident* and *reflected waves* may even generate a late systolic peak with increased p_S[167]; compare late systolic peak in Fig. 2.46 from the aortic wave. Such early reflections

---

itself, a *generalized transfer function* is proposed to describe upper limb amplification at *all ages* and even in a *general nonindividual* way (Nichols and O'Rourke 2005). The transfer function is defined as a ratio of the pressure amplitudes P^k of the sinusoidal components at the *upper limb* to P^k at the *ascending aorta*, with the index k indicating the amplitude's dependency on the frequency; compare sinusoidal decomposition from Footnote 150. The transfer function behaves approximately as a band-pass with an amplification of about 3 at around 4 Hz, decreasing to about 1 at 0 Hz and 10 Hz (O'Rourke 2009). This is certainly a very convenient way to *estimate* the ascending *aortic p* using noninvasive principles for *measuring radial or brachial p* (compare Fig. 2.46) or applying optical plethysmography (Sect. 3.1.3.1). To be more precise, the synthesized aortic waveform allows for determination of ejection duration and separation of the aortic wave into systolic and diastolic intervals.

[166] *Hypertension* is characterized by elevated $\langle p \rangle$ (2.31) which causes an increase in r and thus in the stiffness of elastic arteries (Fig. 2.42), along with an increase in aortic Z_0, aortic v (2.28), and even peripheral R_T. The implication of the stiffening is an increase in pulsatile p in the aorta with early reflections becoming prominent. These reflections yield a disproportionate increase in p_S and a relative reduction of p throughout diastole. In consequence of increased Z_0, v, and R_T, the spectrum of $|\underline{Z}_I|$ (Footnote 153) of the ascending aorta is shifted upward and to the right, similar to the *aging* impact on $|\underline{Z}_I|$ (Nichols and O'Rourke 2005).
Interestingly, already more than a century ago the *diagnostic importance* of hypertension was stressed while examining the radial pulse (Mahomed 1872). In the latter work, the late systolic component was described as "the tidal wave is prolonged and too much sustained."

[167] As shown in Nichols and O'Rourke (2005), the *secondary peak* during systole—representing the reflected wave—constitutes the systolic peak of the *carotid* pressure wave; i.e., the secondary peak becomes dominant and even determines p_S, after the fourth decade of life. In contrast, the secondary peak in the *radial* artery remains lower than the systolic peak (with p_S) even in the eighth decade, as demonstrated in Fig. 2.46.

yield *disadvantageous* effects of overloading[168] the ventricle during blood ejection, reducing V_S and coronary perfusion.

In contrast, *adolescents* with more elastic arteries, lower $\langle p \rangle$, and lower v may exhibit secondary peaks during *diastole*. The peaks represent *late reflections* improving (coronary) blood perfusion during diastole. In contrast to early reflections, late reflections imply a relatively large *travel time* Δt in Fig. 2.48b.

Correspondingly, the *elderly* show smoother waveforms than *adolescents* because of less prominent reflection phenomena within arterial pathways and higher v, causing reflected waves to lose their discrete identity and merge into the incident wave. A figurative comparison between the old and young is given in Fig. 2.51a, b, whereas the waveform from an old person is smoother with a hardly recognizable reflected wave contained within it. In *infants*, early reflections prevail because of their short body length and thus reduced Δt. Generally, early reflections arise in humans of *short body length* or in the case of prolonged ventricular ejection duration, also involving the elderly. Furthermore, early reflections increase with decreasing body length because the accumulated damping is reduced for the reflected wave (2.25).

Besides *early reflections* in the *elderly*, it is interesting to observe that the *reflection level* (or strength) changes *with age* nonlinearly. In particular, the value of R_T increases with age due to progressing vascular rarefaction and decreased arteriolar r; compare Fig. 2.40b. This yields a *progressive mismatch* between the large muscular arteries and the peripheral vascular tree. On the other hand, the aortic v (not peripheral v) increases significantly with age, reducing attenuation of the reflected wave during its propagation (section "Pulse Propagation"). Both effects lead to augmented reflected waves prior to an age of about 60. However, *after 60 years* of age, the level of the *reflection levels off*, possibly due to reduced V_S and increased *aortic* stiffness, i.e., increased aortic v and $|Z_0|$; see (2.28) (O'Rourke 2009). That is, the latter effect yields a proximal Z_0 matching (2.32) between the already stiffened aorta and intrinsically stiffer muscular arteries; compare Fig. 2.40b; it should be recalled that the stiffness of muscular arteries does not change with age (Sect. 2.5.1). The reflection site is moved more distally, which yields larger accumulated damping of reflections (2.25), delays the reflection, and thus reduces the amount of the reflected wave component within the p waveform. In consequence, the waveform of p changes and the pulse amplitude of the forward wave increases in the periphery (Mitchell 2009), which, for instance, may induce microvascular damage (see section "Blood Pressure and Flow").

For a quantitative and *physiologically relevant* assessment of the reflection level in the contour of p, the *augmentation index* I_A (or reflection wave ratio) is defined according to

[168]For instance, aortic $\langle p \rangle$ during systole and end-systolic p can be taken as indices of left ventricular systolic load (Nichols and O'Rourke 2005).

$$I_A = \frac{p_S - p_{IF}}{p_S - p_D}. \tag{2.33}$$

Here p_{IF} is the inflection point or shoulder indicating the beginning of the reflected wave,[169] as illustrated in Figs. 2.48b and 2.49b. Alternatively, the level of I_A can be defined as the height of the secondary systolic peak following p_{IF}, compare Fig. 2.48b, divided either by the height of the primary systolic peak ($= p_S$) or by the pulse amplitude ($= p_S - p_D$) (Nichols and O'Rourke 2005).

Generally, the level of I_A *increases* with advancing arterial stiffness and rising v. It also increases with *aging*, mainly due to augmented reflections, as described above. At an age of 60 years, however, the value of I_A *plateaus* due to a continuously increasing forward wave amplitude [increasing $p_S - p_D$ because of advancing arterial stiffness, (2.33)] but concurrently damped reflections (decreasing $p_S - p_{IF}$). For instance, the value of I_A levels off at about 25% for carotid artery (Mitchell 2009; Nichols and O'Rourke 2005). Even a *classification* scheme *describing the contours* of the ascending aortic p has been established on the basis of the I_A level and the timing between the inflection point and the systolic peak, as described in section "Pulse Waveforms of Pressure and Flow."

It should be noted that I_A is *influenced by* f_C because the total waveform of the constructively interfered incident and reflected waves changes with f_C; a constant round-trip time Δt of the reflected wave is assumed here, i.e., a constant Δt in Fig. 2.48b. In addition, the level of I_A increases with decreasing *body length* because the accumulated damping is reduced in the reflected wave (Nichols and O'Rourke 2005).

According to Footnote 164, *re-reflections* may occur with the waves retraversing the asymmetric arterial tree and thus yielding further *tertiary peaks* or even *oscillations* of p and q in the diastolic phase; for instance, a third diastolic peak may be observed (Nichols and O'Rourke 2005). The re-reflected waves exhibit relatively low amplitudes due to accumulated damping along increased traveled distance (2.25). It should be noted that the discussed phenomena—basically determining the waveform and its changes along the propagation path—are usually based on a *single reflection site* in the periphery. In fact, as noted in Footnote 164, waves travel along *multiple pathways* and their *multiple reflection sites* in vessels of the upper and lower body should be considered separately.

Reflections of the closed-end type (see Figs. 2.48a and 2.54a) between lower-limb arteries and the (closed) aortic valve (compare Footnote 164) may even create

[169] Actually the precise *definition of* I_A, particularly the sign of it, depends on the contour of the ascending aortic p. As elaborated in (Nichols and O'Rourke 2005), (2.33) applies if the peak p_S occurs *after the inflection point* ($I_A > 0$), i.e., a contour of type A or B is typical in the elderly (see section "Pulse Waveforms of Pressure and Flow"); see aortic wave in Fig. 2.46 as an example. Otherwise, $I_A = -(p_S - p_{IF})/(p_S - p_D)$ if the peak p_S *precedes the inflection point* ($I_A < 0$); i.e., a contour of type C is typical in adolescents; see Fig. 2.48b as an example.

a *standing wave*[170] or damped resonance of the entire arterial system. However, there is no anatomical location representing a global pressure node, since nodes of different harmonics of the waveform (Footnote 150) occur at different locations along the aorta (Nichols and O'Rourke 2005).

From an engineering point of view, it is interesting to note that the *reflections* may be artificially[171] *influenced by drugs* and deliberately *reduced by exercise.*[172] Furthermore, *reflections can be neglected* if

- Relatively *long arterial pathways* to reflection sites are considered
- *Short time intervals* are examined before the return of the reflected wave (within Δt in Fig. 2.48b) or
- Only strongly *damped high-frequency components* are present in the pulse wave (section "Pulse Propagation")

An effective approach in the *artificial management* of *reflections* comprises the so-called *Valsalva maneuver*,[173] through which *late reflections* during diastole

---

[170]*Standing waves* arise at the superposition of incident and reflected waves (Fig. 2.48), whereas zeros (=*nodes*) and minima and maxima (=*antinodes*) are located at certain positions along the arterial tree, spaced at $\lambda/2$ for a given *harmonic* with f and v (2.21). For instance, the nodes of p would be at distances $\lambda/4$, $3\lambda/4$, etc. from the closed end (e.g., the lower limb, compare Fig. 2.40 and Footnote 164), whereas the nodes of q at distances 0, $\lambda/2$, etc. This strongly influences the frequency behavior of $\underline{Z}_{\mathrm{I}}$ (compare Footnote 153) and the amplification of p along the arterial tree. However, the exponential damping of the reflected wave along its propagation path (2.25) excludes a complete cancellation of p and q even at the first node from the closed end.

[171]For instance, if *vasodilation* is induced pharmacologically in the *peripheral vascular bed* (compare Footnote 130), R_{T} from (2.20) decreases along with reduced v. There is increased accumulated damping of reflected waves and thus reduced early reflections. In contrast, artificial vasoconstriction is much less effective in increasing R_{T} because of the normally high resting arteriolar tone (Footnote 130).
In analogy, *vasodilation* effects on *peripheral muscular arteries* are accompanied by a decrease in their stiffness (compare Footnote 142), which manifests as a reduction of early wave reflections. The *reflections are delayed and weakened* because of

- Reduced v
- A distal move of the reflection site due to better matching of Z_0 between central elastic arteries and peripheral muscular arteries [Fig. 2.40 and (2.32)]
- Larger accumulated damping of the reflected wave on its way

[172]For instance, leg *exercise* markedly *reduces reflections* in the lower part of the body because of decreased arteriolar tone (relaxed smooth muscles), improved endothelial function, and thus reduced R_{T} (2.32). In addition, muscular arteries are relaxed during exercise yielding reduced wave reflections; for physiological mechanisms involved, see Footnote 171. Thus exercise is associated with a decrease in amplification of the p wave between the aorta and the active limb (Nichols and O'Rourke 2005).

[173]The *Valsalva maneuver*, named after Antonio Maria Valsalva (1666–1723), an Italian physician and anatomist, entails forced expiration while keeping the mouth and nose closed. It yields an increase in intrathoracic and intraabdominal pressure that compresses large veins of the systemic circulation and exerts mechanical force on vessels of the pulmonary circulation; compare the low-pressure part from Fig. 2.39. The return of systemic venous blood decreases, because the driving force for venous blood weakens; i.e., the pressure gradient from peripheral venous vessels

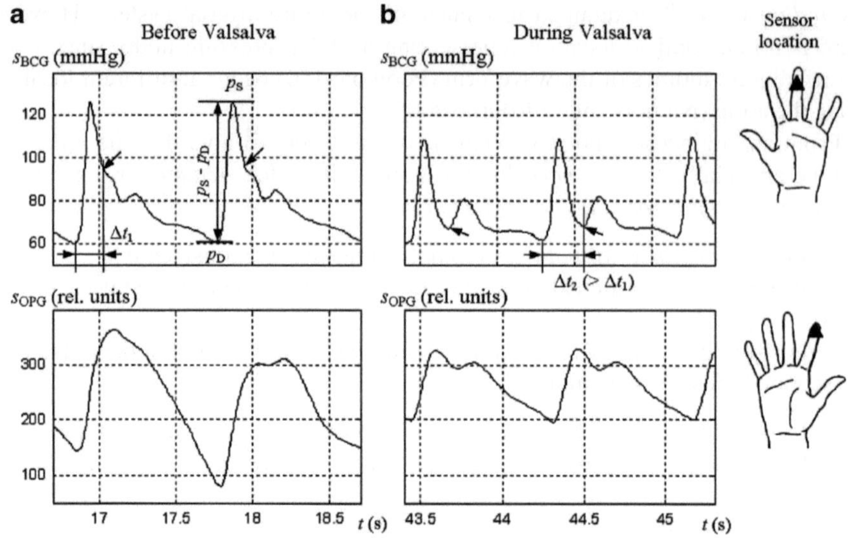

Fig. 2.55 Morphological waveforms of mechanic biosignal barocardiogram s_{BCG} from a finger on the left hand (*upper figures*) and optic biosignal optoplethysmogram s_{OPG} from a finger on the right hand (*lower figures*). (**a**) Before the Valsalva maneuver (Footnote 173). (**b**) During the maneuver. Approximate starts of the reflected waves (inflection points) are indicated by *arrows* (compare Fig. 2.48b)

can be *induced*. During this maneuver the transmural pressure (Footnote 141 and Fig. 2.44b) is decreased along intrathoracic and intraabdominal arteries. The stiffness of the arterial wall decreases (Fig. 2.42) and thus v of reflected waves decreases as well (2.22). In consequence, reflections are postponed, promoting late reflections.

Figure 2.55 *demonstrates* the obvious delay of the reflected wave during the *Valsalva maneuver*. The depicted barocardiogram reflects the behavior of the radial p; see Sect. 3.1.3.1. The reflection onset is moved from the late systole (before Valsalva) to the diastole (during Valsalva), making the diastolic wave more prominent. In consequence, the time Δt from the systolic onset to the reflection onset is subject to change, in line with $\Delta t_2 > \Delta t_1$ (Fig. 2.55b). In addition, the level of p_S is reduced along with the pulse amplitude $p_S - p_D$, as expected from Footnote

to intrathoracic vessels decreases. Consequently, cardiac filling is impeded. If *physiological parameters* are considered during Valsalva, there is an initial increase of p_S because of compression of the thoracic aorta and the forcing of blood from the pulmonary circulation into the left atrium. Later, particularly as a result of the diminished cardiac filling (see Footnote 225), the values of V_S and q drop, as well as p_S; compare Fig. 2.55. The *Baroreflex compensates* for the latter changes with a subsequent peripheral vasoconstriction (increase in R_T) (2.20) and an increase in f_C (Sect. 3.2.2.1). When the air is finally exhaled, the cardiac filling normalizes and the baroreflex balances again p_S by reducing elevated levels of R_T and f_C.

173. In other words, the Valsalva maneuver temporarily "undoes" the aging process of proximal arteries in terms of the waveform morphology.

Lastly, Fig. 2.46 depicts aortic and radial pressure waveforms recorded synchronously to *illustrate* some of the above effects, as related to the *reflection phenomena*. That is, different locations of secondary peaks with different onsets of reflected waves (compare Fig. 2.48b) can be observed, as well as the amplification of the pulse amplitude at the radial site and a steepening of the radial wave front before the systolic peak. At the aortic site, an earlier onset of the reflection on the ascending part of the contour can be observed. This can be attributed to the earlier arrival of the reflected wave at the aortic site from the lower body than at the radial site; compare Fig. 2.40b and Footnote 164.

2.6 Respiratory System

The respiratory system has the complex task of gas exchange, closely interrelated with the circulatory system (Sect. 2.5). In particular, functions of the respiratory system include

- *Mechanical ventilation*, moving air into and out of the lungs.
- *Gas exchange* between air and blood in the lungs and, on the other hand, between blood and bodily tissues.
- *Oxygen utilization* in body tissues within the scope of energy-liberating reactions of cell respiration (Footnote 19).

It should be stressed that only those structures and functions of the respiratory system will be considered, which are relevant for *biosignal generation* and *propagation* (compare Fig. 1.3a).

2.6.1 Structure

The part of the respiratory system that conducts air starts with the *upper breathing airways*. As shown in Fig. 2.56, the upper airways include the *nasal cavity* and *oral cavity*. The oral cavity is bounded by the *soft and hard palate* and encloses the *tongue*. Both cavities merge into each other at the site of the uvula and lead to the *pharynx*, a cavity behind the mouth. At the level of the epiglottis, the pharynx divides into the *trachea*, which leads toward the lungs (for air supply), and the *esophagus*, which leads toward the stomach (for food supply).

As illustrated in Fig. 2.57, the trachea is a windpipe made up of connective tissue, surrounded by smooth muscles. The *trachea* branches into two *primary bronchi* entering the *lungs*. Bronchi successively branch further into even smaller airways, known as *bronchioles*, ending up with a very fine network of airways terminated by thin-walled air sacs, known as *alveoli*.

Fig. 2.56 Pharyngeal airways and surrounding structures of the upper airways

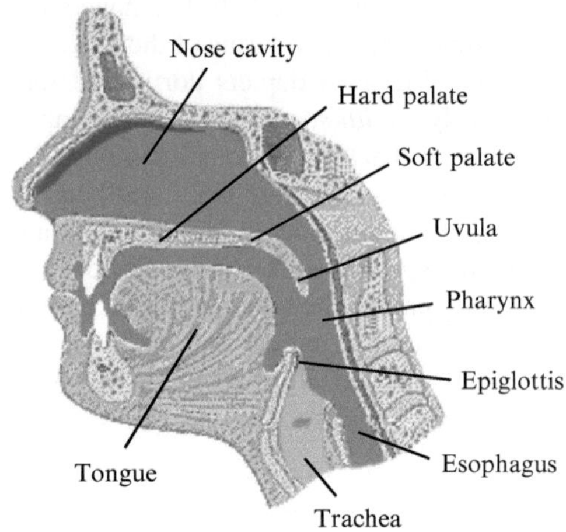

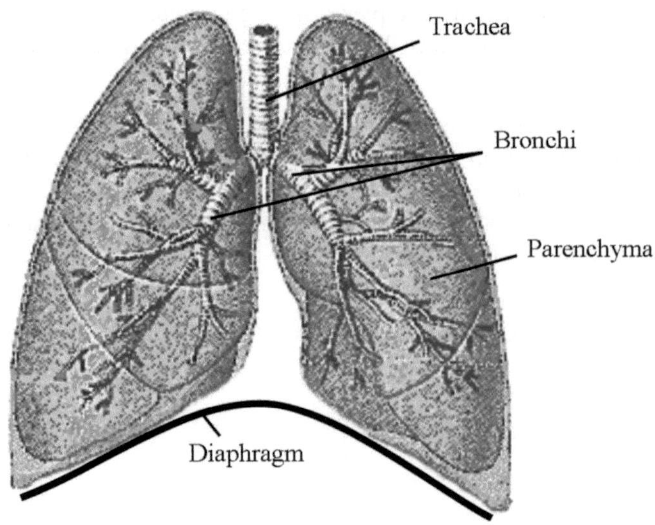

Fig. 2.57 Lungs and adjacent airways

A dense *network of pulmonary capillaries* covers the surface of each alveolus, providing an almost continuous sheet of capillaries *surrounding alveoli* for gas exchange. Alveoli provide an enormous surface area needed for an efficient gas exchange; with about $3 \cdot 10^8$ alveoli, each of which has a diameter of about 0.5 mm, yielding a total surface area of up to 80 m^2 (Fox 2011). Pulmonary *arterioles* supply deoxygenated blood to pulmonary *capillaries* while downstream pulmonary *venules* transport already oxygenated blood back to the heart.

The outer sides of the *lungs* connect to the *thoracic wall*, whereas the lung's lower sides connect to the thoracic *diaphragm*. A special membrane, i.e., a connective layer with a lubricating function, lines the inside of the thoracic wall while another membrane covers the surface of the lungs, with both membranes being known as *pleural membranes*. Under normal conditions, there is only a thin *layer of fluid* between both pleural membranes. These membranes are adhered to each other like two wet glass slides. Consequently, the lungs fill the thoracic cavity and the pleural membranes allow sliding of the lungs relative to the thoracic wall during breathing. The lungs have a high *compliance*[174] and a high *elasticity* to accommodate their actively forced volume increase during *inspiration* and its passive volume decrease due to elastic recoil during *expiration*, respectively. Furthermore, the chest wall exhibits dynamic compliance (Pelosi et al. 2000), usually described in terms of *viscoelastic properties* (Footnote 134).

The *diaphragm* is a dome-shaped skeletal muscle which divides the upper body into *thoracic cavity* and *abdominal cavity* (Fig. 2.58a). The thoracic cavity contains, among others, the heart, lungs, and large blood vessels, whereas the abdominal cavity contains organs of the digestive tract and many others.

2.6.2 Function

The upper airways have to maintain ventilation of the lungs while allowing *food intake*, *speech*, *tasting*, and *smelling*. The ventilation serves not only for *conduction* of the inspired air but also for its *warming* toward the core body temperature, its *humidification* to protect the lungs from desiccation, and its *filtration* through trapping of small particles by secretion of a viscous fluid, known as mucus.

The function of the *respiratory system* is closely related to that of the *pulmonary and circulatory systems*; see Fig. 2.39. That is, the primary function of the *lungs* is to provide *oxygen* to the bloodstream passing the lungs. Oxygen *diffuses passively* from alveoli (filled with air) into *pulmonary capillaries*,[175] to the blood plasma and then to red blood cells where oxygen is finally stored (Sect. 3.1.4). In fact, oxygen diffuses down its concentration gradient from air to blood because the oxygen concentration of alveolar air is higher than that in the blood plasma. In parallel,

---

[174]The *lung's compliance* can be defined in analogy to (2.23) with p as the pressure difference between the intraalveolar pressure and the intrapleural pressure (Sect. 2.6.2).

[175]It is interesting to note that *pulmonary arterioles* constrict when the partial pressure of oxygen p_{O2} in alveoli is low and dilate when p_{O2} is high. Likewise, local *lung ventilation* is *matched* to local *blood perfusion*, e.g., the pulmonary blood flow to poorly ventilated alveoli is decreased. This response is in contrast to that of *systemic arterioles* which constrict when p_{O2} in the tissue is high. For instance, metabolic vasodilation of systemic arterioles helps to supply more blood and oxygen to tissues (such as muscles) which momentarily have an increased metabolic rate, increased oxygen consumption, and thus lower p_{O2}.
From a cellular point of view, a low level of alveolar *oxygen* (low p_{O2} in alveoli) causes depolarization of smooth muscle cells in the walls of *pulmonary arterioles* by inhibiting *outward diffusion of K^+ ions* through gated channel proteins. The *depolarization* yields contraction of the vascular smooth muscle (Footnote 106).

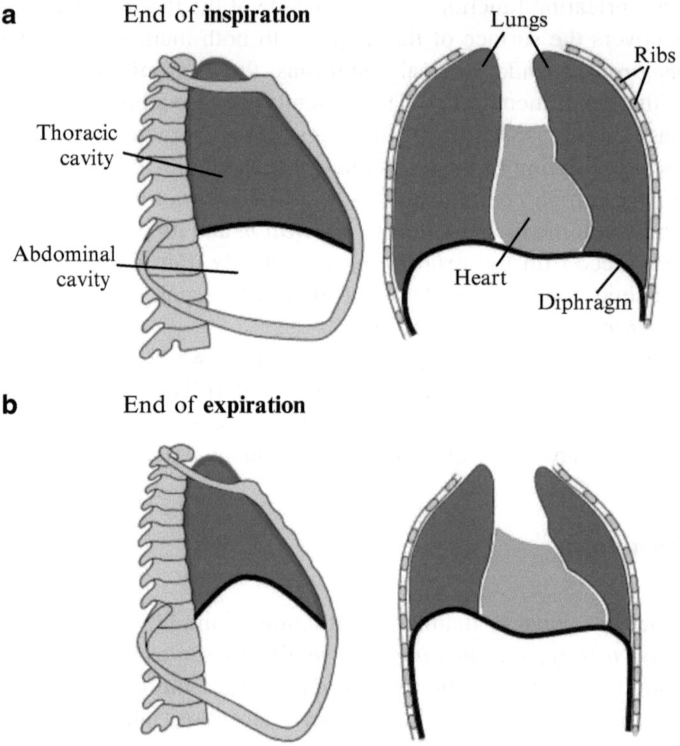

Fig. 2.58 Respiratory movements of the thorax during (**a**) inspiration and (**b**) expiration. Image data partly taken from (Lang 2000)

carbon dioxide is released from blood through its *passive diffusion* from the blood plasma to alveolar air.[176] In both cases, the *exchange of gases* between air and blood occurs across thin walls of both alveoli and capillaries, governed by the

---

[176] *Carbon dioxide* (CO_2) is transported in blood in three forms:

- As *bicarbonate* ion (HCO_3^-) in the *blood plasma* accounting for about 70% of all transported CO_2
- As CO_2 bound to hemoglobin in *red blood cells* accounting for about 20%
- As CO_2 dissolved in the *blood plasma* accounting for about 10% (Fox 2011); see Footnote 219.

Likewise, *CO_2 gas is mainly transported as HCO_3^- ions* when CO_2 combines with water (H_2O) forming *carbonic acid* (H_2CO_3), which can ionize and release a hydrogen ion ($H_2CO_3 \leftrightarrow H^+ + HCO_3^-$) forming a HCO_3^- ion. In *systemic capillaries*, enzymes in red blood cells favor the production of H_2CO_3 while the dissociation product HCO_3^- then diffuses out of red blood cells into the blood plasma. In *pulmonary capillaries*, HCO_3^- and H^+ ions combine to H_2CO_3 which is then converted back to CO_2 gas and H_2O.

concentration gradients of gases.[177] The *amount of gas* dissolved in the blood plasma *depends directly* on the *partial pressure* of this gas in a gaseous mixture; see Footnote 219. Likewise, the *diffusional force* acting on *oxygen* increases with increasing (positive) difference between the partial pressure of oxygen in alveolar air and that in the blood plasma; compare Footnote 24. In analogy, partial pressures of carbon dioxide in alveolar air and the blood plasma determine the *diffusional force* acting on *carbon dioxide*. Consequently, blood leaving the lungs (via pulmonary veins toward the heart) has a higher oxygen level and a lower carbon dioxide level than blood entering the lungs (via pulmonary arteries from the heart).

In *bodily tissues*, the reverse process of gas exchange takes place; i.e., oxygen is delivered from blood to tissue cells while carbon dioxide is extracted from cells and accumulated in blood for its transportation. In *systemic capillaries* (Fig. 2.39), *oxygen* is unloaded from red blood cells and diffuses through the blood plasma into surrounding tissue cells down its concentration gradient (Sect. 3.1.4). In parallel, *carbon dioxide* follows its diffusional force directed from tissue cells to the blood plasma and is then stored in blood (Footnote 176). Consequently, blood leaving bodily tissues (via systemic veins toward the heart) has a lower oxygen level and a higher carbon dioxide level than blood entering tissues (via systemic arteries from the heart).

In addition, respiration helps in maintaining the pH value[178] of the blood plasma by regulating the concentration of carbon dioxide. For instance, *hypoventilation* yields increased levels of carbon dioxide and carbonic acid in the blood plasma (Footnotes 176 and 185), diminishing the pH value (known as respiratory acidosis). On the contrary, *hyperventilation* reduces levels of carbon dioxide and carbonic acid below normal, raising the pH value (known as respiratory alkalosis).[179]

---

[177] *Diffusion* of gases (oxygen and carbon dioxide) between air in alveoli and blood in pulmonary capillaries occurs *very rapidly* because

- Alveolar walls have a large *surface area* for diffusion (up to 80 m^2, Sect. 2.6.1)
- A *tight contact* exists between alveoli and surrounding capillaries
- *Diffusion distance* is very small (about 2 μm only, i.e., diffusion across one alveolar cell and one capillary cell)

Typically, a *diffusional equilibrium* is reached in less than 1 s (Silbernagl and Despopoulos 2007), whereas gas exchange is facilitated by slow blood flow in pulmonary capillaries. By comparison, carbon dioxide diffuses more easily than oxygen.

[178] The *pH value pH*, basically meaning "power of hydrogen," is a dimensionless measure of the acidity or basicity of a solution, i.e., measure of the density of hydrogen ions H^+ within. It is defined as

$$pH = -\log_{10} \frac{\rho_H}{1\,\mathrm{mol/l}}$$

with ρ_H as the molar density of H^+ in mol/l. A solution whose *pH* is 7 is said to be neutral, like distilled water, whereas $pH < 7$ indicates an acid, and $pH > 7$ a base. For instance, blood is slightly basic and has a *pH* of about 7.4 in a normal case.

[179] *Hyperventilation* also raises the pH value of cerebrospinal fluid, inducing cerebral vasoconstriction and thus leading to a reduced blood flow, which can cause *dizziness*.

Mechanical ventilation of the lungs is an integral part of the discussed gas exchange. A rhythmic process of

- *Inspiration*
- *Expiration*

occurs, which moves air into and out of the lungs. Obviously, the inspired air contains more oxygen and less carbon dioxide (with the corresponding partial pressures[180] of 159 mmHg and 0.24 mmHg in dry air) than the expired air (115 mmHg and 33 mmHg in humid air) (Silbernagl and Despopoulos 2007). Likewise, oxygen is supplied during inspiration and carbon dioxide is eliminated during expiration.

While *inspiration* is produced by active muscle contraction, i.e., it is an *active process*, *expiration* is produced by elastic recoil of breathing muscles and is a *passive process*. During normal *inspiration*[181] respiratory skeletal muscles, such as intercostal muscles running between the ribs and the diaphragm, contract to expand the rib cage and thus to *increase the thoracic volume*. As shown in Fig. 2.58, the ribs are lifted up due to contraction of *intercostal muscles* while the thoracic volume increases *laterally*. The contraction of the *diaphragm* flattens and lowers the diaphragm while the thoracic volume increases in the *vertical* direction and the abdominal region becomes compressed (Footnote 117). Since the lungs are stuck to the thoracic wall because of pleural membranes adhered to each other, increasing *thoracic volume* parallels increasing *volume of the compliant lungs*. The increasing lung volume, in turn, decreases air pressure within the lungs.[182] Namely, *intraalveolar pressure* drops below atmospheric pressure which the upper airways are exposed to. The air flows into the lungs along its *pressure gradient*,[183] just as blood flows along vessels governed by blood pressure gradient. The lungs become inflated, whereas the pressure in the intrapleural space between both pleural membranes, known as the *intrapleural pressure*, decreases.[184]

---

[180]For comparison, the respective *partial pressures* of *oxygen* and *carbon dsioxide* are

- About 105 mmHg and 39 mmHg in *alveoli*
- About 100 mmHg and 40 mmHg in the *arterial blood* of the systemic circulation
- About 40 mmHg and 46 mmHg in the *venous blood* of the systemic circulation (compare Sect. 3.1.4)

[181]During *forced inspiration*, *accessory muscles* of respiration also contract, e.g., skeletal muscles situated in the lateral neck or the upper part of the chest, elevating the ribs and stabilizing the rib cage. In addition, *forced exhalation* involves activation of abdominal muscles, forcing abdominal organs up against the diaphragm, in the course of which the thoracic volume decreases further.

[182]According to *Boyle's law*, the absolute pressure of a fixed amount of gas is inversely proportional to its volume at a constant temperature.

[183]For instance, during *quiet inspiration* and *quiet expiration* the *intraalveolar pressure* is about −3 mmHg and 3 mmHg related to atmospheric pressure, respectively (Fox 2011).

[184]*Intrapleural pressure* amounts to about −6 mmHg and −3 mmHg (related to atmospheric pressure) during quiet inspiration and quiet expiration, respectively. Intrapleural pressure is more

In contrast, normal *expiration* is marked by relaxation of all breathing muscles. The relaxation causes the thoracic volume to shrink in size because of a *passive elastic recoil* of the rib cage and the lungs (Footnote 181). Reduced lung volume increases intraalveolar pressure (Footnote 182) and raises it above atmospheric pressure, pressing alveolar air out of the lungs. During expiration, intrapleural pressure increases as well (Footnote 184).

Rhythmic breathing is under voluntary and involuntary *control* of the central nervous system. *Sensorial feedback* to the nervous system is mainly established through stretch-sensitive *mechanical receptors* in the lungs and *chemical receptors* in the brainstem, aorta, and carotid arteries. Chemical receptors are sensitive to pH value as well as levels of oxygen and carbon dioxide in the arterial blood.[185] For instance, decreasing pH value, decreasing level of oxygen, or increasing level of carbon dioxide prompt the central nervous system to strengthen breathing activity and increase air flow, i.e., volume of air per time unit.

In the *resting state*, respiratory rate is in the range from about 10 to 20 per minute while inspiratory volume is about 0.5 l (Silbernagl and Despopoulos 2007), known as the *tidal volume*. Thus, the resulting air flow amounts to about 8 l/min [compare (2.30)]. During *deep breathing*, the maximum volume of exhaled air after a maximum inhalation can significantly increase and reach values up to 5 l, known as the *vital capacity*.

The *tidal volume* was shown to correspond to mechanical *respiratory deformations*[186] in the chest region and abdominal region (e.g., changes in the chest circumference), or rather, correspond to the sum of deformations in both regions (Keenan and Wilhelm 2005). However, *viscoelastic* properties of the chest wall render the relation between tidal volume and respiratory deformations even more nonlinear.

---

negative during inspiration because of a stronger elastic recoil of the lungs than it is during expiration. This pressure is always *subatmospheric* (<0 mmHg) because the elastic lungs pull in one direction and the thoracic wall pulls in the opposite direction during the entire respiratory cycle so that this subatmospheric pressure (or suction) between pleural membranes keeps the lungs and the thoracic wall together. Likewise, intrapleural pressure is always lower than intraalveolar pressure (Footnote 183), the difference of which forces the lungs to stuck to the thoracic wall.

[185]It should be noted that changes in *ventilation affect* the *carbon dioxide* level and *pH value* in the arterial blood more immediately than the *oxygen* level. For instance, during inadequate ventilation (such as *hypoventilation*) the amount of carbon dioxide quickly rises and pH value falls. This is because *carbon dioxide* combines with water forming *carbonic acid* which can release hydrogen ions (Footnote 176) and thus reduce pH value. The level of *oxygen* in the blood plasma decreases much more slowly because oxygen is stored in hemoglobin which maintains a certain partial pressure of oxygen in the blood plasma (Sect. 3.1.4). Likewise, rapid changes in the level of *carbon dioxide and pH value* provide a sensitive measure for *control of breathing* based on *chemical sensorial feedback*.

[186]As shown in Groote et al. (1997), during inspiration at rest in the seated posture, displacement of the chest wall occurred primarily in the cranial and ventral directions and averaged 3 up to 5 mm, while displacement in the lateral outward direction was only 1 up to 2 mm. In parallel, the abdominal wall displayed motions primarily in the ventral direction.

Of particular importance is the difference between chest breathing and abdominal breathing. In the case of *chest breathing* prevailing (thoracic breathing), contraction of respiratory muscles in the thorax dominates inspiratory efforts. In contrast, contraction of abdominal muscles governs *abdominal breathing*. The latter difference influences respiratory biosignals registered on the chest and abdomen, with this difference being relevant for detection of obstructive sleep apneas; see paradoxical respiration in Sect. 3.1.2.

References

J.D. Bronzino: The biomedical engineering handbook. CRC press (1995).

P. Carmeliet, M. Tessier-Lavigne: Common mechanisms of nerve and blood vessel wiring. Nature 436, 193–200 (2005).

C.H. Chen, E. Nevo, B. Fetics, P.H. Pak, F.C. Yin, W.L. Maughan, D.A. Kass: Estimation of central aortic pressure waveform by mathematical transformation of radial tonometry pressure, validation of generalized transfer function. Circulation 95(7), 1827–1836 (1997).

S.Y. Chiu, J.M. Ritchie, R.B. Rogart, D. Stagg: A quantitative description of membrane currents in rabbit myelinated nerve. The Journal of Physiology 292, 149–166 (1979).

R.W. deBoer, J.M. Karemaker, J. Strackee: Hemodynamic fluctuations and baroreflex sensitivity in humans: a beat-to-beat model. American Journal of Physiology, Heart and Circulatory Physiology 253(3), 680–689 (1987).

H.U. Dodt, unpublished microscopic image data, Vienna University of Technology, Austria (2010).

S.I. Fox: Human Physiology. McGraw-Hill Publisher (2011).

B. Gribbin, A. Steptoe, P. Sleight: Pulse wave velocity as a measure of blood pressure change. Psychophysiology 13(1), 86–90 (1976).

A.D. Groote, M. Wantier, G. Cheron, M. Estenne, M. Paiva: Chest wall motion during tidal breathing. Journal of Applied Physiology 83(5), 1531–1537 (1997).

D.K. Hartline, D.R. Colman: Rapid conduction and the evolution of giant axons and myelinated fibers. Current Biology 17(1), R29–35 (2007).

B. Hille: Ionic channels in excitable membranes. Current problems and biophysical approaches. Biophysical Journal 22(2), 283–294 (1978).

A.L. Hodgkin: A note on conduction velocity. The Journal of Physiology 125(1), 221–224 (1954).

A.D. Hughes, K.H. Parker: Forward and backward waves in the arterial system: impedance or wave intensity analysis? Medical and Biological Engineering and Computing 47(2), 207–210 (2009).

E.R. Kandel, J.H. Schwartz, T.M. Jessell: Principles of Neural Science. McGraw-Hill Medical Publisher (2000).

E. Kaniusas, H. Pfützner, L. Mehnen, J. Kosel, J.C. Téllez-Blanco, G. Varoneckas, A. Alonderis, T. Meydan, M. Vázquez, M. Rohn, A.M. Merlo, B. Marquard: Method for continuous non-disturbing monitoring of blood pressure by magnetoelastic skin curvature sensor and ECG. IEEE Sensors Journal 6(3), 819–828 (2006).

D.B. Keenan, F.H. Wilhelm: Adaptive and wavelet filtering methods for improving accuracy of respiratory measurement. Biomedical Sciences Instrumentation 41, 37–42 (2005).

F. Lang: Basic knowledge in physiology (in German: Basiswissen Physiologie). Springer Publisher (2000).

F.A. Mahomed: The physiology and clinical use of the sphygmograph. Medical Times and Gazette 1, 62–64 (1872).

J. Malmivuo, R. Plonsey: Bioelectromagnetism, Principles and Applications of Bioelectric and Biomagnetic Fields. Oxford University Press (1995).

B. McMillan: Human body: A visual guide. The Reader's Digest Association Limited (2006).

W.R. Milnor: Hemodynamics. Williams & Wilkins Publisher (1989).

G.F. Mitchell: Clinical achievements of impedance analysis. Medical and Biological Engineering and Computing 47(2), 153–163 (2009).

M. Moser, M. Frühwirth, T. Kenner: The symphony of life. Importance, interaction, and visualization of biological rhythms. IEEE Engineering in Medicine and Biology Magazine 27(1), 29–37 (2008).

W.W. Nichols, M.F. O'Rourke: McDonald's blood flow in arteries: Theoretical, experimental and clinical principles. Hodder Arnold Publication Publisher (2005).

M.F. O'Rourke: Time domain analysis of the arterial pulse in clinical medicine. Medical and Biological Engineering and Computing 47(2), 119–129 (2009).

T.J. Pedley: The fluid mechanics of large blood vessels. Cambridge University Press (1980).

P. Pelosi, A. Aliverti, R. Dellaca: Chest wall mechanics in normal subjects and in critically ill patients. Intensivmedizin und Notfallmedizin 37(4), 341–351 (2000).

H. Pfützner: Applied biophysics (in German: Angewandte Biophysik). Springer Publisher (2003).

F. Rattay, M. Aberham: Modeling axon membranes for functional electrical stimulation. IEEE Transactions on Biomedical Engineering 40(12), 1201–1209 (1993).

Y.W. Shau, C.L. Wang, J.Y. Shieh, T.C. Hsu: Noninvasive assessment of the viscoelasticity of peripheral arteries. Ultrasound in Medicine & Biology 25(9), 1377–1388 (1999).

S. Silbernagl, A. Despopoulos: Pocket-atlas of physiology (in German: Taschenatlas Physiologie). Georg Thieme Publisher (2007).

D.U. Silverthorn: Physiology (in German: Physiologie). Pearson Studium (2009).

J. Sola, S.F. Rimoldi, Y. Allemann: Ambulatory monitoring of the cardiovascular system: the role of pulse wave velocity. In Book New developments in biomedical engineering (editor D. Campolo), 391–424 (2010).

M. Sugawara, K. Niki, H. Furuhata, S. Ohnishi, S. Suzuki: Relationship between the pressure and diameter of the carotid artery in humans. Heart Vessels 15, 49–51 (2000).

G. Wang, R. Sougrat, unpublished microscopic image data, Imaging and Characterization Lab, King Abdullah University of Science and Technology, Kingdom of Saudi Arabia (2011).

J.P. Wijngaard, M. Siebes, B.E. Westerhof: Comparison of arterial waves derived by classical wave separation and wave intensity analysis in a model of aortic coarctation. Medical and Biological Engineering and Computing 47(2), 211–220 (2009).

Wikipedia, free encyclopedia, en.wikipedia.org (2010).

D. Zemaityte: Autonomic regulation of heart rhythm: mechanisms, registration, clinical value (in Lithuanian: Sirdies ritmo autonominis reguliavimas: mechanismai, vertinimas, klinikine reiksme). Kaunas Medical Academy Publisher (1997).

D. Zheng, J.N. Amoore, S. Mieke, A. Murray: Estimation of mean arterial pressure from the oscillometric cuff pressure: comparison of different techniques. Medical and Biological Engineering and Computing 49, 33–39 (2011).

C.M. Zierhofer: Analysis of a linear model for electrical stimulation of axons - critical remarks on the "activating function concept", IEEE Transactions on Biomedical Engineering 48(2), 173–184 (2001).

Chapter 3
Physiological Phenomena and Biosignals

Since a biosignal can be defined as a description of a physiological phenomenon (Sect. 1), it is obvious that *biosignal parameters* reflect *physiological parameters*. For proper *diagnostic* or *therapeutic* interpretation of the derived biosignal's parameters (or results of the biomedical signal analysis), an adequate understanding of their *physiological causes* is necessary. Here, definitions and basics of biosignal parameters are relevant, as well as their time-related or event-related behavior along with their *mutual interdependencies*.

Since there are numerous biosignals as well as numerous physiological phenomena, we concentrate on a few *vital phenomena* which are (nearly) always of interest in clinical praxis: *heartbeat*, *respiration*, *blood circulation*, *blood oxygenation*, and *body temperature*. We focus on the typical behavior of the corresponding physiological parameters along with their typical interrelations. In addition, different *biological rhythms* and *sleep* patterns are considered, which represent common physiological states showing highly sophisticated but physically very reasonable interrelations of vital physiological parameters. The inevitable[187] influence of *healthy aging* is considered throughout the chapter to fortify understanding of biosignal parameters with respect to (aging) physiological phenomena.

It should be noted that there is a clear difference between recorded *biosignals* and physiological *parameters* of interest, as shown in Fig. 3.1. While the biosignals are raw signals given by different sensors, the parameters are usually hidden in the biosignals and are extracted with various signal processing approaches. For instance, radically different biosignals (acoustical, mechanical, etc.) may contain information about the same physiological parameter (heart rate, respiratory rate, etc.), as will be shown in manifold cases discussed below. On the other hand, multiple parameters may be extracted from a single biosignal using multiparametric

[187]"Aging seems to be the only available way to live a long live," according to the words of Daniel Francois Esprit Auber (1782–1871).

E. Kaniusas, *Biomedical Signals and Sensors I*, Biological and Medical Physics, Biomedical Engineering, DOI 10.1007/978-3-642-24843-6_3,

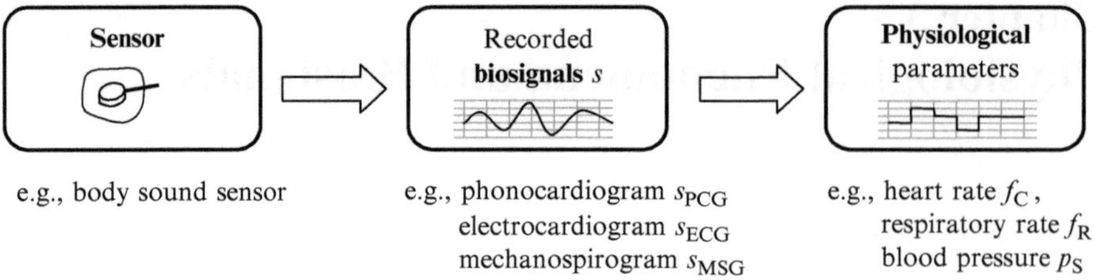

Fig. 3.1 From sensor, to biosignals s to physiological parameters of interest (compare Fig. 1.17)

monitoring (Sect. 1.4). In the following sections, all biosignals will be denoted as s with a *subscript* indicating the specific type of biosignal, compare Fig. 3.1 and "Symbols of Biosignals."

3.1 Vital Phenomena and Their Parameters

3.1.1 Heartbeat

The *heart* is responsible for pumping blood in the circulatory system with its *rhythmic* contractions (Sects. 2.4 and 2.5), creating pulsatile waves of blood pressure and blood flow (Sects. 2.5 and 3.1.3). An easy and reliable assessment of the resulting *cardiac cycle* with heart rate f_C is of utmost importance for diagnosis and therapy. There are numerous approaches to register the heartbeat, as discussed below.

Figure 3.2 demonstrates three widely used *methods* to register cardiac activity. The electrocardiogram in Fig. 3.2a, an *electric biosignal,* shows rhythmic waves and peaks which correspond to electrical heart muscle excitation with f_C, for details see Sect. 4. The optoplethysmogram in Fig. 3.2b, an *optic biosignal,* depicts a smoother waveform reflecting pulsatile blood absorption of incident artificial light (Sect. 6). Last, the phonocardiogram in Fig. 3.2c, an *acoustic biosignal,* demonstrates two consecutive temporary Signal deflections—first and second heart sound—induced by consecutive closures of heart valves (Sect. 5).

Obviously, the intrinsic *interrelations* in between these *cardiac biosignals* are of high interest for a proper understanding of the physiological phenomena involved, whereas these biosignal's relationship with respiration—vital activity encompassing the whole body—will be described in Sect. 3.2 in more detail. Before the biosignals can be compared with each other in relation to time, the *time delay* of the above biosignals with respect to a *reference site*, e.g., the heart, should be considered. While the electric biosignal arrives nearly instantaneously at the corresponding sensor (electrodes) location, as depicted in Fig. 3.2 to the right, the acoustic biosignal arrives marginally later (with sound velocity in tissue

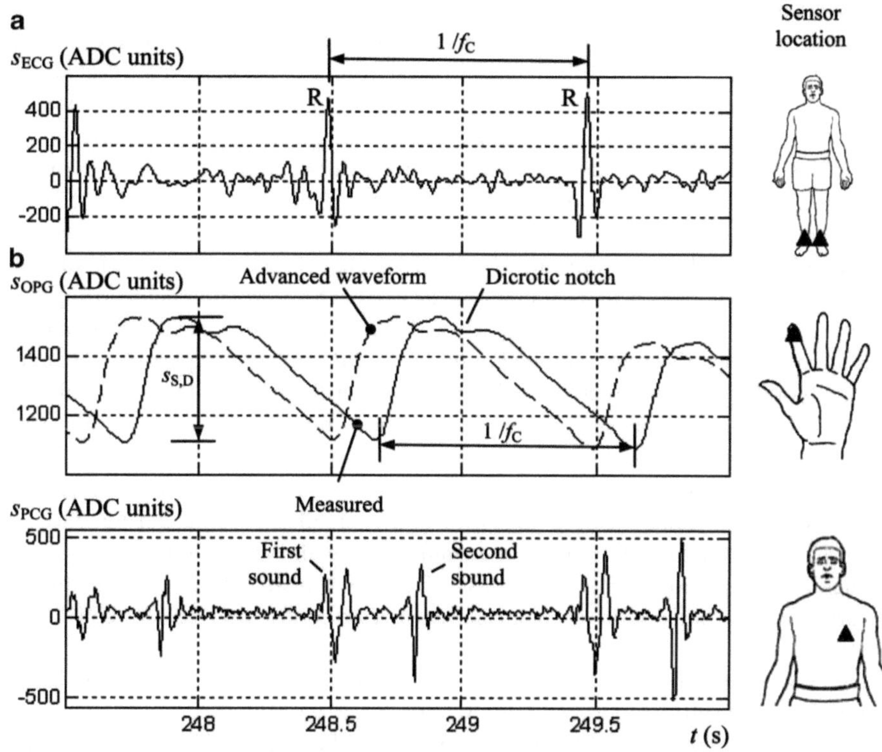

Fig. 3.2 Heartbeats as given by (**a**) electric biosignal electrocardiogram s_{ECG} (lead from the left and right leg with indicated R peak as fiducial point), (**b**) optic biosignal optoplethysmogram s_{OPG} (from finger) with its advanced version (by 160 ms) and indicated systolic–diastolic deflection $s_{S,D}$, (**c**) acoustic biosignal phonocardiogram s_{PCG} (from the heart region on the chest) with indicated heart rate f_C

of about 1,500 m/s, Sect. 5), and the optic biosignal is considerably delayed with respect to the heart (pulse wave propagation velocity v of about 5 m/s, Fig. 2.41). In other words, the *optic biosignal* should be *advanced* by about 160 ms (Fig. 3.2b), i.e., the *arrival time* $l/\langle v \rangle$ (3.6) calculated with $l \approx 80$ cm estimated as the distance from the heart to the finger and an assumed v of about 5 m/s, to allow for its (approximate) time correct comparison with the electric and acoustic biosignals.

When the *electrocardiogram and heart sounds* are compared (Fig. 3.2a vs. c), it can be observed that the first heart sounds correspond to R peaks. From a physiological point of view, it confirms that *atrioventricular valves* close at the onset of ventricular systole (Fig. 2.38c, d). The second heart sounds occur at the end of T waves, i.e., at the end of ventricular systole when *semilunar valves* close (compare also Fig. 3.10b, c).

Considering the *electrocardiogram and optoplethysmogram* (Fig. 3.2a vs. advanced waveform in Fig. 3.2b), it can be observed that R peaks initiate a steep increase in the light absorption (or local blood volume). Namely, the onset of the ventricular systole initiates ejection of blood (pulse) out of the heart. On the other hand, the *heart sounds and optoplethysmogram* (Fig. 3.2c vs. b) reveal that the second heart sounds coincide with dicrotic notches (Footnote 158), i.e., coincide with the consecutive closures of the semilunar (aortic) valve.

Other methods to be mentioned for the registration of the cardiac activity include electroplethysmogram, an *electric* biosignal, based on detection of pulsatile changes of electrical tissue impedance (see Sect. 4); magnetocardiogram, a *magnetic* biosignal, giving magnetic fields emitted by electrical heart muscle excitation; barocardiogram, a *mechanic biosignal*, reflecting periodic changes of the blood pressure (Sect. 3.1.3.1); mechanocardiogram, a *mechanic* biosignal, found through local skin deformations and vibrations because of pulsating blood vessels underneath (Sect. 7); another mechanocardiogram, showing pulsatile changes in the volume of the capillary bed, e.g., a larger finger volume can be observed during blood ejection (systole) and a smaller volume during diastole (Sect. 7); and ballistocardiogram, a *mechanic* biosignal, based on the recording of accelerated mass movements, i.e., mass of both the heart and circulating blood, imparted to the body with each atrial/ventricular contraction and forced blood ejection.

Other more sophisticated but *less direct methods* include ultrasound Doppler that reflects pulsatile blood velocity within large vessels (Sect. 3.1.3.2); laser Doppler in combination with interferometric methods quantifying minute skin displacements because of pulsating blood (Ul'Yanov and Tuchin 1993); microwave Doppler radar technology that measures cardiac displacements of the chest region with respect to an ambient antenna (Greneker 1997); or high-resolution thermal imaging of superficial blood vessels yielding pulsatile blood flow (Garbey et al. 2007).

Each of the aforementioned biosignals shows a *unique waveform and behavior* in normal and pathologic cardiac cases, as could be expected from their cardinally different waveforms in Fig. 3.2, see also Sect. 3.1.3. Here *formation methods* of the respective biosignals, i.e., the interface in between the physiological mechanism forming biosignals and the registered biosignal, are of crucial importance, as discussed in Sect. 4–7. In particular, the phenomena of *pulsatile* blood *pressure*, blood *flow*, and blood *velocity* in vessels are covered by Sects. 2.5 and 3.1.3 in some depth. Since numerous cardiac-related cases of different biosignals will be discussed in the Sect. 4–7, we focus on the physiological *origin of the heartbeat*, i.e., the initiation of the heartbeat, described by its *instantaneous* f_C.

Although spontaneous cardiac activity is intrinsic to various pacemaker tissues in the heart, the level of f_C and its change are *largely under the control* of the *autonomic nervous system* mediated via the *sinoatrial node*, the main pacemaker in the heart (Sect. 2.4). That is, efferent *sympathetic*[188] and *parasympathetic*[189] (=vagal) activities directed to the sinoatrial node are usually *reciprocal*[190] and characterized by discharges (predominantly) synchronous with each cardiac cycle. Both activities, especially their *balance* determining the *instantaneous* f_C, can be *modulated* via

- *Central oscillators* in the central nervous system, e.g., vasomotor and respiratory centres
- *Peripheral oscillators*, e.g., respiratory movements (Sect. 3.2.1) and arterial pressure fluctuations (Sect. 3.2.2). In terms of control engineering, *inhibitory* and *excitatory* peripheral reflexes act as control loops with *negative* and *positive feedback*, respectively (Zemaityte 1997)

These oscillators generate seemingly noisy *fluctuations in the heart period* or in the corresponding instantaneous f_C. However, a similar rhythmic behavior of these fluctuations can be observed over different time scales, from short-term changes up to long-term changes (see below). In addition, the *level of* f_C provides one of the most efficient measures of estimating *energy expenditure* in the body; for instance, f_C increases with increasing *oxygen consumption* (Keytel et al. 2005). The heart period ($= 1/f_C$) is usually referred to as the *RR interval*, i.e., the time interval between two consecutive R peaks (fiducial points in Fig. 3.2) of the

---

[188]The activity of the *sympathetic nervous system* (SNS) is due to a disruption of the autonomic balance because of the body's physical and mental activity (body accelerator); a mobilization of metabolic resources and *increased energy expenditure* are given during times of stress, arousal, or other external challenges, supported by elevated blood pressure and redirected blood from the intestinal reservoir toward skeletal muscles. The SNS is mediated by the release of hormones (e.g., adrenalin) induced via splanchnic nerves, the neuronal activation of beta receptors in the heart (sinoatrial node, atrioventricular node, and heart muscles), the acceleration of the slow diastolic depolarization (Footnote 111), and a decrease in cardiac refractory period, typically resulting in an *increased* f_C and increased contractile force of heart muscles.

[189]The activity of the *parasympathetic nervous system* (PNS) is concerned with the regulations of routine functions of the body during rest or digestion (body brake), promoting restoration and conservation of bodily energy, i.e., *reducing energy expenditure*. The PNS influence is mediated via release of acetylcholine by the *vagus nerve* toward the heart (sinoatrial node and atrioventricular node) which mainly slows down the diastolic depolarization, increases cardiac (ventricular) refractory period (Footnote 56), and thus typically *decreases* f_C and diminishes blood pressure (Footnote 71).

[190]It is worth noting that *reciprocal behavior* of the sympathetic and parasympathetic activities is given for *tonic* (basic) control of f_C while both reciprocal and nonreciprocal behaviors arise for *reflex* (stimulus related) control of f_C (Zemaityte 1997). Interestingly, in the case of concurrent sympathetic and parasympathetic activities, *parasympathetic* vagal stimulation seems to *override sympathetic* stimulation in the heart. For instance, a relative *decrease in* f_C due to PNS stimulation (e.g., due to artificial electrical vagal stimulation) is largest when the baseline level of f_C is relatively high and SNS activity is well expressed (Levy and Zieske 1969). By contrast, a relative *increase in* f_C due to SNS stimulation is smallest when prevailing PNS activity (or level of vagal tone) is high.

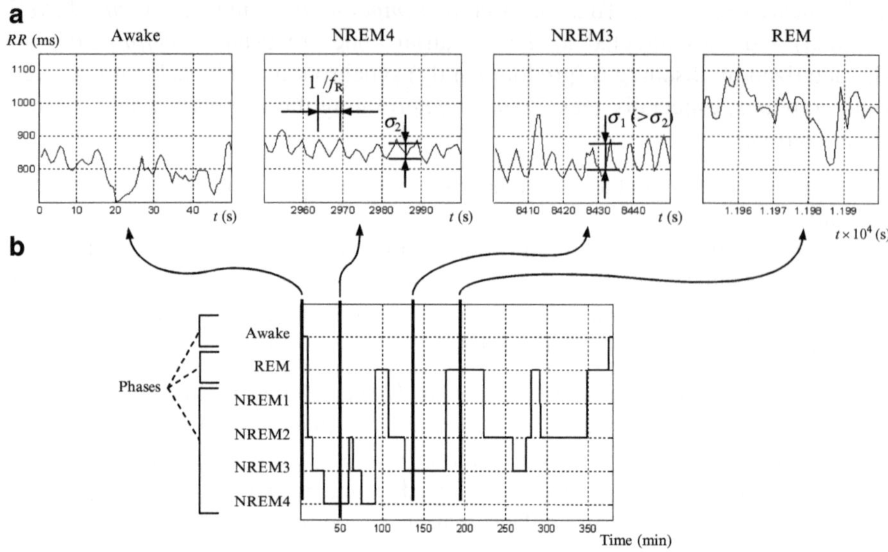

Fig. 3.3 Typical behaviors of interbeat intervals RR ($= 1/f_C$ with f_C as instantaneous heart rate) over time t as derived from R peaks of electrocardiogram during different awake/sleep phases and stages. (**a**) The RR intervals during, from left to right, awake phase, deepest sleep [nonrapid eye movement (NREM) phase, stage 4], deep sleep (NREM, stage 3), and paradoxical sleep, i.e., rapid eye movement (REM) phase, with an identical resolution of RR, indicated respiratory rate f_R and standard deviation σ. (**b**) The corresponding somnogram of the subject showing successive awake/sleep phases and stages

electrocardiogram (Sect. 4); consider limitations of the RR interval (from Footnote 196) to assess the activity of the sinoatrial node.

The *heart rate variability* (HRV) has become the standard term to describe both the oscillation in the heart period and the oscillation between consecutive instantaneous values of f_C (less common). To exemplify the HRV, Fig. 3.3a demonstrates the specific behaviors of interbeat intervals, the so-called *rhythmograms*, during different sleep phases; see Sect. 3.2.4 for definitions of the sleep phases. It can be observed that the most obvious phenomenon in the rhythmograms is the interval modulation with respiration, as discussed later.

Thus, the HRV is closely interrelated with regulatory mechanisms of the autonomic nervous system which responds immediately to any *physiological state* including *respiration* phase (Elstad et al. 2001), *exercise* and rest (HRV task force 1996), wake and sleep (Penzel et al. 2007), *sleep* phases and stages (Schmitt et al. 2009), *circadian rhythms* (see Sect. 3.2.3), or even *emotional activity* (Bigger 2006; Varoneckas 2003; Clifford et al. 2006). Furthermore, it is a good indicator of the *functional integrity* of physiological processes (thermal, hormonal, neural, autonomic, etc.) reflecting the physiological state of interest and general well-being. Thus, the assessment of HRV allows for the registration of the impact of any *physiologic perturbation* or the early signs of *pathological developments*, e.g., cardiovascular diseases (HRV task force 1996). The HRV is one of the most promising robust[191] markers measuring fluctuations (rather than mean levels) of autonomic inputs of the heart.

In particular, a continuous interplay[192] (or balance) between the activity of the SNS and PNS can be measured by the HRV. As a matter of fact, the level of

- The *SNS* (and *partly the PNS*, which is a controversial issue; HRV task force 1996) is interrelated with the power of the *low-frequency* (LF) band 0.04–0.15 Hz of the power spectral density[193] of interbeat intervals, compare Fig. 3.4a. In other words, the relatively *slow oscillations* in the rhythmogram (related to the LF band), known to contribute to Mayer wave arrhythmia (Sect. 3.2.2.1), exhibit a modulation period of about 10 s [=1/(0.1 Hz)] and indicate *mid-term regulatory mechanisms* (see below). In particular, the changes of interbeat intervals are related to arterial *blood pressure* oscillations in terms of baroreflex, i.e., blood pressure control is tightly linked with f_C adjustment. In short, the modulation can be explained as a resonance phenomenon in the control loop of the baroreflex involving SNS and PNS. This resonance occurs because of delays between hemodynamic and neural pathways, i.e., the compensatory effect of the control loop is not immediate but comes up with a latency of 2–5 s (Seydnejad and Kitney 2001).
- The *PNS* is interrelated with the power of the *high-frequency* (HF) band 0.15–0.4 Hz, compare Fig. 3.4a. It is mainly characterized by relatively *rapid respiratory modulation* of interbeat intervals (normally) at about 0.2 Hz, known as *respiratory sinus arrhythmia* (Sect. 3.2.1) since respiration is mediated by the PNS. That is, *short-term regulatory mechanisms* are involved with a modulation period of respiratory cycle duration. Figure 3.3a demonstrates this most evident modulation of the rhythmogram by respiration.

---

[191] Advantageously, f_C is a *frequency-related parameter* but not amplitude related, in contrast to, for instance, chest circumference changes induced by airflow or respiration. Thus, the influence of unavoidable body movements or external noise can be expected to be lower.

[192] In general, an increased activity of the PNS represents overall healthier people while decreased values reflect temporal dysfunction. An increased *activation of the SNS* can be attained by a 90° tilt, standing, physical activity, or stress, while an *activation of the PNS* can be attained by controlled respiration, compare Sect. 3.2.1 (HRV task force 1996). Furthermore, a *positive stress* could be assessed by an increase in both PNS and SNS, while *distress* (or physical exhaustion, nervous tension) is indicated by an increase in the SNS with a simultaneous decrease in the PNS (Riftine 2006). Furthermore, there is significant evidence that the SNS plays an important role in the genesis of arrhythmias, while the PNS has a protective role, decreasing the probability of arrhythmias (Bigger 2006).

[193] The *power spectral density* provides information of how signal power (i.e., variance) is distributed as a function of frequency, compare Footnote 150. In the case of the interbeat intervals, which have a physical unit of 1 ms, the spectral density has the unit of 1 ms^2/Hz. Thus, the power in a particular frequency range is an integral of the power spectral density over the frequency range (compare Fig. 3.4a). It is important to note that an estimation of the power spectral density assumes a *stationary* sequence of interbeat intervals; obviously, this requirement is more valid for short sequences than long because different short-term, mid-term, and long-term *regulatory mechanisms* impact the stationarity of the sequence (compare Fig. 3.3a).

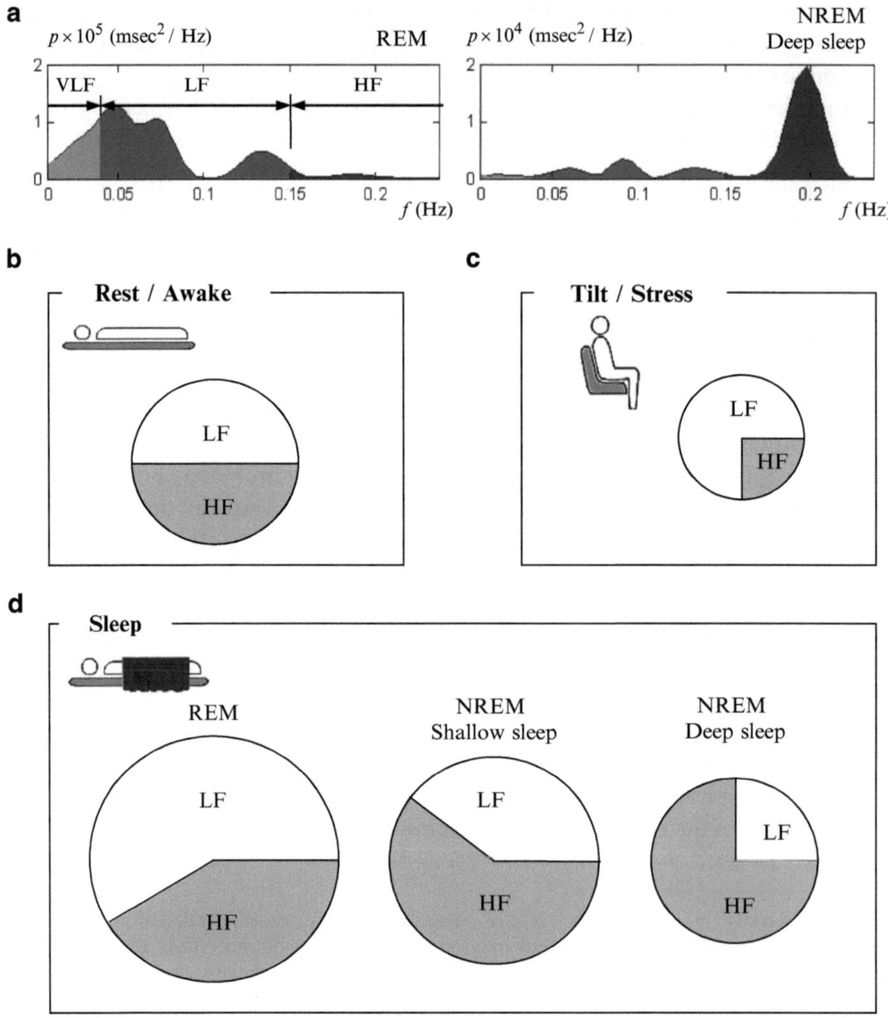

Fig. 3.4 The rationale of the HRV analysis with (**a**) power spectral density p over frequency f of interbeat intervals during different sleep phases, rapid eye movement (REM) phase, and non-REM (NREM) phase (or deep sleep). The spectral ranges of very-low-frequency (VLF) components, low-frequency (LF) components, and high-frequency (HF) components are depicted with different fill colors. Below, the *pie charts* show the relative distribution of power of LF and HF components in healthy subjects during (**b**) rest/awake, (**c**) tilt/stress, and (**d**) different sleep phases and stages. The absolute power of the two components is approximately represented by the area on the same scale. Data aggregated from HRV task force (1996), Tsunoda et al. (2008), Villa et al. (2000), Busek et al. (2005), Varoneckas and Zemaityte (2002), and others

Here, it should be noted that the *ratio of the LF/HF* components is often used as an indicator for *sympathovagal balance*, since the LF component is influenced by both the SNS and PNS.

As can be inferred from Fig. 3.4a, in addition to the LF and HF band there is a *very-low-frequency* (VLF) band <0.04 Hz. It is related to *very slow* oscillations in the rhythmogram, whereas the *long-term regulatory mechanisms* involved and their physiological origins are much less defined. The most likely candidates are

- *Thermoregulation* (compare Sect. 3.1.5)
- *Humoral* and *metabolic* regulation
- *Circadian* variations (Sect. 3.2.3)
- *Blood pressure* regulation (Sect. 3.2.2)

For instance, the mechanism of the blood pressure regulation is mainly due to autorhythmicity of the vascular smooth muscle and baroreflex response, thereby modulating the whole cardiovascular system including the interbeat interval; it is also known to contribute to Mayer wave arrhythmia (Sect. 3.2.2.1).

It is important to note that during

- *SNS activation* (increasing expenditure of bodily energy) the resulting relatively *slow increase in* f_C (or slow shortening of cardiac intervals, interrelated with the LF band) is usually accompanied by a marked *reduction in HRV* (or total power of cardiac intervals). This is because the range of possible variability of the cardiac interval is reduced. Consequently, the response in f_C is relatively slow with the *response time* >5 s.

The reverse occurs during

- *PNS activation* (diminishing expenditure of bodily energy) which results in a relatively *rapid decrease in* f_C (interrelated with the HF band) accompanied by an *increase in HRV*. Consequently, the response in f_C is relatively rapid with a *response time* within one to two heartbeats; in other words, the slowing of the heartbeat happens very quickly given the discrete nature of the heartbeat

This basic behavior is exemplified in Fig. 3.4b, c by comparing *rest and tilt*[194]; in the tilt position the total power of the HRV is reduced while the LF fraction is

[194]The subject's transition *from supine to upright* reflects the dynamic response of the autonomic nervous system and is typically used as a standard test in the HRV analysis (=*orthostatic test*). When standing, the gravitational force yields a widening of compliant venous vessels of the legs, which reduces the available blood volume, venous return, and thus filling of the left side of the heart (compare Footnote 225). In order to counteract a decrease in all three *stroke volume, cardiac output, blood pressure* (see baroreflex, Sect. 3.2.2.1)

- The *peripheral resistance* increases (through vasoconstriction)
- The *heart contractility* increases (through neuronal control)
- The *level of* f_C increases (through neuronal control); compare (2.31)
- When standing the *total power of the HRV* is reduced (Fig. 3.4b, c)

The lower is the base level of f_C (i.e., the higher is the *prevailing PNS activity*), the more pronounced and more rapid is the f_C rise in response to standing up. By contrast, the higher is the base level of f_C, i.e., the higher is the *prevailing SNS activity*, the less pronounced and more inert is the f_C rise.

being increased. That is, a high HRV is usually associated with an increased HF fraction, while a low HRV with an increased LF fraction. It is important to note that an *increase in* f_C affects *diastole* more than *systole*, with the duration of systole increasing *relative* to the diastolic duration and decreasing when the heart slows.

According to the examples in Figs. 3.3a and 3.4, the *quantitative level of HRV* can be assessed by numerous *time domain* measures such as the standard deviation σ (=nonstationarity) of the consecutive interbeat intervals or numerous *frequency domain* measures such as spectral power in the aforementioned frequency ranges, e.g., power of HF, LF components or their ratio. From a general point of view, there are linear and nonlinear measures of the HRV since linear and nonlinear phenomena[195] are certainly involved in the genesis of the HRV (HRV task force 1996). The *linear measures* include (nonexclusively) the above time and frequency domain measures. *Nonlinear measures* that have been employed include the detrended fluctuation analysis, Lyapunov exponents, point process framework, and others (HRV task force 1996; Chen et al. 2010; Penzel et al. 2007).

For instance, the HRV *across sleep/awake phases* (Sect. 3.2.4) reveals a varying influence of sleep regulation on autonomic cardiac activity (Tsunoda et al. 2008; Villa et al. 2000; Busek et al. 2005; Varoneckas and Zemaityte 2002; Baharav et al. 1995; Schmitt et al. 2009; Zemaityte et al. 1984). *Nonrapid eye movement* (NREM) phase ("slow" sleep) *PNS activity* dominates compared to the awake phase. This is demonstrated in Fig. 3.3a by the amount of the HF component in the NREM3 and NREM4, which show dominating and clear oscillations of interbeat interval duration with respiratory rate. The prevalence of the HF component in the NREM is also shown in Fig. 3.4d depicting absolute and relative power of LF and HF components. In contrast to the NREM, during *rapid eye movement* (REM) phase ("fast" sleep or dreaming) and wake *SNS activity* prevails. Figures 3.3a and 3.4b, d demonstrate prevailing LF components during wake and REM compared to NREM. Obviously this varying interplay of nervous systems leads to very different morphologies in the time series of the heartbeat intervals, as demonstrated by Fig. 3.3a.

In particular, the HRV is increased during *sleep* as compared with *wake*, indicating increased restorative effects of the body. This can be observed in Fig. 3.4b–d as the area of circles, i.e., total HRV level, is higher in sleep than during wake. The HRV is highest in the REM and it decreases with increasing *sleep depth*, which is also indicated by the oscillation amplitude of the rhythmograms in Fig. 3.3a, compare σ_1 for deep sleep in NREM3 and $\sigma_2(<\sigma_1)$ for deepest sleep in NREM4. The PNS fraction increases with increasing sleep depth with its absolute level remaining nearly the same, while the SNS fraction and its absolute level strongly decrease (Fig. 3.4d). For instance, the value of σ is reduced by about 0.02 s from

[195] A *linear* phenomena or linear system satisfies the principle of additivity and homogeneity. That is, the net response of the linear system caused by multiple stimuli is the sum of the responses caused by each stimulus separately, according to the so-called *superposition principle*. In other words, *nonlinear measures* (or features) describe nonlinear (output) signals produced by nonlinear (biological) systems.

shallow sleep to deep sleep (Schmitt et al. 2009), compare Fig. 3.3a. In analogy with the sympathovagal balance across sleep/awake phases, only the *short-range correlations* (over a few seconds) of the subsequent interbeat intervals remain with increasing sleep depth, as disclosed by the nonlinear detrended fluctuation analysis (Penzel et al. 2000, 2007). The short-range correlations indicate a more random behavior and the presence of only HF components (Figs. 3.3a and 3.4d). By contrast, during the REM and wake, *long-range correlations* exist proving the presence of LF components (Figs. 3.3a and 3.4d). Finally, it should be noted that the particular mechanisms of interaction between autonomic control and sleep structure are currently not well understood (Schmitt et al. 2009).

To demonstrate the *diagnostic value* of the HRV (HRV task force 1996), the effect of *congestive heart failure* should be shortly discussed. In comparing healthy subjects and subjects with congestive heart failure, this pathology was associated with reduced power density along all frequencies, reduced tone of the PNS, predominance of the SNS, and thus increased ratio LF/HF. Besides the latter *linear measures*, the *nonlinear measures* change as well. The dynamics of the HRV from healthy subjects are more nonlinear, with a prevalent nonlinearity due to increased SNS and reduced PNS activity (Chen et al. 2010).

Healthy aging is marked by reduced HRV level, e.g., the value of σ is reduced by about 0.02 s from young persons with average age 33 years to elderly persons with average age 78 years, consistently across all sleep stages (Schmitt et al. 2009). This shows that σ decreases very gradually with aging, reaching 60% of its baseline (second-decade values of age) by the tenth decade (Umetani et al. 1998; Villa et al. 2000). In particular, the PNS tone is reduced in the elderly, e.g., by about 50% when comparing age 33 with age 78, as can be derived from Schmitt et al. (2009).

It is interesting to note that the influence of *sleep regulation* on the cardiac autonomic regulation does not break down with *healthy aging* despite a reduction in the HRV and strong sleep fragmentation (Sect. 3.2.4), i.e., both young and elderly exhibit similar aforementioned stratification patterns in the HRV across sleep/awake phases. That is, the effect of sleep regulation is even stronger than that of aging; according to Schmitt et al. (2009), the difference in σ between wake and deep sleep (≈ 0.04 s) is roughly twice that of the difference between the young and elderly (≈ 0.02 s).

The apparently easy derivation of the HRV has popularized its use leading to misinterpretation of the results. The recording conditions and preprocessing should actually fulfill numerous *requirements* in order to obtain a reliable HRV estimation.[196] On the other hand, the meaning of the many different HRV *measures depend on recording* conditions, e.g., σ increases with the length of analyzed

[196]The most important *requirements for a reliable HRV analysis* are summarized below:

- The *sampling rate* of the biosignal (compare Fig. 3.2) should be relatively high (>500 Hz) for a precise estimation of the instantaneous f_C; i.e., significantly higher than used during standard electrocardiogram recordings (≤ 256 Hz)

recording, as the contribution of the long-term regulatory mechanisms rises (HRV task force 1996).

3.1.2 Respiration

The *lung* is responsible for oxygen delivery to the bloodstream and release of carbon dioxide from the blood through its *rhythmic* expansion and contraction (Sect. 2.6). An assessment of the resulting *respiratory cycle* with the respiratory rate f_R is a vital issue from a diagnostic and therapeutic point of view. Numerous biosignals can be used to register the respiration.

Figure 3.5 shows three established *methods* to register respiratory activity. The mechanorespirogram in Fig. 3.5a is a *mechanic biosignal* resulting from circumference changes of the abdomen (and chest) during breathing.[197] A periodic waveform with f_R can be observed during *normal breathing*, while it disappears when holding breath. *Snoring* is provoked, which induces an increase in amplitude deflection; this is due to intensified respiratory efforts in order to increase tidal volume and to overcome an (artificially) increased respiratory resistance. If the biosignals from the abdomen and chest are compared (Fig. 3.5a), differing amplitudes can be observed, which are due to different strengths of abdominal and chest breathing. Not only the amplitude but also the phase of the latter biosignals may differ, for the abdominal and chest breathing are usually slightly asynchronous, i.e., the waveform of the abdominal breathing is delayed with respect to chest breathing.

In a first approximation, the phonocardiogram in Fig. 3.5b, an *acoustic biosignal*, mainly visualises the heart sounds arising from consecutive closures of heart valves. However, the lung sounds are also present during *normal breathing* due to air turbulences in the branching airways of the lung. These sounds cannot be easily distinguished in the time course due to their much lower amplitude, for details see Sect. 5. Only when *snoring*, an additional overlapping signal component can be

---

- *Abnormal events* as *ectopic beats*, arrhythmic events, and noise should be appropriately eliminated, because they are not governed by the sinoatrial node (Footnote 120)
- The choice of *fiducial point* to asses interbeat intervals may be critical,
  - Either the *P wave* in the electrocardiogram (compare Fig. 1.15a) is the most appropriate fiducial point to assess the activity of the *sinoatrial node* but low in amplitude and difficult to detect
  - The prominent maximum during *QRS complex* after the P wave (Fig. 3.2a), or even
  - The centre of area under the QRS complex

It should be noted that clinical and scientific practice relies mostly on the fiducial points based on the QRS complex.

[197]In the given case of Fig. 3.5a, the *body circumference* changes due to respiratory muscle activity are detected by *respiratory belts* positioned around the subject's abdomen and chest. The extension and contraction of an elastic, tightly applied belt, relative to its initial length, is detected during inspiration and expiration, respectively. A piezoelectric transducer in the belt converts its mechanical tension into voltage which is amplified and shown in Fig. 3.5a.

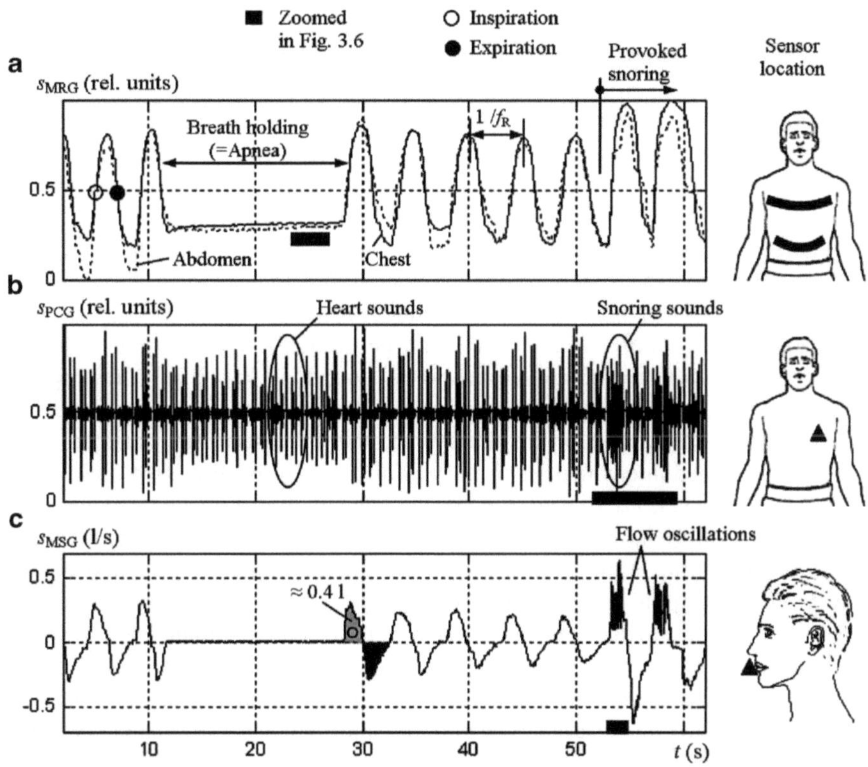

Fig. 3.5 Respiration as given by (**a**) mechanic biosignal mechanorespirogram s_{MRG} (from chest and abdominal circumference changes), (**b**) acoustic biosignal phonocardiogram s_{PCG} (from the heart region on the chest), and (**c**) mechanic biosignal mechanospirogram s_{MSG} (from mouth airflow) with indicated respiratory rate f_R and respective sensor locations. For zoomed versions see Fig. 3.6

recognized during the inspiration phase, which corresponds to the snoring sounds (Fig. 3.5b). In short, snoring sounds are generated from elastic oscillations of the pharyngeal walls which may even lead to a temporal closure of the airway. It should also be noted that the amplitude of the heart sounds (=envelope) remains nearly constant when holding breath, while it is strongly amplitude modulated during breathing; this behavior already indicates obvious cardiorespiratory interrelations to be discussed later (Sect. 3.2.1).

Lastly, another *mechanic biosignal* reflecting respiration is depicted in Fig. 3.5c, which is quite common in clinical practice. It is given by a recording of the *air flow* through the mouth, while (eventual) nasal airflow is stopped by a clip.[198]

---

[198]In the given case of Fig. 3.5c, the *air flow* through the mouth is registered by a mouthpiece connected to an *airflow transducer*. The transducer basically consists of a tube with a woven screen inside which acts as a flow-resisting object (=flow resistance). During inspiration, the outer

During *normal breathing*, the air flow is positive during inspiration and negative during expiration, compare with inspiratory increase in the circumference (Fig. 3.5a) and the following expiratory decrease. In other words, the air flow is zero (or flow reversal occurs) at extreme values of the thoracic circumference, compare Fig. 3.5a, c at the time instant of 30 s (see also Fig. 3.33b, c from Sect. 3.2.1.2). If the flow waveform is integrated over time for a single inspiration (or expiration) event, it gives the volume of the inspired (or expired) air, compare Fig. 3.5c with an indicated inspiratory air volume of about 0.4 l (tidal volume) after the breath holding. Obviously, the flow is zero when holding breath.

While intentionally making snoring sounds (referred to as provoked *snoring*), high-frequency *oscillations of the flow* can be recognized during the inspiratory phase (compare with the concomitant snoring sounds in Fig. 3.5b); the amplitude of the flow increases during both inspiration and expiration. These oscillations are due to intermittent closures of the airways and the increased amplitudes are due to the aforementioned intensified respiratory efforts, as will be discussed in Sect. 5.

The *zoomed versions* of Fig. 3.5 are depicted in Fig. 3.6, which demonstrate waveforms of particular interest in specific time intervals. When holding breath, the mechanorespirogram in Fig. 3.6a discloses a *residual cardiac component* oscillating with f_C. This component reflects periodic (global) vibrations of the chest circumference due to (local) vibrations of the chest wall, which are initiated by mechanical pumping action of the heart. That is, the vibrations themselves are imparted to the body with each heartbeat during the heart's ventricular contraction when the blood is forcefully ejected into arteries. The phonocardiogram in Fig. 3.6b depicts *snoring sounds* during the inspiration phase which overlap the heart sounds. The air *flow oscillations* are shown in Fig. 3.6c, whereas the oscillation rate ($\approx$ 40 Hz) of the intermittent airway closures during provoked snoring can be easily derived.

Another biosignal is depicted in Fig. 3.7b which reflects respiration. It is a thermorespirogram, a *thermal biosignal*, resulting from variations of the air temperature in front of the nostrils during breathing.[199] Since the thermal biosignal is depicted in parallel to the aforementioned mechanic biosignal (from Fig. 3.5a), it is obvious from the figure that the temperature varies with f_R by about 1°C. An inverse relationship between the thermal and mechanic biosignal can be observed, i.e., the temperature increases during expiration and decreases during inspiration.

---

side of the flow resistance has a higher air pressure than the inner side (toward the mouth) while during expiration the reverse is true. The resulting pressure difference across the flow resistance reflects the nonzero air flow (=pressure difference divided by flow resistance). The measured pressure difference is converted to voltage with the amplified and calibrated version being shown in Fig. 3.5c.

[199] In the given case of Fig. 3.7b, the *temperature* of the *airflow* is detected by a miniature *thermistor* close to the nostrils. During inspiration, the relatively cool air (compared to the subject's body temperature) enters the nostrils and is warmed up in the body. During expiration, the warm air leaves the respiratory airways. The thermistor converts the air's temperature into voltage with the amplified and calibrated version given in Fig. 3.7b.

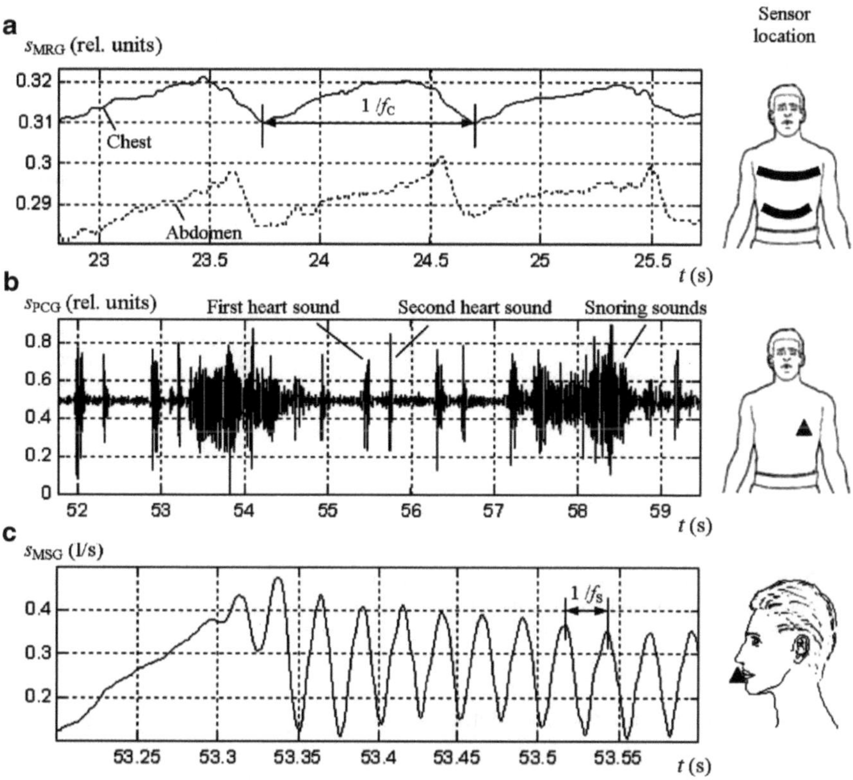

Fig. 3.6 Respiration and cardiac effects as given by zoomed versions of Fig. 3.5: (**a**) cardiac component in the mechanorespirogram s_{MRG} when holding breath, (**b**) provoked snoring sounds and heart sounds in the phonocardiogram s_{PCG}, and (**c**) flow oscillations with the rate f_S ($\approx$40 Hz) during provoked snoring in the mechanospirogram s_{MSG} with indicated heart rate f_C and respective sensor locations

Other methods to be mentioned for the registration of the respiratory activity include electroplethysmogram, an *electric* biosignal, based on monitoring the thoracic electrical impedance changes due to cyclically inflated and deflated lungs (Sect. 4). In analogy to the mechanorespirogram from Fig. 3.5a (Footnote 197), a *mechanic* biosignal, respiratory circumference changes of the abdomen (and chest) can be established by inductance plethysmography. Here, a loop of wire is placed around the abdomen (or chest) and the resulting electrical inductance of the wire is measured; namely, the inductance increases with the area the loop encloses and thus with the inspiration phase of the respiratory cycle.

More indirect methods to detect respiration include optoplethysmogram, an optic biosignal, which reflects peripheral blood volume changes over the respiratory cycle. Electrocardiogram, an electric biosignal, echoes periodic and mechanic heart displacements due to varying lung volume in the inner thorax; see Sect. 3.2.1. In addition, flexible pads (e.g., piezoelectric foils) sensitive to pressure can be used

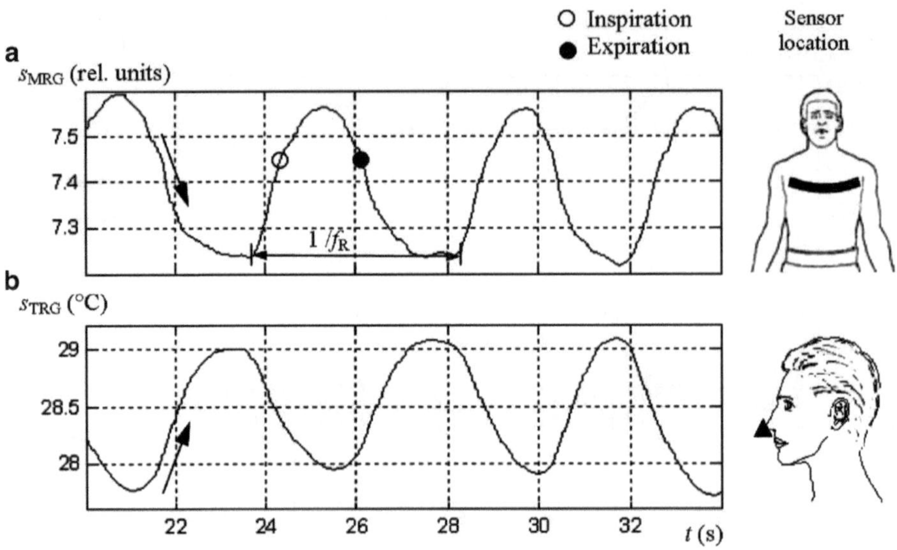

Fig. 3.7 Respiration as given by (**a**) mechanic biosignal mechanorespirogram s_{MRG} (from chest circumference changes), and (**b**) thermal biosignal thermorespirogram s_{TRG} (from nose airflow) with indicated respiratory rate f_R and respective sensor locations

under the abdomen and chest to detect shifts of subject's centre of mass during respiration.

As already indicated in Fig. 3.5 by a simulated breath holding, an *apnea* (greek *apnia* breathlessness) deserves an extended description. An apnea is a *respiratory disturbance* characterized by a complete (or partial) cessation of effective respiration. Sleep apnea which occurs[200] during sleep at night is well known and is usually detected by the so-called polysomnography (described in Sect. 3.2.4). There are several types of sleep apneas, but the following three are the most important:

- *Obstructive sleep apnea* (OSA) is characterized by occlusion of the upper airways, as demonstrated in Fig. 3.8, with maintenance of rhythmic contractions of inspiratory pump muscles yielding thoraco-abdominal excursions, for at least 10 s and with a (delayed) fall in oxyhemoglobin saturation (Sects. 3.1.4 and in section "Ceased Respiration" under Sect. 3.2.1.1) of more than 4%.

[200] *Sleep apnea* affects the sleep of millions of individuals, with a minimum *prevalence* of about 1%, dominating in men (Saletu and Saletu-Zyhlarz 2001; Peter et al. 1995). For instance, 2% of women and 4% of men between 30 and 60 years have more than five *obstructive* apneas and hypopneas per sleep hour and accompanying daytime sleepiness (Lee-Chiong 2006), the latter drastically reducing quality of live (the definitions of apnea types follow in the text). *Central* apneas are reported to be less common than obstructive, i.e., they occur at about 10% the rate of obstructive (De Backer 1995).

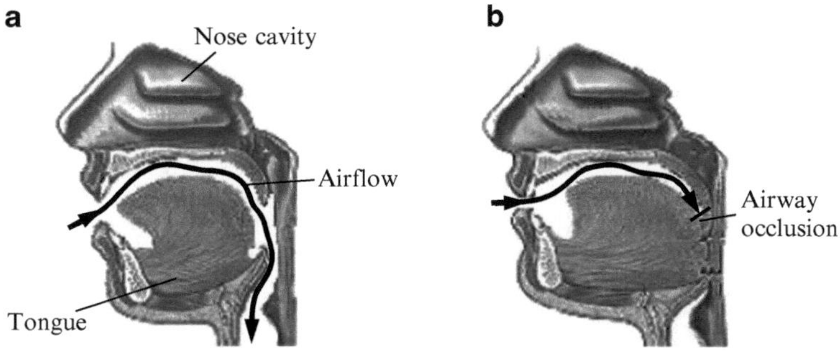

Fig. 3.8 (**a**) Normal (open) airway. (**b**) Obstructive (occluded) airway yielding obstructive sleep apneas

The OSA is surrounded by obstructive snoring,[201] i.e., intermittent, loud, and irregular snoring (Sect. 5). The OSAs are mainly due to anatomic and physiologic abnormalities which yield a decreased stability and increased compliance of the upper airway. Contributing factors include pathologic narrowing of the upper airways, redundant soft tissue within, obesity (a high body mass index[202]), increased neck circumference, and loss of upper airway muscle tone.

- *Central sleep apnea* (CSA) shows temporal failure in breathing rhythm generation due to a lack of neural stimulation lasting at least 10 s. The CSAs arise because of a deteriorated metabolic control of breathing (i.e., the metabolic control responsible for maintaining (nearly) constant oxygen, carbon dioxide, and hydrogen levels in arterial blood), a decreased activity of the respiratory muscles, and inhibited upper airway reflexes.
- *Mixed sleep apnea* (MSA) consists of successive CSA and OSA segments, i.e., showing an initial period of decreased central drive followed by an obstructive breath.

There are also other types of sleep apnea syndrome, characterized by a mere reduction in effective respiration, namely, obstructive *sleep hypopnea* (OHA), central sleep hypopnea (CHA), and mixed sleep hypopnea (MHA). During hypopneas, the airflow is reduced by at least 50% for more than 10 s with a noticeable fall in oxyhemoglobin saturation. Proper classification is complicated by the fact that it is difficult to distinguish in between OHAs and CHAs, with both OSAs and CSAs being commonly present together in the same subject (Lee-Chiong 2006).

Figure 3.9 demonstrates consecutive *MSA* and *OSA* by the use of various *biosignals*, as being applied in a polysomnographic investigation (Sect. 3.2.4). The

[201]For instance, epidemiological studies have shown that 37% of males and 19% of females were *snorers* while 10% and 7%, respectively, represent potential apnea patients (Saletu and Saletu-Zyhlarz 2001).

[202]The *body mass index* is an anthropometric measure defined as human weight in kilograms divided by the square of height in meters. Usually index values over 30 kg/m^2 indicate obesity.

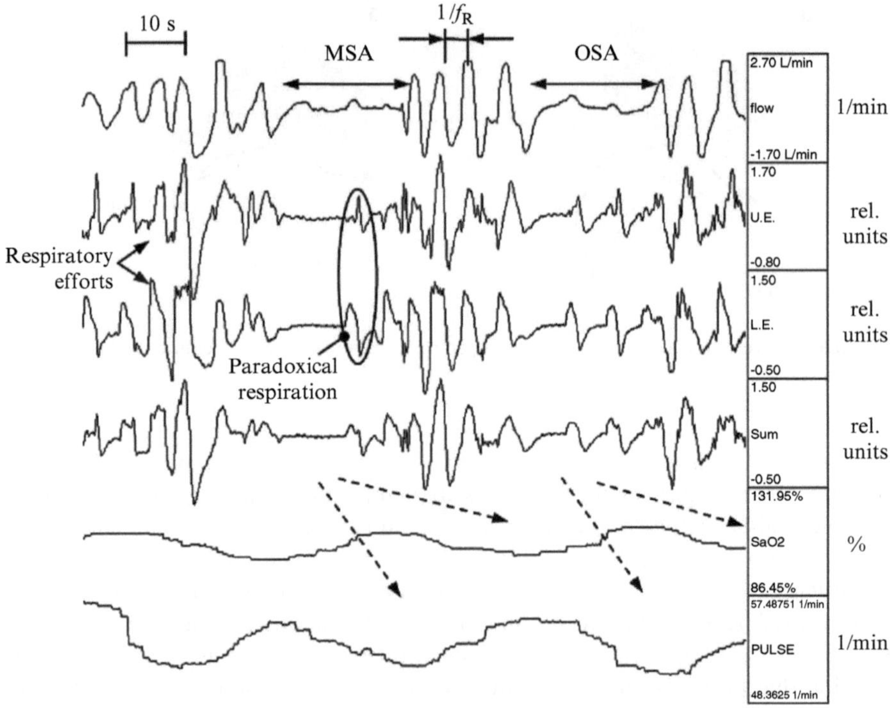

Fig. 3.9 Multichannel monitoring of sleep within the scope of polysomnography, showing consecutive mixed sleep apnea (MSA) and obstructive sleep apnea (OSA). The depicted signals include, from top to bottom, airflow, chest extension, abdominal extension, sum of the chest and abdominal extensions, blood oxygenation (hemoglobin oxygen saturation), and heart rate. *Dashed arrows* indicate temporal desaturations and drops of the heart rate which correspond to the depicted MSA and OSA

MSA begins with a central segment characterized by a zero line in the extensions of the thorax and abdomen, which indicates an absence of respiratory efforts. By contrast, the second half of the MSA corresponds to an obstructive segment in which respiratory efforts prevail while the airflow is still zero. The *OSA* shows the respiratory efforts throughout the entire period of ceased breathing. In addition, both MSA and OSA yield desaturations and reductions in the heart rate that are temporal and delayed. *Changes* in vital physiological *parameters* during apnea are described in section "Ceased Respiration" under Sect. 3.2.1.1 and exemplified in section "Ceased Respiration" under Sect. 3.2.1.2; compare also the voluntary breath holding in Sect. 3.1.4 (Fig. 3.20).

As the obstruction impedes pressure equalization between the lungs and atmosphere, the thoraco-abdominal excursions persist. That is, an inspiratory *expansion of the abdomen* happens in parallel with a *compression of the chest*, whereas for the expiration the reverse is true. This is called *paradoxical respiration*, as demonstrated in Fig. 3.9. By contrast, the abdomen and chest move in phase during normal (nonobstructed) breathing.

Isolated apneas do not represent any threat to health. However, as soon as the *apnea arousal frequency*, i.e., the total sum of apneas and hypopneas divided by total sleep time in hours (known as *respiratory disturbance index*), is *higher than five*, it is considered a *respiratory disturbance* (compare Footnote 200). This may cause severe *deterioration of life quality*, arterial hypertension, cardiac arrhythmias, decreased life expectancy, and other chronic and adverse cardiovascular effects (Peter et al. 1995). *Therapy* is mainly based on two concepts: *reduction of the obstruction*, e.g., with lifestyle changes to reduce obesity, avoiding supine sleep position or with surgical interventions aimed at removing tissues in the obstruction, and *bypassing the obstruction*, e.g., with the use of passive and active breathing appliances keeping the occluded airway open or with medications stiffening the flabby tissues (Lee-Chiong 2006).

Lung function becomes less efficient *with age*. The efficiency of oxygen delivery to the blood stream decreases. The amplitude of the air flow, the vital capacity, and maximal force of inspiration and expiration decline slowly after the third decade. The airways become more collapsible in the elderly, especially during shallow breathing, which increases the risk of lung infections. Breathing control by the brain is reduced with age; that is, the response of the rate and depth of breathing is reduced with respect to (low) oxygen and (high) carbon dioxide levels in the blood. The *prevalence of apneas* increases with age and saturates by the sixth to seventh decade. On the other hand, there are some indications that infants show relatively more CSAs due to immature respiratory control, while in adults OSAs prevail because of excessive tissues (obstructions) in the upper airways (compare Footnote 200).

3.1.3 Blood Circulation

The *blood circulation* is related to systemic and pulmonary circulation (Sect. 2.5). In short, the systemic circulation comprises *rhythmic* transport of the oxygenated blood to the body and of the deoxygenated blood back to the heart, whereas the pulmonary circulation comprises transport of the deoxygenated blood to the lungs and of the oxygenated blood back to the heart.

Besides the assessment of the *pumping action* in terms of the cardiac cycle with the duration $1/f_C$, as discussed in Sect. 3.1.1, a (simultaneous) registration of the *circulatory parameters* as

- *Blood pressure*
- *Blood flow*
- *Arterial radius*

is highly relevant for diagnosis and therapy. The pumping action of the heart yields periodic and *pulsatile waveforms* of the above circulatory parameters. The resultant *pulses*[203] carry information not only on f_C and its relatively *short-term rhythms* but also on the waveforms of the above-listed circulatory parameters, as detailed in Sect. 2.5. In addition, the circulatory parameters exhibit *long-term rhythms* including the so-called circadian rhythms, see Sect. 3.2.3 for the biological rhythms in general. In particular, the blood pressure, arterial diameter, and arterial wall stiffness show *circadian* variations (Nichols and O'Rourke 2005); that is, the blood pressure is reduced during sleep while the diameter and stiffness are increased.

In particular, the *blood pressure p* is physiologically assessed by its *characteristics* such as systolic value p_S, diastolic value p_D, and the pressure pulse *waveform*, as described in section "Pulse Waveforms of Pressure and Flow." The systemic *blood flow q* is mainly defined by the left ventricular stroke volume V_S, blood flow velocity u, and obviously the flow pulse *waveform*, as described in Sect. 2.5.2.2 and section "Blood Pressure and Flow." Lastly, the *arterial radius r* is usually defined by its mean value and the *pulsatile waveform* of r.

Before experimental issues related to the parameters p, q, and r are discussed, *aging effects* on the latter parameters should be quickly *summarized*. Most of the effects were already described in Sect. 2.5 within the scope of *distributed* and specific contexts. In the aging community, *arteries stiffen* and *dilate* progressively while the *arterial wall* becomes thicker, see Sect. 2.5.1. The *reflected wave* amplitude increases nonlinearly, see section "Reflected Pulse Propagation." These changes are the most important determinants of increased p_S, e.g., p_S increases by about 40 mm Hg from 20 to 80 years (Nichols and O'Rourke 2005), and also of increased pressure pulse deflection $p_S - p_D$. The level of p_D changes little with age, increasing by only about 10% up to age 50 and then progressively declining (Nichols and O'Rourke 2005). The latter changes of the pressure characteristics come with multiple cardiovascular complications such as hypertension (Footnote 166). The *amplification* and *waveform* changes of the pressure pulse from proximal to distal arteries are diminished with increasing age due to reflections and increasing stiffness, as discussed in section "Reflected Pulse Propagation."

The level of the *pulse wave velocity v* for old (and obese) patients has been found to be larger than for young (and thin) ones (Franchi et al. 1996). This is because of both normal *age-related* arterial stiffening (=senile arteriosclerosis) and abnormal arterial stiffening (=atherosclerosis), see Footnote 137 and section "Pulse Propagation." In addition to blood pressure and age, the propagation velocity is affected by other physiological parameters such as gender and serum lipids in blood (Okada 1988).

Besides the aforementioned age impact on the *arterial* system, functional changes in the *ventricular* function should be mentioned. With *aging*, slowed and delayed left ventricular *filling* occurs during diastole. Ventricular *relaxation* during diastole is also delayed on account of progressively developing left ventricular

[203]The relevance of the pulse can be best illustrated in words "...The pulse ranks first among our guides; no surgeon can despise its counsel, no physician shut his ears to its appeal..." (Mahomed 1872).

hypertrophy. The ventricular *contraction* is prolonged and the left ventricular *load* is increased because of aortic stiffening and early reflections, see section "Reflected Pulse Propagation." Thus, the *stroke volume* is reduced with age, whereas *progressive stiffening* of heart tissues and negative reflections of the blood flow wave also contribute to a decrease in the stroke volume.

Lastly, the ongoing deterioration of the autonomic *regulatory functions* should be exemplified with *age*. In elderly persons, blood circulation and heart rhythm respond weakly to stress or physical load. For instance, the blood pressure tends to drop if they stand up or change body position too suddenly, with the person becoming dizzy. This may be explained by reduced sensitivity of the baroreceptors with aging, see Sect. 3.2.2.

It can be expected that there are *numerous methods* to register the *vital circulatory parameters* p, q, and r and their respective characteristics; thus, there are numerous *biosignals* of interest. A short overview will be given below about the most popular methods to obtain a faithful reproduction of the circulatory parameters (Sects. 3.1.3.1–3.1.3.3), whereas both *invasive* and *noninvasive* approaches will be considered. Obviously the invasive methods tend to be more precise, while the noninvasive are more comfortable but often require a calibration on-site and on patient. Most methods will be exemplified with the corresponding biosignals to facilitate their understanding.

3.1.3.1 Blood Pressure

The *recording* of *blood pressure* has always been a *challenge*.[204] In particular, unobtrusive and long-term monitoring is difficult to establish, aiming at diagnostically relevant blood pressure values free of artifacts. Basically,

- Invasive and direct methods
- Noninvasive and indirect methods

can be distinguished. Before the most typical methods will be dealt with, an introductory illustration of the blood pressure monitoring will be given.

Figure 3.10 *illustrates* continuous and unobtrusive *blood pressure monitoring*, performed in *synchrony* with two other vital biosignals. The barocardiogram in Fig. 3.10a, a *mechanic biosignal*, is based on the volume clamp method (see below) and represents p waveform in the radial artery with the corresponding p_S and p_D. The biosignal exhibits a periodic waveform oscillating with fundamental frequency f_C; the waveform reveals primary and secondary peaks within the cardiac cycle, as discussed in section "Pulse Waveforms of Pressure and Flow."

[204]The *first measurement* of the *blood pressure* is ascribed to Stephen Hales (1677–1761), an English physiologist. He pioneered quantitative measurements of the blood pressure in animals by a vertical glass tube with one end inserted into a horse artery. The rise in the column of blood in the tube was an estimate for the instantaneous blood pressure level (compare Footnote 223).

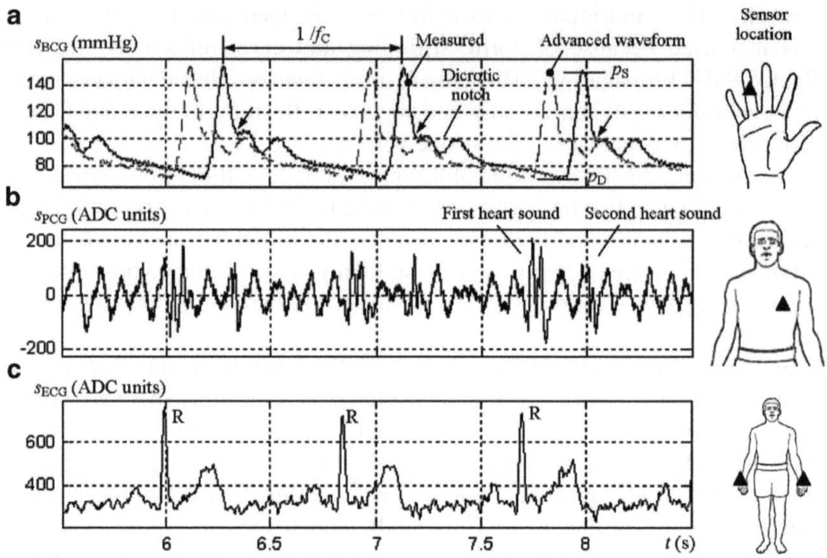

Fig. 3.10 Simultaneous recording of (**a**) mechanic biosignal barocardiogram s_{BCG} (from a finger on the right hand) with its advanced version (by 160 ms). Approximate starts of the reflected waves (inflection points) are indicated by *arrows* on the measured waveform. (**b**) Acoustic biosignal phonocardiogram s_{PCG} (from the heart region on the chest), and (**c**) electric biosignal electrocardiogram s_{ECG} (lead I Einthoven) with indicated R peaks

The phonocardiogram in Fig. 3.10b, an *acoustic biosignal*, demonstrates the heart sounds (Sect. 5), whereas the electrocardiogram in Fig. 3.10c, an *electric biosignal*, shows electrical heart muscle excitation (Sect. 4).

A close *connection* between the *p waveform* and, on the other hand, the *heart sounds* and *electrocardiogram* is highly conclusive about the origin of the respective biosignals; compare the *mutual interrelations* of the (primarily) cardiac biosignals from Sect. 3.1.1 including electrocardiogram, optoplethysmogram, and heart sounds. Here, it should be noted that the *respiratory impact* on *p* will be discussed in Sect. 3.2, for it offers valuable insights.

As in the case of Fig. 3.2, before the recorded signals in Fig. 3.10 can be *synchronously compared*, the relatively low propagation velocity of the pressure pulse waves should be considered. If the heart is chosen as a reference site for comparison (cardiac origin), the *delay* of the *p waveform* amounts to about 160 ms (=80 cm/5 m/s), as derived in Sect. 3.1.1. That is, the *p* waveform should be advanced by this amount of time to facilitate its reasonable comparison[205] with the heart sounds and electrocardiogram, see Fig. 3.10a.

[205]It should be stressed that the *advancement* alone does not yield the waveform of the aortic pressure at the heart, for the pressure pulse waveform changes strongly during pulse propagation toward the periphery (section "Reflected Pulse Propagation" in Sect. 2); *only the running time* from the heart to the finger is compensated through the deliberate advancement in Fig. 3.10a.

The comparison with the advanced waveform of p yields a demonstrative overlap between the onset of *systolic increase in p* (Fig. 3.10a), the *first heart sound* (Fig. 3.10b), and the *R peak* (Fig. 3.10c). In physiological terms, the onset of ventricular systole (R peak) initiates an increase in arterial pressure (systolic increase) and precedes blood ejection while atrioventricular valves close (heart sounds). Moreover, the time instant of the second heart sound coincides with the *dicrotic notch* (Footnote 158) of the advanced p waveform, i.e., coincides with the closure of the semilunar (aortic) valve. Lastly, the above verification of the dicrotic notch position by the second heart sound allows for a reliable identification of the wave—residing between the primary systolic peak and dicrotic notch—as a *reflected wave*, see corresponding arrows in Fig. 3.10a.

The *blood pressure* can be recorded in an *invasive* way directly in the vessel by inserting a catheter with a mounted *internal pressure sensor* or by inserting a fluid-filled and rigid catheter[206] for transmitting the blood pressure waveform (including p_S and p_D) to an *external pressure sensor*. However, such invasive methods, though precise and direct, lack popularity because of their invasiveness and related complications for routine use.

The most popular *noninvasive* methods for the determination of p_S, p_D, and (ideally) *blood pressure* waveform include the auscultatory method, oscillometric method, volume clamp method, and tonometric method. These methods will be described below while *experimental examples* follow the descriptions:

- *Auscultatory method* utilizes *Korotkoff sounds*[207] detected by a *stethoscope* head to determine p_S and p_D. In particular, an inflatable cuff encircles an extremity

[206] Different *application and technical issues* have to be considered if *catheters* are used. For instance, the inserted end of the catheter should be kept at the same vertical level as an external pressure sensor so that the hydrostatic pressure does not impact the blood pressure being measured. From a technical point of view, the catheter behaves as a low-pass filter with a resonant frequency (Neuman 2011). That is, a smaller diameter of the catheter [increasing catheter resistance to flow, (2.19)] or a larger length of the catheter (increasing fluid mass and inertia) would damp high-frequency components of the blood pressure waveform or, in engineering terms, would decrease the cut-off frequency of the low-pass filter.

[207] Nikolai Sergeyevich Korotkoff (1874–1920) was a Russian military surgeon who described the sounds heard over an artery below a compression cuff. The *Korotkoff sounds* arise at each *blood pressure pulse* and are due to the *local turbulence* in the blood flow passing through *constricted opening of the artery*, as long as the cuff pressure is in between p_D and p_S. That is, the artery temporally opens during systole and close during diastole. The *turbulent blood flow* moves not only in the axial direction of the vessel but also in other directions, including its radial direction, which causes mechanical vibrations of the vessel wall and thus generates *acoustical sounds*. When the cuff pressure is lower than p_D, the artery remains open, the *blood flow is nearly laminar* (Sect. 2.5.2.2), i.e., all lamina of the blood move parallel to the axis of the vessel, radial vibrations of the wall are absent, and the *vessel is silent*.

From the *historical perspective*, it is interesting to observe that Korotkoff was actually not looking for a method to measure the blood pressure (Geddes and Roeder 2009). He was rather interested in *collateral blood circulation*, which he evaluated by feeling the pulsations of a stenosed artery while he pressed down on the artery. The pulsations meant pulsatile flow of blood. Later, as a byproduct of this investigation, he introduced auscultated sounds from beyond the cuff on the upper arm, named as Korotkoff sounds, to estimate the blood pressure. Harold Nathan Segall (1897–1990), (upper arm) and the cuff pressure is increased until a complete cessation of the blood circulation downstream the (brachial) artery (the cuff must expand against the extremity). During subsequent release of the cuff pressure, the first released Korotkoff sound indicates the time instant when the upper (systolic) part of the *blood pressure pulse* wave passes under the cuff and the cuff pressure is equal to p_S. The following transition from muffling to silence indicates the time instant when the lower (diastolic) part of the pulse wave passes and the cuff pressure is equal to p_D.

- *Oscillometric method*, successor of the ancient mercury sphygmomanometer[208] (Fig. 1.13), uses the principle that pulsatile blood flow produces *radial oscillations* of the arterial vessels wall, which are transmitted to the cuff encircling an extremity and thus to a *pressure sensor* within. As soon as the intra-arterial blood pressure exceeds the cuff pressure (during its deflation), the oscillations of the vessels walls are strengthened due to turbulent blood flow and progressing arterial decompression. The cuff pressure at the time of the initial increase in the oscillations amplitude corresponds to p_S and that during a subsequent rapid decrease in the oscillations corresponds to p_D.[209] In fact, the maximal amplitude of the oscillations (of both vessels walls and cuff pressure) is observed when the cuff pressure passes the *mean arterial pressure*; i.e., when the systolic vessel radius has nearly reached its final value because of the actual decompression but the diastolic radius is still nearly zero.
- *Volume clamp method* (or vascular unloading method) uses a miniaturized cuff encircling a finger, equipped with an optical transmission sensor (Sect. 6) inside the cuff. When the radius (volume) of the finger artery tends to increase during the blood pressure (volume) pulse, as detected by the *transmitted light intensity*, the cuff pressure (volume) is increased just enough to keep the radius and thus the *transmural pressure* p_T (Footnote 141) constant. The resulting cuff pressure waveform compares favorably with the *blood pressure waveform*, because at constant p_T the cuff pressure (outside of the blood vessel) follows the intraarterial

---

a Canadian physician, went on and showed that the auscultation can be substituted by palpation just beyond the cuff.

[208]It should be noted that Scipione Riva-Rocci (1863–1937), an Italian physician, significantly improved the mercury *sphygmomanometer* (Fig. 1.13) in terms of its easy and general use. He used a simple rubber tube as an inflatable cuff on the upper arm to constrict the brachial artery, a bulb to inflate the cuff, a glass manometer filled with mercury to measure the cuff pressure, and manual palpation of the radial pulse. The obliteration of the palpated pulse corresponded to mercury pressure equal to p_S.

[209]In practice, the *pressure under the cuff is not homogenous*, with lower values at the edges of the cuff (Neuman 2011). Thus, while the pressure under the centre of the cuff is above p_S, the pressure near the edges could be even lower than p_S. In the latter case, the blood pressure pulse would open the artery under the edge region, increase the local *limb volume*, decrease the *cuff volume* to a small amount, and thus slightly increase the *cuff pressure*. In other words, cardiac pulsations in the cuff pressure begin at cuff pressures *higher than* p_S. In analogy, the pulsations in the cuff pressure do not disappear when the cuff pressure is *lower than* p_D because the limb volume changes by a small amount over the cardiac cycle. The volume and pressure of the cuff must follow these changes.

pressure up to a constant factor. In addition, a pneumatic feedback system is established for cuff pressure (volume) control so that p_T approaches zero and *pulsatile changes of the vessel radius* become a maximum. Actually the deflection of the vessel radius is maximal when p_T approaches zero since the arterial compliance (section "Pulse Propagation") becomes maximal with $p_T = 0$ (Asada et al. 2003); the *artery* is said to be *unloaded*. In consequence, the cuff pressure pulsations roughly equal those of the intraarterial pressure (at $p_T = 0$), as, for instance, recorded invasively in the *radial artery*. Advantageously, no previous calibration is necessary within the patient to attain *absolute blood pressure* values.

- *Tonometric method*, successor of the ancient sphygmograph (Fig. 1.14), applies a rounded probe over a superficial (radial or carotid) artery which can be pressed against the bone (or another backside support of the artery) allowing the artery to be flattened in a reproducible way. The flattening eliminates the tangential forces in the arterial wall and the probe is exposed only to the *pressure within the artery* (normal to the flattened arterial wall). Keeping the artery flattened, the force applied by the probe is opposite and equal to the pulsatile force that the blood pressure exerts on the flattened arterial wall. The probe is connected to a *pressure sensor* whose output reflects arterial *blood pressure waveform*. Usually an initial calibration is necessary within the patient to cancel out changes in arterial mechano-elastic function (e.g., its stiffness) among patients in order to attain absolute blood pressure values.

Figure 3.11 *demonstrates* the working principle of the *auscultatory method*. The barogram in Fig. 3.11a, a mechanic biosignal, corresponds to the *cuff pressure* which is depicted in the state of its release from 140 to 60 mm Hg. The synchronous recording of the phonogram in Fig. 3.11b, an acoustic biosignal, yields *Korotkoff sounds* from the brachial artery. For comparison, the *radial waveform of p* oscillating with f_C is depicted in Fig. 3.11a, recorded as the barocardiogram, a mechanic biosignal, by the use of the *volume clamp method*. As expected from the basic principle of the auscultatory method, the Korotkoff sounds start to arise when the cuff pressure is as low as p_S shortly before 115 s, as can be nicely verified by the waveform of p. Later, at the time instant of about 154 s, the Korotkoff sounds disappear, when the cuff pressure crosses p_D.

The *zoomed* version of Fig. 3.11 is given in Fig. 3.12, illustrating the time relations between the above biosignals in more detail. It proves that the Korotkoff sounds arise during the time interval when the peak of p exceeds the cuff pressure and a bolus of blood passes down the arm. Actually, the *radial p* seems to be *delayed* with respect to the sounds, recorded at the brachial site. Thus, the time delay of the recorded radial p with respect to the brachial p, which in fact determines the timing of the sounds, should be considered. In analogy to considerations in section "Blood Pressure and Flow" with respect to Fig. 2.49, the time delay or the estimated propagation time is about 70 ms. Advancing of the waveform of p in Fig. 3.12a by about 70 ms would lead to an evident overlap of the systolic peak and the peak of the Korotkoff sounds.

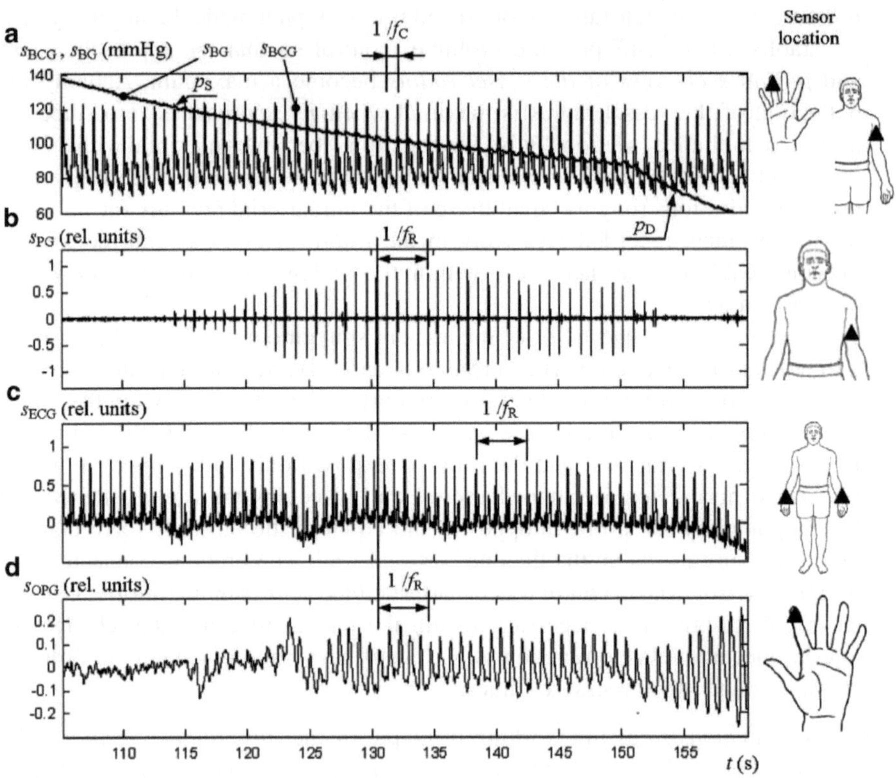

Fig. 3.11 The auscultatory method in comparison with the volume clamp method for blood pressure monitoring (compare with the zoomed version in Fig. 3.12 from 115 to 117 s). (**a**) Mechanic biosignals barocardiogram s_{BCG} (from a finger on the right hand) and barogram s_{BG} (from the upper arm on the left), (**b**) acoustic biosignal phonogram s_{PG} (from the upper arm on the left), (**c**) electric biosignal electrocardiogram s_{ECG} (lead I Einthoven), and (**d**) optic biosignal optoplethysmogram s_{OPG} (from a finger on the left hand)

Figures 3.11 and 3.12 also show the electrocardiogram, the *electric biosignal*, and the optoplethysmogram, an *optic biosignal* (Sect. 6), to illustrate their *synchronous behavior* during the *release of the brachial blood flow* to the *left arm*. As long as the blood flow is completely ceased by the inflated cuff, there are no pulsations in the optic biosignal due to the absence of pulsatile blood absorption in the finger on the *left hand*, see Fig. 3.11d. The pulsations are formed during the time when the cuff pressure comes within the systolic part of the p waveform (Fig. 2.48b), as can be observed in Figs. 3.11d and 3.12d; in the diastolic part the deflection amplitude of the pulsations remains nearly constant.

To be more precise, in the diastolic part of the p waveform, the *deflection amplitude* of the pulsatile blood absorption is *modulated by breathing* with the respiratory rate f_R, as clearly seen in Fig. 3.11d and will be discussed in Sect. 3.2. The respiratory modulation is also obvious in the radial p waveform (Fig. 3.11a), the

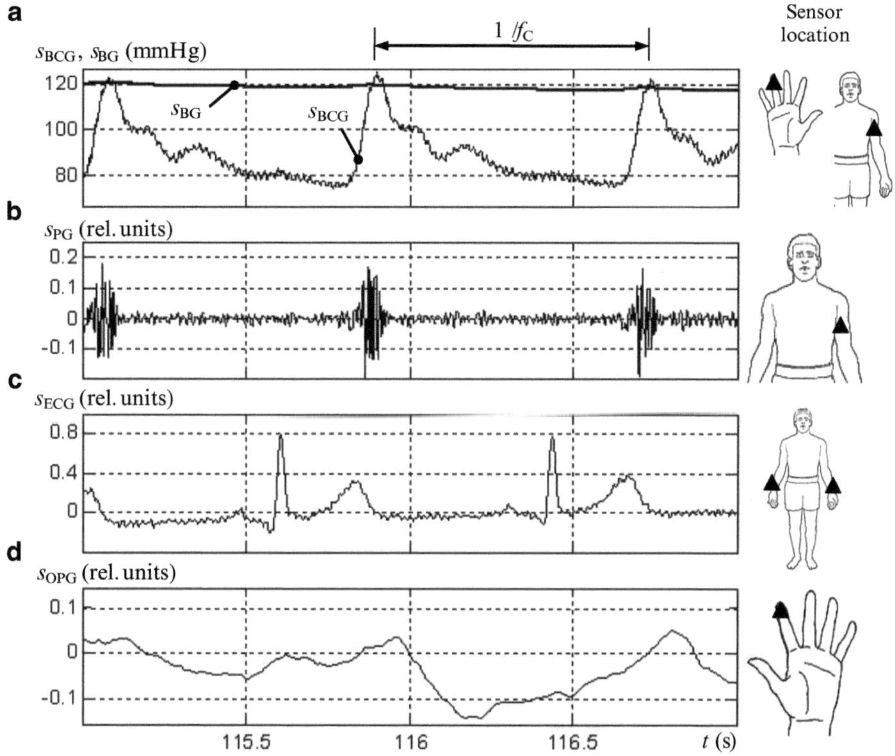

Fig. 3.12 The auscultatory method in comparison with the volume clamp method for the blood pressure monitoring considering an extract from Fig. 3.11. (**a**) Mechanic biosignals barocardiogram s_{BCG} (from a finger on the right hand) and barogram s_{BG} (from the upper arm on the left), (**b**) acoustic biosignal phonogram s_{PG} (from the upper arm on the left), (**c**) electric biosignal electrocardiogram s_{ECG} (lead I Einthoven), and (**d**) optic biosignal optoplethysmogram s_{OPG} (from a finger on the left hand)

amplitude of the Korotkoff sounds (Fig. 3.11b), and the amplitude of the R peaks of the electrocardiogram (Fig. 3.11c).

The principle of the *oscillometric method* is depicted in Fig. 3.13, demonstrating an excerpt from Fig. 3.11a, d. That is, the contour of the *cuff pressure* yields *pulsatile oscillations* as long as the cuff pressure is between p_S and p_D (Fig. 3.13a), with the oscillation strength serving as a basis for the estimation of p_S and p_D. The synchronous pulses in the optic biosignal from the finger exemplify the cardiac origin of the cuff pressure oscillations.

Lastly, Fig. 3.14 exemplifies the *tonometric method*. The barocardiogram, a *mechanic biosignal*, yields the waveform of p from the radial artery. The typical waveform features can be observed, as discussed in section "Pulse Waveforms of Pressure and Flow." There is a noticeable *influence of breathing* on the *secondary peaks* after the systolic peak. During *inspiration*, the secondary peak is more

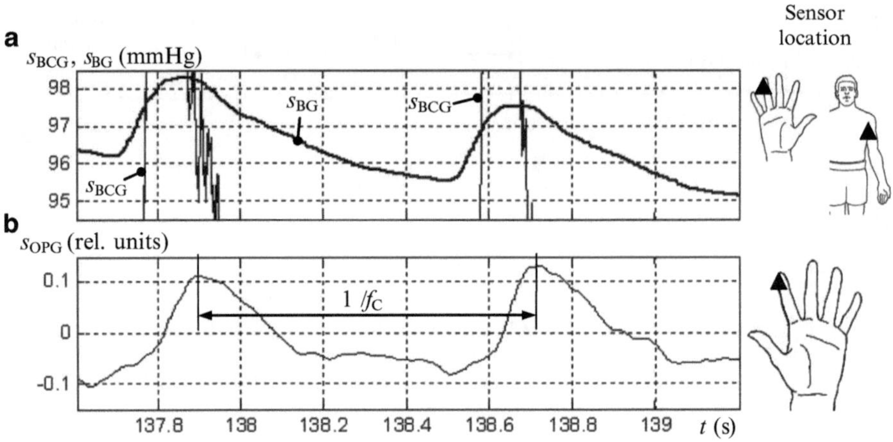

Fig. 3.13 Visualization of oscillometric method for the blood pressure monitoring. The zoomed versions from Fig. 3.11a, d demonstrate pulsatile oscillations of (**a**) mechanic biosignals barogram s_{BG} (from the upper arm on the left) and barocardiogram s_{BCG} (from a finger on the right hand), and (**b**) optic biosignal optoplethysmogram s_{OPG} (from a finger on the left hand)

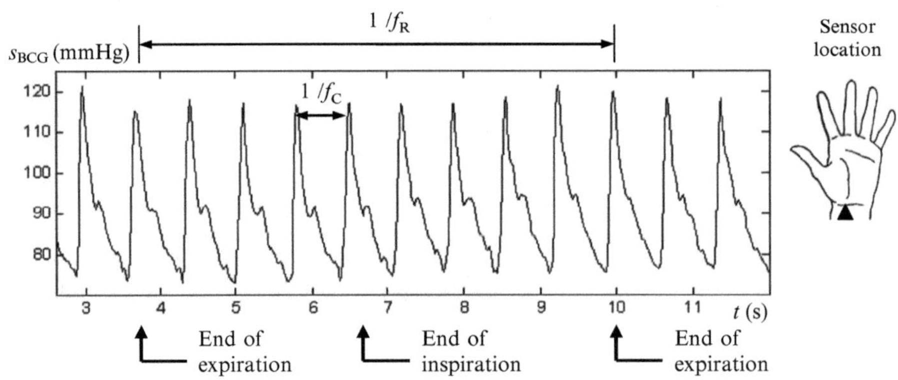

Fig. 3.14 The tonometric method yielding mechanic biosignal barocardiogram s_{BCG} from the left wrist

prominent, for p_S and thus the arterial stiffness tends to decrease. In consequence, the pulse wave velocity v of the secondary (reflected) wave decreases leading to a later (delayed) interference of the incident and reflected waves and thus to a more prominent wave after the systolic peak, as described in Sect. 3.2. At the end of *expiration*, the (early) reflected wave merges with the incident wave so that the secondary peak becomes less prominent.

Estimation from Arterial Radius

The numerous invasive and noninvasive methods discussed above are devised specifically for the estimation of the p waveform and its characteristics. It is interesting to pursue the question of whether the characteristics of *p can be estimated* from a *single parameter* such as the arterial radius r or, alternatively, from *r as the main parameter* and v as an auxiliary parameter. For this aim, *numerous models* have been introduced, an *excerpt* of which will be given below.

The depicted *nonlinear behavior* of the stiffness in Fig. 2.42 already indicates the difficulties in establishing models for the estimation of the characteristics of p, if either of the parameters r and v or even both of them are experimentally available. A major requirement for such *models* is that they should account for *increasing stiffness* of the arterial wall *as r increases*, as described in Sect. 2.5.2.1.

An easy relationship is suggested by Hardy and Collins (1982), which fits experimental data well for several types of circulatory elements including arterial data:

$$\frac{1}{\kappa} \cdot (A_{\mathrm{M}} - A) = \frac{\mathrm{d}A}{\mathrm{d}p}, \tag{3.1}$$

where A_{M} is the limiting maximum value of A, i.e., $A_{\mathrm{M}} = \pi r_{\mathrm{M}}{}^2$ with r_{M} as the maximum value of r. According to the above equation, $\mathrm{d}A/\mathrm{d}p$ tends to zero as p increases or (and) the difference $A_{\mathrm{M}} - A$ decreases, i.e., the arterial stiffness increases with increasing p, compare (2.23). If the value of κ is assumed to be constant since the varying stiffness of the artery is already considered through introduction of A_{M}, then the *pressure–radius* relationship can be simply obtained by integration of (3.1):

$$p = p_0 - \kappa \cdot \ln \frac{A_{\mathrm{M}} - A}{A_{\mathrm{M}} - A_0} = p_0 - \rho \cdot v^2 \cdot \ln \frac{r_{\mathrm{M}}^2 - r^2}{r_{\mathrm{M}}^2 - r_0^2}, \tag{3.2}$$

with p_0, A_0, and r_0 being the values of p, A, and r, respectively, at an arbitrary point on the *pressure–cross section* (or pressure–radius) curve. For instance, Fig. 2.42 compares experimental data and the correspondingly modeled behavior with $2 \cdot r_0 = 5.4$ cm, $2 \cdot r_{\mathrm{M}} = 5.8$ cm, $p_0 = 78$ mm Hg, and $\kappa = 8.1$ mm Hg. An obvious *fit* between the *experiment* and the *model* from (3.1) can be observed.

Another simple function according to Pedley (1980), in which the artery becomes less distensible as r increases, is

$$p = \frac{1}{2} \cdot \rho \cdot v_0^2 \cdot \frac{r^2}{r_0^2} + c, \tag{3.3}$$

with c being a constant and v_0, r_0 the values of v, r, respectively, at an arbitrary point on the *velocity–radius* curve.

The applied models in Wibmer (2004) assume axisymmetric *radial loading* and deformation of the artery wall, as shown in Fig. 2.44b. Furthermore, the relation

$h \ll r$ is assumed [compare assumptions for (2.24)], reducing the external forces to stresses acting only in the radial direction of the artery wall. Applying *Laplace law*[210] (Pedley 1980; Milnor 1989), which states that the circumferentially directed force per unit vessel length is related to the transmural pressure p_T (Footnote 141) and r (Fig. 2.44b), the circumferential tensile stress σ results to

$$\sigma = \frac{r \cdot p_\mathrm{T}}{h} = \frac{r \cdot (p - p_\mathrm{E})}{h} = \frac{E}{1 - \psi_x \cdot \psi_\theta} \cdot \frac{r - r_\mathrm{T}}{r_\mathrm{T}}. \tag{3.4}$$

Here, p_T is given by $p - p_\mathrm{E}$, i.e., the excess pressure or the difference between intraarterial blood pressure p and external pressure p_E (Fig. 2.44b). The radius r_T is the arterial radius at $p_\mathrm{T} = 0$. Furthermore, the symbols ψ_x, ψ_θ are the *Poisson ratios*[211] in the longitudinal and circumferential direction, respectively; for the incompressible artery wall $\psi_x = \psi_\theta = 0.5$ holds. Equations (2.24) and (3.4) yield an estimation for p, to give

$$p = \frac{8 \cdot \rho \cdot v^2}{3 \cdot r_\mathrm{T}} \cdot (r - r_\mathrm{T}) + p_\mathrm{E}. \tag{3.5}$$

It is important to note that the derivation of (3.4) and (3.5) is based on purely elastic theory; nevertheless, the velocity v in (3.5) accounts for the pressure dependence of the stiffness (Fig. 2.42). As a restriction, (3.4) assumes a constant σ throughout the vessels wall while σ actually decreases through the wall thickness to the outer surface, as illustrated in Fig. 2.44b.

Estimation from Pulse Running Time

Besides model based estimation of the waveform of p with r as the main parameter (section "Estimation from Arterial Radius"), (2.22) and (2.23) suggest the possibility of *estimating p* out of v as the main parameter, as utilized by many authors (Franchi et al. 1996; Gribbin et al. 1976). The main *rationale* behind this approach is based on the phenomenon that the level of v tends to increase with increasing p because of increasing arterial stiffness, see Fig. 2.42.

From an experimental point of view, the estimation of v can be based on the determination of the *running time* τ of the pulse wave caused by every heartbeat (section "Pulse Propagation"), as demonstrated in Fig. 3.15. Generally, the pulses could be of *different origins*, including *pressure pulses*, *flow pulses*, or even *arterial*

[210] Pierre-Simon Laplace (1749–1827) was a French mathematician, physicist, and astronomer who contributed significantly to mechanics, statistics, mathematics, and astronomy.

[211] Simeon Denis Poisson (1781–1840) was a French mathematician and physicist after whom the ratio was named that describes the quotient of the transverse strain and the axial strain of material under axial stress. Incompressible materials yield a *Poisson's ratio* of 0.5.

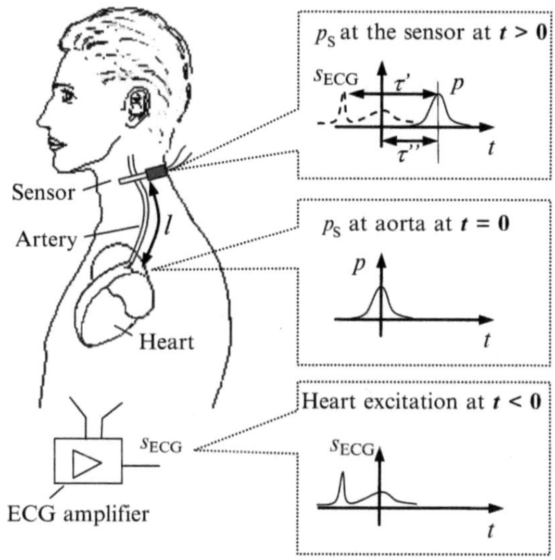

Fig. 3.15 Time relationships between the *electrical* heart excitation as the pulse triggering event, monitored by the electrocardiogram s_{ECG}, the following *mechanical* blood pressure p wave at the aorta as the pulse origin, and the delayed mechanical propagating p wave on the neck over the carotid artery at the distance l from the heart. The pulse *arrival time* τ' from the aorta to the neck is approximately given by the elapsed time between the R peak in s_{ECG} and the systolic blood pressure peak p_S at the neck. By contrast, the pulse *transit time* $\tau''(<\tau')$ from the aorta to the neck is approximated by the time delay between p_S peak at the aorta and p_S peak at the neck

diameter pulses, see Sect. 2.5.2.3. Thus, the estimation of v is simply based on

$$v = \frac{l}{\tau}. \tag{3.6}$$

Here, l is the propagating distance of the pulse wave; in Fig. 3.15 it is given as the distance between the heart and neck. The time τ could be basically estimated in two ways, as

- Pulse *arrival time* or
- Pulse *transit time*

In the case of *pulse arrival time*, the time τ' in Fig. 3.15 could be the *time difference* measured between the R peak of the electrocardiogram, i.e., the *pulse triggering* event released by *electrical* heart excitation, and the pulse wave maximum at a peripheral artery such as the carotid artery on the neck, i.e., the pulse arrival time conveyed by *mechanical* pulse wave. The quantity l is the distance between the heart and the peripheral artery.

Alternatively, the *pulse transit time*, the time τ'' in Fig. 3.15, could be the *time difference* detected between two *mechanical pulse waves* from two different locations along the arterial tree with *different distances to the heart.* In the case of Fig. 3.15, one location is the aorta while the other is the carotid artery on the neck, with the obvious relationship $\tau'' < \tau'$. In a general case, the quantity l is the distance between the two arterial sites.

In most situations, peripheral arteries are considered as candidates for the estimation of v because they are accessible in an *unobtrusive way*, e.g., the carotid artery and radial artery. The sensors used usually apply plethysmographic noninvasive principles (Sect. 4, 6, 7). Furthermore, it is important to note that the *accuracy*[212] of the estimation of v is higher if the pulse transit time is used as τ in (3.6).

Numerous *models* have been suggested for the *estimation* of p_S and p_D by the use of the single parameter τ from (3.6) (Pedley 1980; Franchi et al. 1996; Gribbin et al. 1976; Kaniusas et al. 2006a). Here, it should be stressed that physiologically the *level of* τ as well as that of v primarily depends on $\langle p \rangle$ and p_D, compare section "Pulse Propagation"; for instance, a nearly linear relationship between v and $\langle p \rangle$ was observed with a correlation coefficient larger than 0.9.

Franchi et al. (1996) and Barschdorff and Erig (1998) suggest the following model:

$$p_S = c_{S,1} + c_{S,2} \cdot \frac{1}{\tau} \text{ and } p_D = c_{D,1} + c_{D,2} \cdot \frac{1}{\tau}, \tag{3.7}$$

where c_S and c_D are constants which can be calculated using calibration. However, the authors conclude that a better estimation is given for p_D which supports the physiological background.

A nonlinear relationship between p_S, p_D, and τ is proposed by Kolluri et al. (2003) with two different τ related to different time instants within the period of the pulse wave. That is, the time τ_S is the pulse *transit time during systole* and τ_D the pulse *transit time during diastole*; to give

$$p_S = c_{S,1} + c_{S,2} \cdot \left(\frac{1}{\tau_S}\right)^{\alpha} \text{ and } p_D = c_{D,1} + c_{D,2} \cdot \left(\frac{1}{\tau_D}\right)^{\alpha}. \tag{3.8}$$

[212] In the case of the *pulse arrival time* as τ, the use of the electrocardiogram wave as a *timing signal* provides data with much more scatter than are obtained when using an arterial pulse for timing (Geddes et al. 1981a, b). The reason for this is the *large variability* of the pulse arrival time between the R peak and the pulse wave arrival. This *pulse arrival time* consists of a portion of the time required for

- *Propagation of excitation* over the ventricles
- Period of *isovolumetric contraction*
- *Pulse transit time* from the aortic valve to the artery at the measuring site, e.g., at the neck in Fig. 3.15

Of these times, the *isovolumetric period* is the most variable, showing a significant variability during changes of the blood pressure and the same order of magnitude if compared, for instance, to the pulse transit time between the aorta and the ear lobe (Franchi et al. 1996). Using the arterial pulse as a timing reference, in the case of the *pulse transit time* as τ, eliminates this variable (Geddes et al. 1981a); it implies also $\tau'' < \tau'$ in Fig. 3.15. In addition, if more *distal arteries*, e.g., femoral instead of carotid, are used in connection with the electrocardiogram as a timing reference, then the isovolumetric period builds a smaller fraction of the arrival time and the present variations in the isovolumetric period are less significant (Geddes et al. 1981b; Sugo et al. 1999).

Here, α is a positive number. The values of τ_S and τ_D can be experimentally assessed as the time difference between the systolic peaks of the pulse waves (compare τ'' in Fig. 3.15) and the corresponding diastolic troughs, respectively.

An additional *consideration of* f_C is proposed by Golub (1998):

$$p_S = c_{S,1} + c_{S,2} \cdot \left(\frac{1}{\tau}\right)^2 \text{ and } p_D = c_{D,1} + c_{D,2} \cdot \left(\frac{1}{\tau_D}\right)^2 + c_{D,3} \cdot f_C. \qquad (3.9)$$

The estimation of p_D includes a term proportional to f_C, for p_D is proportional to f_C in physiological terms, see section "Pulse Waveforms of Pressure and Flow." In addition to f_C, Lin et al. (2004) suggest considering the *width* τ_{PW} *of the pulse wave* for the estimation of p_S, to give

$$p_S = c_{S,1} + c_{S,2} \cdot \tau + c_{S,3} \cdot \tau_{PW} + c_{S,4} \cdot \tau_{PW} \cdot f_C. \qquad (3.10)$$

Here, τ_{PW} is considered to reflect V_S which is proportional to p_S (section "Pulse Waveforms of Pressure and Flow"). In addition, the product $\tau_{PW} \cdot f_C$ or the ratio $\tau_{PW}/(1/f_C)$ accounts for the fractional change of V_S, being roughly proportional to $\mathrm{d}A/A$ from (2.23).

Some authors suggest *lookup tables* or *calibration curves* for p_D (Chen et al. 2003) and a subsequent derivation of p_S using the pressure–volume relationship (2.23) of the artery at the pulse wave detection site (Hardy and Collins 1982; Greubel et al. 1993). For instance, Greubel et al. (1993) suggest

$$p_S = p_{S,D} + p_D \text{ with } p_D = f(\tau) \text{ and } p_{S,D} \propto s_{S,D}, \qquad (3.11)$$

where $p_{S,D} (= p_S - p_D)$ is the systolic–diastolic deflection of p during the cardiac cycle, $f(\tau)$ the function of the lookup table for the estimation of p_D out of τ, and $s_{S,D}$ the systolic–diastolic deflection of the applied biosignal s for the pulse detection. Figure 3.2b illustrates $s_{S,D}$ based on the optoplethysmogram s_{OPG}.

3.1.3.2 Blood Flow

Analogous to the methods pertaining to monitoring of the blood pressure (Sect. 3.1.3.1), the blood flow can be recorded in an *invasive* and *noninvasive* way. The *parameters of interest* are the stroke volume V_S, blood flow velocity u, and obviously the pulsatile flow *waveform* (Sect. 2.5.2.2). Generally, the applied methods are more sophisticated than in the case of the blood pressure.

Invasive approaches for blood flow monitoring enjoy limited acceptance because of their invasiveness and related complications. The following methods are in use:

- *Indicator method*, in which an indicator, e.g., *oxygen*, is introduced into the stream of blood flow and the resulting arterial and venous concentrations of the indicator are measured, according to the so-called Fick principle (Footnote 23).

Alternatively, a bolus of *ice-cold saline* is introduced into the right atrium by a catheter and the resulting drop in *temperature* of the blood in the pulmonary artery is detected by a thermistor catheter. The cardiac output is related to the amount of indicator injected divided by the area underneath the blood temperature dilution curve.

- *Electromagnetic method* is focused at an exposed blood vessel with flowing blood (as moving conductor), which is placed into a transverse magnetic field. A potential difference is induced in the conductor, measured perpendicular to both flow direction and magnetic field direction. The potential difference is directly related to the internal diameter $2r$ of the vessel and the mean blood flow velocity $\langle u \rangle$; thus (2.17) can be used here to estimate the blood flow q.
- *Transit-time ultrasonic method* that uses an ultrasound[213] beam (waves) passing through the blood vessel. Two ultrasound transceivers are *diagonally* placed on either side of the vessel. That is, the waveform of the flow velocity is obtained from the difference in time taken for the ultrasound to pass in one direction as opposed to the other.

The most popular *noninvasive methods* for the determination of the aforementioned *blood flow* parameters comprise the following three approaches:

- *Echocardiographic method*, based on ultrasonic *Doppler*,[214] uses an ultrasound beam in the frequency range of a few MHz, which is backscattered from the moving blood cells. The *frequency shift* of the backscattered sound is proportional to the velocity of the blood; basically, the frequency increases if the blood moves toward the ultrasound probe. The volumetric blood flow can be estimated from the velocity profile (or its average) over the cross-sectional area of the vessel combined (multiplied) with the cross-sectional dimensions, compare (2.17). The cross-sectional dimensions are usually measured by ultrasound as well (Sect. 3.1.3.3).
- *Impedance cardiography method*, also known as electric field plethysmography (Sect. 4), introduces an electric current and measures the resulting voltage in the axial direction of the thorax. While the current seeks the path of least resistance in the thorax, i.e., seeks the blood filled aorta, the measured voltage represents

[213]*Ultrasound* is given by a mechanical vibration of particles of elastic media with a frequency above the range of human hearing (>20 kHz). When the particles are displaced from their equilibrium positions, internal restoration (electrostatic) forces arise between the particles, which lead to the oscillatory local motions of the particles. For instance, local oscillations in the density of tissue may be induced in the direction of ultrasound propagation, yielding the so-called compressional or *longitudinal waves*. It is important to note that the ultrasound waves are *reflected* and scattered on anatomical inhomogenities or whenever the waves encounter different acoustic impedance; specifically, anywhere there are density changes in the body.

[214]The *Doppler effect* was named after Christian Andreas Doppler (1803–1853), an Austrian mathematician and physicist, who first described the change in frequency [and wavelength, compare (2.21)] of a propagating wave as perceived by an observer moving relative to the wave's source.

the thoracic impedance. In consequence, the volumetric changes of the aorta during the cardiac cycle induce thoracic impedance changes allowing for the determination of the cardiac *stroke volume* (even) in absolute units.

- *Pressure pulse contour method* derives the aortic flow waveform from the arterial pressure waveform, e.g., from the radial pressure waveform assessed in a noninvasive way (Sect. 3.1.3.1). In a first step, a generalized transfer function (Footnote 165) is used to derive the aortic pressure from the radial pressure and then an aortic impedance model is applied to derive the aortic flow; the latter model describes the ratio between the aortic pressure and aortic flow (section "Blood Pressure and Flow"). The stroke volume is the integral of the flow waveform over a single cardiac cycle. However, this method requires a previous calibration to attain absolute blood flow values.

Figure 3.16 illustrates *experimental data* showing *u waveforms* in different regions (sample volumes) of the carotid artery (Fig. 2.40a). The radial distance of the *sample volume* from the arterial axis was varied. In the demonstrated case, the echocardiographic method was applied for both imaging[215] of the anatomical structure of the artery (Fig. 3.16a) and measuring u. The miniature sample volumes within the artery, as shown in Fig. 3.16a by overlaid artificial structures, were made possible by the *pulsed Doppler*[216] approach. Maximal values of u appear along the axis of the artery, with a tendency to decrease toward the wall (Fig. 3.16b), as expected from Fig. 2.44b and section "Blood Pressure and Flow." Multiple *secondary peaks* can be observed in u due to reflections, compare Fig. 2.49a and section "Pulse Waveforms of Pressure and Flow." Actually, the smeared curves in Fig. 3.16b result from the limitations of Doppler technology and local turbulences in the blood flow.

As already mentioned, a diagnostic assessment of instantaneous arterial q is more difficult than that of p, whereas both parameters are clinically significant. Thus, a reasonable *experimental derivation of q out of p* is of high practical importance and should be quickly addressed.

The most obvious *estimation of q* relies on a *model* for the input impedance $\underline{Z}_I$ according to (2.29), with reflections being considered (section "Reflected Pulse Propagation"). Then the waveform of q can be derived from the measured p

---

[215]*Ultrasound imaging* uses echoes from inhomogenities in the tissue, since part of the sound wave is reflected back to the ultrasound probe and detected as an echo, compare Footnote 213. The amplitude of the echo is related to the *dominance* of the corresponding inhomogeneity and is used to modulate the *pixel* brightness in the image, compare Fig. 3.16a. The time delay of the echo reflects the *depth* (or distance) of the inhomogeneity from the ultrasound probe and is given on the *y-axis*. Lastly, the *location* of the sound beam determines the *x axis* and the beam is swept over an area of interest to vary x. This type of display is often referred to as a "*B-mode*" display.

[216]The *pulsed Doppler* approach yields Doppler information (Footnote 214) from only a small blood sample volume, with a longitudinal size usually < 1 mm, compare Fig. 3.16a. The depth (or distance) of the sample volume from the ultrasound probe is determined by the propagation time of a short ultrasound burst (pulse), i.e., by the time from pulse emission until its reception after reflection.

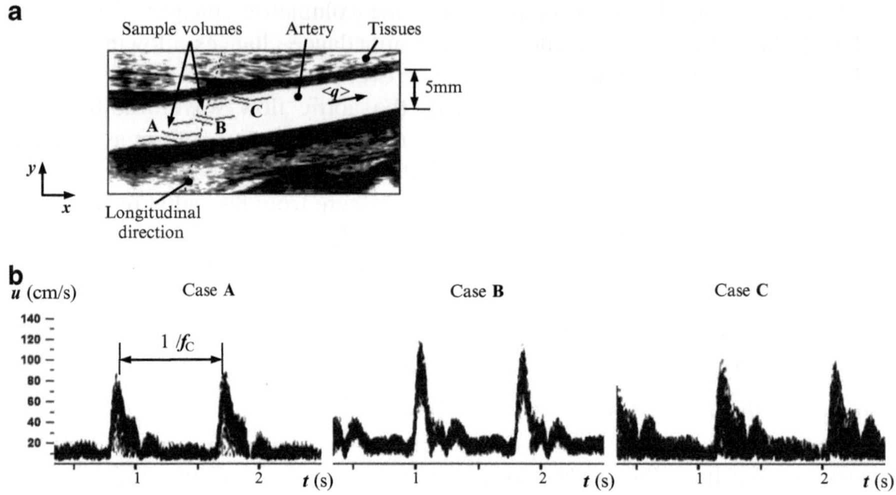

Fig. 3.16 Blood flow q velocity u in different regions of the carotid artery. (**a**) Ultrasonic image (Footnote 215) of the carotid artery with indicated sample volumes of the longitudinal size of 0.7 mm. (**b**) The waveforms of u for the sample volumes A, B, and C from (**a**), as derived by the pulsed Doppler (Footnote 216)

waveform which inevitably includes reflections. The models are usually *nonlinear* and *time varying*, including parameters such as the total peripheral resistance R_T [from (2.20)], arterial compliance V/κ (2.23), and characteristic impedance $\underline{Z}_0$ (2.28). In general, such models show a good agreement with measured q waveforms (Wesseling et al. 1993).

Another simple empirical approach to *derive* q is based on the *time derivative* $\mathrm{d}p/\mathrm{d}t$ rather than on the p waveform itself (Nichols and O'Rourke 2005). In a first approximation, it is based on the decrease in $|\underline{Z}_I|$ over f (Footnote 153), implying $\underline{Q} \propto j\omega \cdot \underline{P}$ in the *frequency domain* [compare (2.28)] or $q \propto \mathrm{d}p/\mathrm{d}t$ in the *time domain*. An experimental demonstration of the possibility of using $\mathrm{d}p/\mathrm{d}t$ to estimate q is given in Fig. 2.50b, c. It can be observed that the spatial derivative $\mathrm{d}p/\mathrm{d}x$ determining q, as discussed in section "Blood Pressure and Flow," and the time derivative $\mathrm{d}p/\mathrm{d}t$ exhibit very similar waveforms. In addition, the primary peak of $\mathrm{d}p/\mathrm{d}t$, i.e., the primary peak of the estimated waveform of q, precedes that of p, as expected from section "Blood Pressure and Flow."

3.1.3.3 Arterial Radius

Common methods to monitor blood pressure within the arteries were discussed in Sect. 3.1.3.1, while those to monitor blood flow were discussed in Sect. 3.1.3.2. Here, *invasive* and *noninvasive* methods for *arterial radius* monitoring will be briefly reviewed. Naturally the mean value of the radius and its *pulsatile waveform* are of physiological interest.

The waveform of the *arterial radius* shows strong *similarities* to that of the *blood pressure*, as could be extracted from Fig. 2.52. That is, both waveforms exhibit a steep systolic rise, a secondary peak during systole, a dicrotic notch, and a slow diastolic decrease. However, the differences between the waveforms are determined by the *nonlinear relationship* between the *radius and blood pressure* (Fig. 2.42), and, on the other hand, by the *viscoelastic* behavior of vessels, see Sect. 2.5.1.

The methodology used to measure the radius is somewhat *similar* to that used to measure the blood pressure. However, *measurements* of the arterial *radius* should generally be *more sensitive* than those of the blood pressure. This is because the radius changes by about 10% while the pressure may change by as much as 50% within a cardiac cycle (Nichols and O'Rourke 2005; Shau et al. 1999), compare the relative deflection amplitude in Fig. 2.52a with that in Fig. 2.52b.

Invasive methods for radius monitoring comprise mainly the following approaches:

- *Resistance/inductive strain gauges* fixed directly to the outer wall of an (exposed) artery or even inserted by a catheter to measure internal radius changes.
- *Photoelectric devices* are applied, with the pulsating artery casting a shadow on a photocell.
- *Transit-time ultrasonic approach* is in use, as was mentioned in Sect. 3.1.3.2, that comprises two ultrasound transceivers positioned *opposite* to each other on the outer sides of the arterial wall. The time that elapses between the emission of an impulse (or burst) and its reception on the opposite site is proportional to the arterial radius.

The most popular *noninvasive methods* to asses the *arterial radius* are based on

- *Ultrasonic beams*, in which reflections of the ultrasound waves are utilized (Footnote 213). The time that elapses between emission of an impulse (or burst) from the ultrasound probe on the skin and reception of the reflected impulse from both arterial walls yields the arterial radius, compare Footnote 215.
- *Optical plethysmography*, aiming to assess *indirectly* the local pulsatile volume of the transilluminated artery (Sect. 6). With each blood pulse or surge, the arterial radius increases and the transilluminated region now encloses an increased ratio of blood, which strongly absorbs the light, compared to surrounding tissue which weakly absorbs the light. As a consequence of the local absorption changes, the intensity of the transmitted light decreases for increased arterial radius or during systole. To be more precise, not only the pulsatile changes of the local blood volume are assessed, which are actually related to (pulsatile) blood pressure (Fig. 2.42), but also the basic level of blood absorption related to (inert) blood oxygenation (Sect. 3.1.4). The recorded traces bear a strong *similarity* with *blood pressure from the carotid* artery and some similarity with the pressure from the ascending aorta, but not from the radial artery (Nichols and O'Rourke 2005).
- *Mechanical plethysmography*, targeting local skin curvature assessment over a superficial artery, e.g., over the carotid artery on the neck. During cardiac

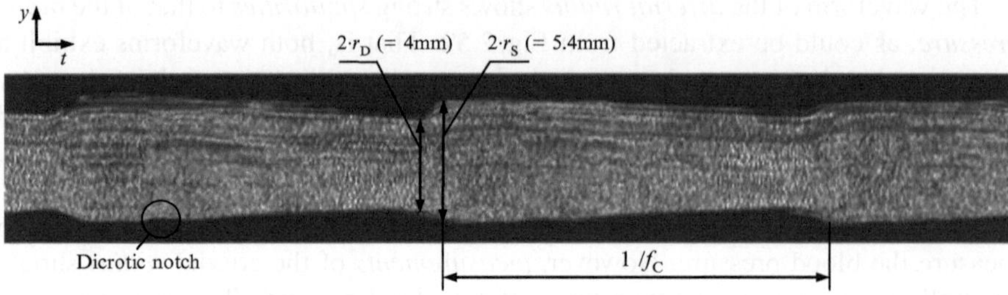

Fig. 3.17 Ultrasonic image (Footnote 215) of the carotid artery in motion mode (Footnote 217). The periodic deflection of the arterial diameter can be observed from its (minimum) diastolic value $2 \cdot r_D$ to its (maximum) systolic value $2 \cdot r_S$, compare Fig. 2.52b

deflections of the artery, the local skin curvature changes can be assessed by a skin curvature sensor (Kaniusas et al. 2006a, b), see Sect. 7.

Figure 3.17 demonstrates the *ultrasonic approach* applied to the noninvasive measurement of the *arterial radius*. Here the course of the ultrasound reflections is monitored over time in *motion mode,*[217] while the direction of the sound beam was kept constant, i.e., the coordinate x from Fig. 3.16a and Footnote 215 is kept constant. It can be observed that the arterial radius changes periodically from its lowest diastolic values to its highest systolic values with a period duration of $1/f_C$. As expected, the course of the arterial width (or the arterial diameter) over a single cardiac cycle strongly resembles the typical pressure waveforms, as exemplified in Fig. 2.48b or 2.52; even the location of the dicrotic notch can be recognized.

Figure 2.51a demonstrates the output from the *optical plethysmography*, comparing a *young* with an *elderly* person. Once more, the obtained waveforms compare favorably with the blood pressure waveforms. In particular, the waveform from the young person is much wavier than from the elderly person and yields a pronounced dicrotic notch; the contour resembles a *Type C* wave type, according to the nomenclature in section "Pulse Waveforms of Pressure and Flow." The waveform from the elderly person mainly exhibits a single systolic peak which is wider than that of the young person, suggesting an early fusion of the incident and reflected wave (section "Reflected Pulse Propagation"); *Type D* could be the appropriate wave type in this case.

In the case of the *optoplethysmogram* from Fig. 2.51, the *second differential* of the *waveform* is often considered as being conclusive in terms of diagnosis. As depicted in Fig. 2.51c, d, the waves "*a*"–"*d*" can be distinguished in the second time

[217]In motion mode, the so-called "*M-mode*," the temporal motions of the inhomogenities within the tissue are revealed along the propagation direction of the sound beam, i.e., the motions relative to the ultrasound probe. In other words, the temporal changes of echo amplitudes are monitored along the coordinate y while x is frozen, compare Fig. 3.16a and Footnote 215. In the image of M-mode, see Fig. 3.17, the echo *amplitude* corresponds to the *pixel brightness*, with the physical *depth* of the inhomogenities displayed along the *y-axis* and the *time* along the *horizontal t-axis*.

derivatives. The positive "*a*" wave during early systole represents a rapid upstroke of the pulse, while the negative "*d*" wave during late systole represents downstroke of the systolic pulse and precedes the diastolic period. The *ratio d/a* actually decreases with aging and hypertension (Footnote 166), and it also decreases with increasing aortic augmentation index from (2.33).

Another example of the *optoplethysmogram* is depicted in Fig. 2.55 (lower figures) demonstrating its changes in response to the *Valsalva maneuver* (Footnote 173). During this maneuver, the diastolic wave becomes more prominent due to the delayed reflection, as expected from section "Reflected Pulse Propagation." The pulse amplitude of the optoplethysmogram decreases, which corresponds to a decrease in the blood pressure deflection $p_S - p_D$, compare the upper figures in Fig. 2.55. Generally, it can be observed that the Valsava maneuver impacts rather differently the course of the blood pressure and that of the optoplethysmogram because the involved biosignals are of different origins, see also Sect. 3.1.3.1.

3.1.4 Blood Oxygenation

Blood circulation implies rhythmic transport of the oxygenated blood to the cells and the deoxygenated blood back to the lung (Sect. 2.5.1). Thus, the *oxygenation level* of arterial blood is a vital *physiological parameter*, usually derived from optic biosignals (Sect. 6). The oxygenation level is normally maintained at a fairly constant value; thus, its monitoring is important in diagnosing cardiac and vascular anomalies, especially in the field of anesthesiology to prevent an inadequate oxygen supply.

In fact, *oxygen* in the arterial blood is carried in two ways,

- *Bound* to hemoglobin[218] within red blood cells
- *Dissolved* in the blood plasma[219]

However, the quantity of oxygen in the blood is mainly determined by the degree of oxygenation of hemoglobin as the *blood plasma carries a negligible amount*

---

[218] A single *red blood cell* has about $280 \cdot 10^6$ hemoglobin molecules while each molecule can bind *four molecules of oxygen*, see Fig. 3.18. Hemoglobin molecule includes four *iron atoms* in the reduced form Fe^{2+}, each of which combines reversibly with one oxygen molecule, forming the so-called *oxyhemoglobin*. Hemoglobin with released oxygen molecules is known as *deoxyhemoglobin*. It should also be noted that hemoglobin contributes to the transport of carbon dioxide, see Footnote 176.

[219] When *oxygen* in a gaseous state comes into contact with the blood plasma, there is a tendency for the gas to *dissolve* in it. At *equilibrium*, the resulting concentration of oxygen is equal to its *partial pressure* p_{O_2} in the gas times its *solubility* in the plasma. Likewise, the plasma is saturated with oxygen at equilibrium, whereas the amount of dissolved oxygen depends on p_{O_2} and the solubility. The level of p_{O_2} corresponds to the pressure that oxygen of a gaseous mixture would have if it alone occupied the same volume. A low amount of oxygen dissolved in the plasma is already indicated by a *low solubility coefficient* for *oxygen* in the plasma, e.g., $10\,kPa^{-1} \cdot \mu Mol/l$ for oxygen vs. $225\,kPa^{-1} \cdot \mu Mol/l$ for carbon dioxide (Silbernagl and Despopoulos 2007).

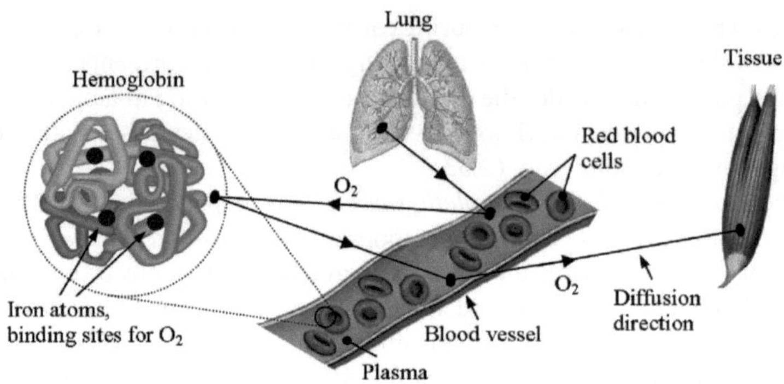

Fig. 3.18 The way of oxygen O_2 transportation from pulmonary capillaries in the lungs, over blood plasma, to hemoglobin molecules for O_2 storing. In analogy, the delivery of the buffered O_2 to the tissues goes over blood plasma as an intermediate medium

of oxygen, e.g., 200 ml of oxygen is carried by hemoglobin while only 3 ml is carried by plasma per 1,000 ml blood and at $p_{O_2} = 100$ mm Hg (Silbernagl and Despopoulos 2007).

Thus, the *oxygenated hemoglobin* (=oxyhemoglobin) acts as a local *oxygen buffer* to maintain p_{O_2} in the plasma. In particular, oxygen is unloaded from oxyhemoglobin in the *systemic capillaries* and then extracted by the surrounding tissue cells for their *cellular respiration* (Footnote 112). On the other hand, the *reduced hemoglobin* (=deoxyhemoglobin) *stores oxygen* in the *pulmonary capillaries* by depleting p_{O_2} in the plasma and forms oxyhemoglobin.

As demonstrated in Fig. 3.18, the dissolved amount of oxygen, though negligible, plays a major role in oxygen delivery to the tissues and the storing of oxygen by hemoglobin in the pulmonary capillaries. *Oxygen delivery* is carried out by *diffusion* which is driven by the positive pressure difference between p_{O_2} in the plasma of the blood capillaries and p_{O_2} in the intracellular fluid. Unloading of oxyhemoglobin is favored by low p_{O_2} in the plasma of the systemic capillaries. On the other hand, the *storing of oxygen* through its *binding to hemoglobin* needs diffusion of oxygen molecules from alveoli into the blood plasma and a subsequent diffusion of dissolved oxygen toward hemoglobin as intermediate processes. *Loading of deoxyhemoglobin* with oxygen is favored by high p_{O_2} in the plasma of the pulmonary capillaries.

Because of the low level of dissolved oxygen, the total quantity of oxygen in the arterial blood is expressed as the *hemoglobin oxygen saturation S* only, as given by

$$S = \frac{\rho_{\mathrm{HbO}}}{\rho_{\mathrm{HbO}} + \rho_{\mathrm{Hb}}}. \tag{3.12}$$

Here, ρ_{HbO} and ρ_{Hb} are the number densities (m^{-3}) of oxyhemoglobin and deoxyhemoglobin, respectively. Thus, the level of S gives the percentage of hemoglobin binding sites already occupied by oxygen; likewise, the level of S approximates the

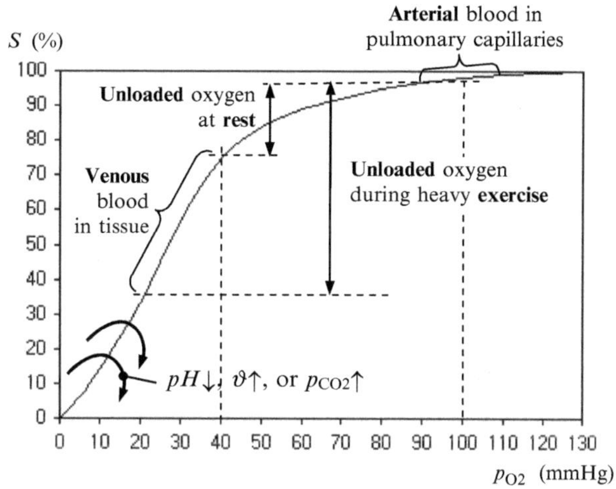

Fig. 3.19 Oxygen–hemoglobin dissociation curve showing hemoglobin oxygen saturation S over the partial pressure p_{O_2} of oxygen in blood plasma. The qualitative shift of the curve is indicated for decreasing pH value *pH*, increasing temperature ϑ, and increasing partial pressure p_{CO_2} of carbon dioxide, all three yielding a reduced affinity to oxygen binding

amount of oxygen in the blood related to the potential capacity of oxygen. Usually S is between 97 and 99% in the *arterial blood*, which indicates that almost all hemoglobin molecules are combined with oxygen, compare Fig. 3.19. In the *venous blood*, after oxygen delivery to the tissue, the level of S is usually about 75% at rest, i.e., there is still a large *reserve* of bound oxygen; however, in intensive working organs such as muscles the saturation S may go down to 35% increasing the amount of unloaded oxygen (Silverthorn 2009). It is important to note that *total hemoglobin* includes—besides oxyhemoglobin and deoxyhemoglobin—also nonfunctional hemoglobins[220] which are not capable of carrying oxygen.

In fact, the level of S increases with increasing p_{O_2} in the plasma to which the hemoglobin is exposed. Figure 3.19 shows the corresponding *oxygen–hemoglobin dissociation curve*. In the terminal part of *pulmonary* capillaries, p_{O_2} has a relatively high value of about 100 mm Hg implying that oxygen binds readily to hemoglobin (compare Footnote 180). The level of S saturates because nearly all hemoglobin molecules bind oxygen; i.e., oxygen is stored in the lungs and the *arterial blood* is released with $S \approx 100\%$.

In the terminal part of *systemic capillaries*, p_{O_2} is less and amounts to about 40 mm Hg that corresponds to $S \approx 75\%$ according to Fig. 3.19. Low p_{O_2} facilitates the depletion of bound oxygen buffer, i.e., the deoxidized *venous blood* leaves the cells with significantly lower values of S (<100%) but over a much larger range, see Fig. 3.19. The released oxygen molecules diffuse through the plasma to the cells

[220] *Nonfunctional hemoglobins* (compare Footnote 218) include, for instance, *carboxyhemoglobin* (combines with carbon monoxide instead of oxygen) and *methemoglobin* (includes iron in the Fe^{3+} state unable to bind oxygen), amounts of which increase in abnormal situations, e.g., in poisoning with carbon monoxide. It is important to note that these hemoglobins do not influence the results of spectrometry or oximetry, see Sect. 6, as long as their amount is relatively small (Kamat 2002; Wukitsch et al. 1988).

while the level of dissolved (available) oxygen in the plasma is maintained by the release of oxygen from the buffer, compare Fig. 3.18.

The *nonlinear behavior* of S over p_{O_2} is obvious from Fig. 3.19. The curve shows a sigmoid shape which indicates an accelerated binding of oxygen by hemoglobin at medium p_{O_2} values. This is because the initial binding of oxygen yields structural changes within the hemoglobin molecule, which in turn increases the affinity to oxygen, compare charge and shape complementarity from Sect. 2.1.1.

The dissociation curve or the *hemoglobin's affinity for oxygen* is also strongly *influenced* by other *factors*, comprising important physiological mechanisms to address regulatory tasks. For instance, as indicated in Fig. 3.19, a reduced pH value (Footnote 178) in the blood plasma or an increased body temperature (e.g., during *physical exercise*) leads to a reduced *affinity for oxygen* (i.e., a greater unloading of oxygen) and consequently to its accelerated release into the (exhausted) tissue. Also an increased partial pressure of carbon dioxide—decreasing the pH value through formation of carbonic acid (Footnote 185)—leads to a strengthened oxygenation of the tissue undergoing a *fast metabolism*. Generally, an increased partial pressure of carbon dioxide in *systemic capillaries* favors unloading of oxygen while decreased partial pressure of carbon dioxide in *pulmonary capillaries* favors loading of oxygen (Fig. 3.19).

Figure 3.20 depicts a typical course of the *physiologic parameter* S during voluntary *breath holding*. The reference signal for *respiration* is a *mechanic biosignal* obtained from local changes of the chest curvature (Sect. 7). The level of S was derived from *optic biosignals* (Sect. 6); that is, the light absorption due to pulsatile arterial blood was measured at two wavelengths and interrelated by an algorithm.

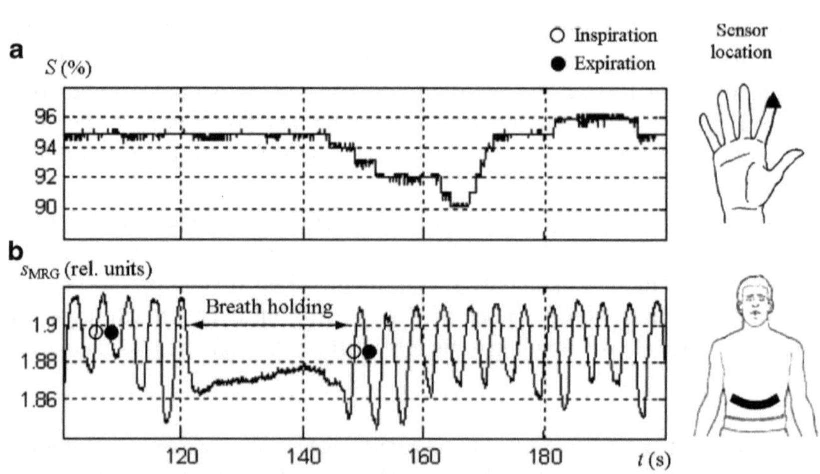

Fig. 3.20 Temporal changes of (**a**) hemoglobin oxygen saturation S derived from optic biosignals (from a finger on the right hand) during voluntary breath holding, recorded synchronously with (**b**) mechanic biosignal mechanorespirogram s_{MRG} (from local chest curvature changes)

A strongly *delayed decrease in S* can be observed in Fig. 3.20 when compared with the breath holding interval, corresponding to suppressed oxygen supply from the lungs. The resumption of breathing is followed by a relatively steep increase in S and thus normalization of the oxygen level in blood; even a temporal overshoot can be observed for the time >180 s. The *time delay* is in the range of 20 s and may be ascribed to both *physiological phenomena* and *applied methodology*.

Inertness in oxygen storage and delivery with hemoglobin as an intermediate buffer determines the observed delay from a physiological point of view. In terms of the methodology used, the *distal location* of the finger—where the optical sensor is applied (Fig. 3.20a)—predisposes a time delay in the response of S. In addition, an averaging algorithm to estimate S out of the optic biosignals contributes to the time delay, whereas the averaging time is usually between 5 and 20 s to reduce motion artifacts (Hill and Stoneham 2000). Particularly in cases of vasoconstriction and poor perfusion, the delay increases from proximal to distal locations[221] (Bebout et al. 2001).

Another example of the *course of S* is given in Fig. 3.9 during consecutive *sleep apneas* (Sect. 3.1.2). The oscillations (temporal reductions) of S correspond to intermittent apneas, whereas the aforementioned time delay can also be observed. Generally, the oscillation amplitude of S reflects the severity of apneas, see definition of the obstructive apnea from Sect. 3.1.2.

Healthy aging is marked by a decreased efficiency of oxygen delivery to blood in the lungs because of reduced diffusion capacity across the alveolar walls. The number of red blood cells is reduced and thus the amount of available hemoglobin for binding oxygen; the alveolar sacs become shallower yielding a reduced alveolar surface area. These effects may contribute to a reduced amount of oxygen in the blood and thus to a fast fatigue in the elderly. Lastly, it should be noted that noninvasive assessment of S by the optical approach is faced with progressing accuracy problems in the elderly (Hill and Stoneham 2000; Kaniusas 2006b); advancing hypoperfusion and vasoconstriction impede the slope and offset of the calibration curve to estimate S, see Sect. 6.

3.1.5 Body Temperature

The temperature of the human body is clearly given by the *balance* between

- *Heat production*
- *Heat loss*

Under resting conditions, *heat production* Is mainly carried out by inner organs such as the liver, kidneys, heart, intestines, and brain within the scope of metabolic activity. Resting metabolic activity accounts for almost 50–70% of all daily energy

---

[221]To give an example, *hypoxemia* manifests approximately 90s later for the *finger* vs. the *forehead* in the case of peripheral *vasoconstriction* (Bebout et al. 2001).

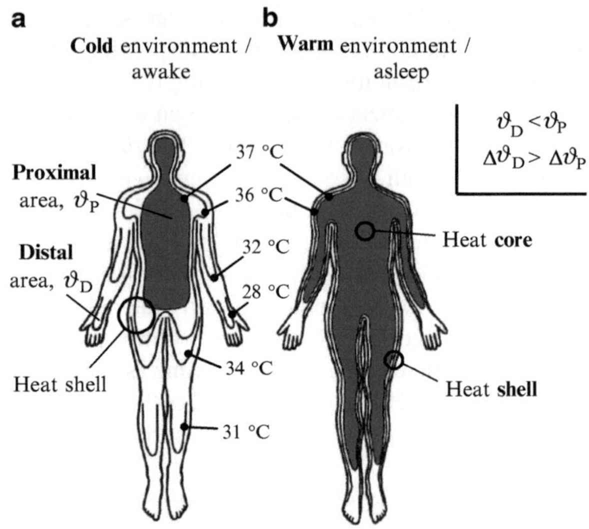

Fig. 3.21 Isothermal lines of the human body at different ambient temperatures ϑ or different postures/activities. (**a**) Cold environment ($\vartheta = 20°C$) or standing/awake with proximal temperature ϑ_P and distal temperature ϑ_D. (**b**) Warm environment ($\vartheta = 35°C$) or lying/sleeping [modified from Aschoff (1971) and Kräuchi (2007)]

expenditure (Kräuchi 2007). Normally, more than 50% of thermal energy is produced by inner organs and about 20% by muscles and skin, whereas under physical work the contribution of muscle and skin may reach 90% (Silbernagl and Despopoulos 2007).

In general, *heat loss* is governed by heat radiation, heat convection, and evaporation (i.e., sweating), whereas radiation prevails at room temperature and evaporation in warm environments. Efficient heat loss cannot be realized by the *proximal* skin surface, for its shape is too flat for efficient heat transfer to the environment. Thus, heat has to be transferred to *distal* body parts such as fingers and toes, which have optimal surface shapes, i.e., an increased surface to volume ratio, to efficiently conduct heat to the environment.

In other words, the body consists of the heat producing *core* (inner body) which is homeostatically regulated around 37°C and the heat-loss regulating *shell* (skin) which is poikilothermic, meaning that its temperature varies along with that of the ambient environment, compare Fig. 3.21. Typically, the *core body temperature*[222] shows a circadian variation of about ±0.6°C with a maximum in the early evening (at about 6 p.m.) and a minimum in the second half of the night (at about 3 a.m.), for details see Sects. 3.2.3 and 3.2.4.

The *regulating mechanisms* involved comprise a readjustment of a *target value of about* 37°C over daytime, according to the endogenous clock (=circadian inner clock) and exogenous clocks (=solar and social clocks), see Sect. 3.2.3. The *target value* is set by the central nervous system in the brain (hypothalamus region), while

[222]From a *practical point* of view, the *core body temperature* can be estimated by measuring axillary, oral, tympanic membrane, (superficial) temporal artery, or even rectal temperature; the latter often being considered the closest to the true core body temperature (Neuman 2010).

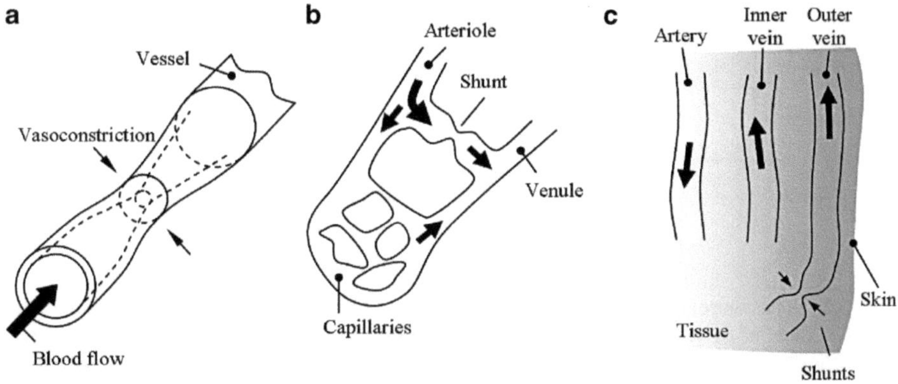

Fig. 3.22 The regulating mechanisms of heat loss in the periphery. (**a**) Dilation/constriction of peripherial vessels. (**b**) Opening/closure of arterio-venous anastomoses, i.e., blood flow shunts. (**c**) Redirection of venous blood return

the registration of the *actual value* of the core body temperature by thermal receptors is also performed within the hypothalamus.

In particular, the regulating mechanisms are of autonomic, *mainly sympathetic*, nature, as depicted in Fig. 3.22, and encompass (Kräuchi 2007)

- Dilation and constriction of *peripheral vessels* (Fig. 3.22a)
- Opening and closure of *shunts* between arterioles and venules, the so-called *arterio-venous anastomoses* bypassing superficial capillary loops (Fig. 3.22b). These anastomoses are structured as arterioles with an unusually thick layer of *smooth muscle* for a shunt (switching) function and are found exclusively in *distal skin* regions.
- In addition redirection of *venous blood return* in the extremities in combination with the arterio-venous anastomoses (Fig. 3.22c) contributes to the regulating mechanisms

In particular, the *dilation* of *vessels* (by smooth muscle control) increases *blood flow in the skin* and thus assists heat exchange with the environment, whereas with the *constriction* of vessels the reverse is true. In analogy, *opening of the shunts* supports a more rapid blood flow from the arterioles to venules, if compared to capillary blood flow; thus *enhancing the heat loss*. The third mechanism of venous *blood return* originates from the so-called *counter-current heat exchange*. In order to protect the body from cooling in a cold environment, the venous blood is redirected by the arterio-venous anastomoses into inner blood vessels where the back stream of venous blood is pre-warmed by the forward stream of arterial blood. In a warm environment, the venous blood is redirected to outer blood vessels near the skin surface, which enhances heat loss.

Figure 3.23 demonstrates schematically the above *regulatory mechanisms* embedded within a *control loop* to control the core body temperature. As an

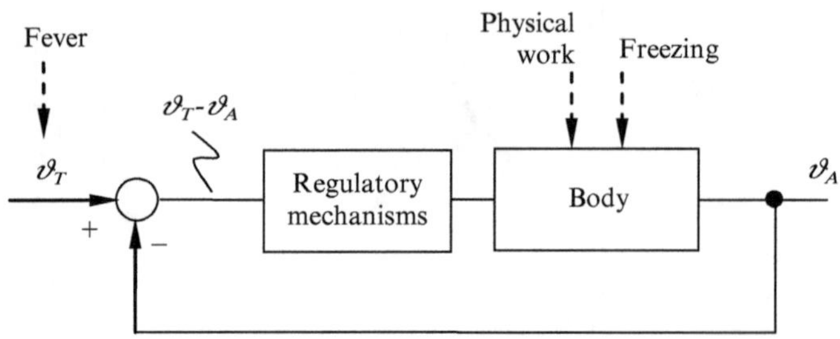

Fig. 3.23 The control loop of the core body temperature ϑ with its target value ϑ_T and actual value ϑ_A. The value of ϑ_T remains constant during, e.g., physical work and freezing whereas it changes during fever (compare Fig. 3.24). In terms of control engineering, the regulatory mechanisms comprise the *controller* while the body represents the *controlled system*. The *command variable* ϑ is a function of fever, whereas physical work and freezing represent disturbances

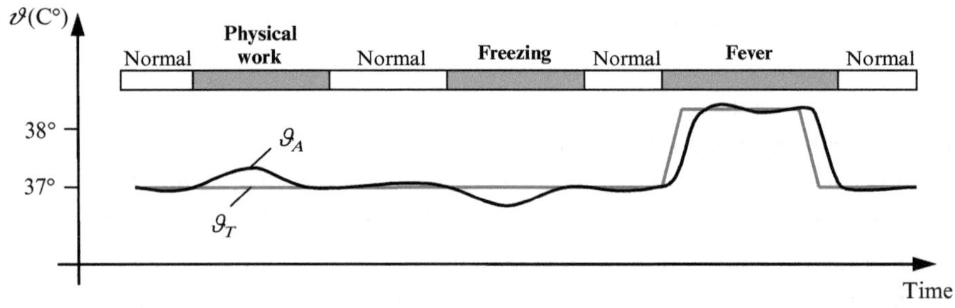

Fig. 3.24 Schematic course of the actual core body temperature ϑ_A for its constant target value ϑ_T (during physical work and freezing) and variable ϑ_T (at fever)

approximation, the mechanisms are activated as soon as the target of the core body temperature deviates from the actual core body temperature.

The efficiency of the control loop (and of the *regulatory mechanisms*) can be demonstrated in the thermally opposing cases of physical work and freezing, as shown in Fig. 3.24. During *physical work*—can be seen as a disturbance in the control loop (Fig. 3.23)—the core body temperature tends to increase due to inner heat production. However, the body counteracts this increase with dilation of peripheral vessels (compare Fig. 3.21), opening of the aforementioned shunts between arterioles and venules, redirection of venous blood to superficial veins, and perspiration. All these measures cool the skin and thus create a necessary temperature gradient between the body core and the skin, yielding an enhanced heat loss. The delayed equilibrium in heat production and heat loss yields only a slight and transient increase in the actual body core temperature and thus its normalization toward a *constant target value*, as indicated in Fig. 3.24.

During *freezing*—a disturbance in the control loop (Fig. 3.23)—the core body temperature tends to decrease because of increased heat loss (Fig. 3.24). However, the body not only restricts the heat loss by vasoconstriction (compare Fig. 3.21) but also generates additional heat with muscle tremors. Both measures normalize the actual core body temperature with respect to its *constant target value*, see Fig. 3.24.

In contrast to physical work and freezing, the *target value* of the core body temperature *changes* during *fever*, compare Fig. 3.23. During the *onset of fever*, the target value is increased (Fig. 3.24). For the actual core body temperature to reach its higher target value, the heat dissipation is reduced by vasoconstriction and reduced blood circulation in the skin. On the other hand, the heat production is increased by shivering. Usually, the reduced blood circulation in the skin yields *sensation of cold*. Before the *end of fever*, the target value normalizes, i.e., decreases toward 37°C. Thus, the actual core body temperature has to decrease, which is attained by increased heat loss due to increased blood circulation and strong perspiration on the skin. In contrast to the onset of fever, the end of fever is usually marked by *sensation of heat*.

Figure 3.25 illustrates the timely behavior of the *distal* and *proximal skin* temperature from *evening to morning* time. The thermograms, the *thermal biosignals*, result from miniature thermistors applied tightly to the skin. As expected, the proximal temperature remains larger than the distal one. In accordance with Sects. 3.2.3 and 3.2.4, during the *onset of sleep,* there is an increase in the temperature of proximal and distal skin regions (heat sink) to downregulate the temperature of the core body (heat source). Actually, the distal temperature increases to a larger extent. During *morning* hours, by contrast, both temperatures decrease, compare later Fig. 3.42e.

In *elderly* persons, peripheral blood perfusion is reduced, not only related to aging processes per se, e.g., enhanced baseline vasoconstriction and loss of dynamic change in vasoconstriction in response to changing ambience (e.g., to cold), but also related to a decreased level of fitness and physical activity (Van Someren et al. 2002). It restricts the thermal regulatory mechanisms described above and thus the heat exchange with the environment. The thermal perception is attenuated

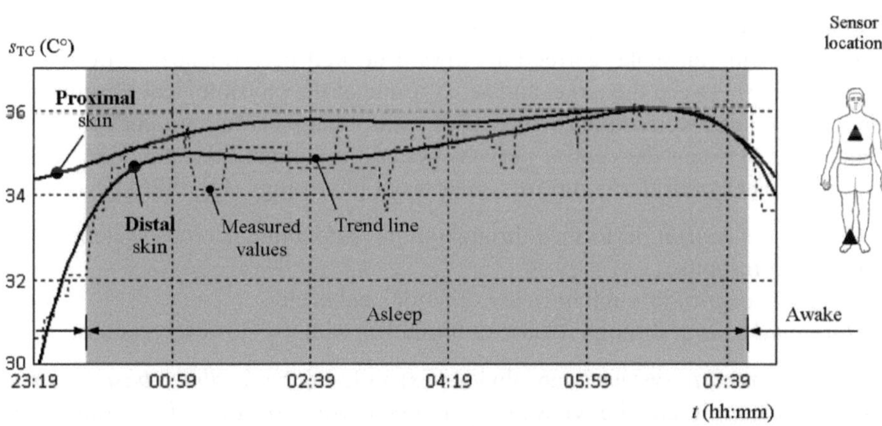

Fig. 3.25 Distal skin temperature throughout the night as given by thermal biosignal thermogram s_{TG} (from sole of the foot) along the corresponding proximal skin temperature as given by another s_{TG} (from chest). Both measured trend values and calculated trend values (based on a polynomial) are given for the distal skin region, whereas only trend values are given for the proximal region

as well as the amplitude and stability of the circadian rhythm of the core body temperature (Sect. 3.2.3). In addition, the loss of subcutaneous fat in the elderly makes it harder to maintain body heat.

3.2 Parameter Behavior

The discussed *vital phenomena* of the heartbeat (Sect. 3.1.1), respiration (Sect. 3.1.2), blood circulation (Sect. 3.1.3), blood oxygenation (Sect. 3.1.4), and body temperature (Sect. 3.1.5) show specific changes in their tonic and reflexive behavior. Moreover, these physiological phenomena along with their parameters are highly interrelated in order to *coordinate and integrate body functions*, as will be discussed throughout this chapter.

The *behavior* of the physiological parameters and their *mutual coordination* facilitate

- Vital physiological *functions*
- Limited resources of body *energy*
- Limited *space* and *time* in cells and organs to cope with life-supporting functions e.g., sufficient and timely blood supply
- *Accommodation* to environmental *challenges* (e.g., winter time)
- *Accommodation* to physical and mental *stress* (e.g., exercise)
- Body *regeneration* tasks (e.g., sleep)

Both behavior and coordination of the physiological parameters are usually based on *control loops* which are exemplified in Fig. 3.26. The *rationale* behind the *control loop* is that the performance of the physiological phenomenon (or function) is controlled via a *quantitative feedback* to the central nervous system, with the feedback usually given by pressure, thermal, or chemical receptors. The central nervous system can make corrections toward desired performance by evaluating the difference between the target and actual value of the physiological parameter to control. The latter corrections, in terms of control engineering, are performed via a controller between the central nervous system and target body functions being controlled. The *controller* comprises, in terms of physiology (Fig. 3.26),

- *Neurogenic control* performed throughout the autonomic nervous system yielding a fast response
- *Myogenic control* through muscle excitation, and even
- *Hormonal control* through release of hormones with the slowest response

The *cardiovascular system* is an obvious example of the *feedback-based control* system because it can be viewed as a *blood pressure-controlled system*, see Sect. 3.2.2. If the *blood pressure drops* below normal, i.e., control parameter is diminished, arterial stretch-sensitive receptors (baroreceptors) start to signal this imbalance to the brain. Then the difference between the target and actual values of the blood pressure becomes significant, see Fig. 3.26. Efferent (output) *neurogenic*

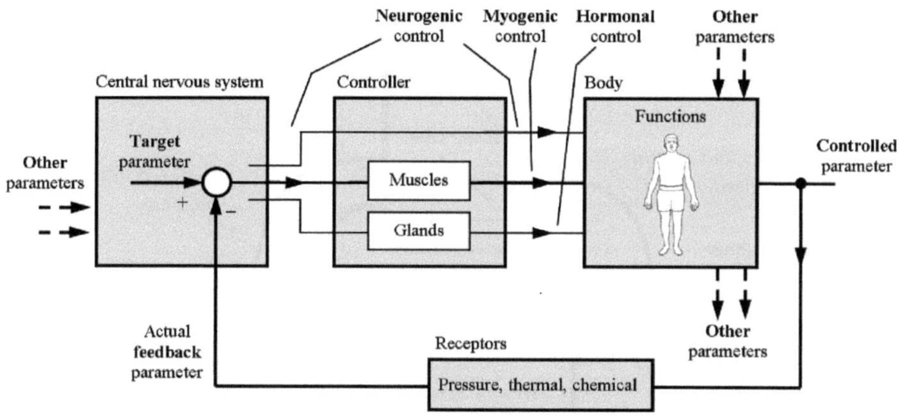

Fig. 3.26 Feedback-based control of vital physiological parameters

control inhibits the parasympathetic activity of the vagus nerve connected to the sinoatrial node (pacemaker) of the heart; in response, the *heart rate* f_C and *contractility* of the heart muscles increase. In parallel, *myogenic control* forces vasoconstriction to increase the *total peripheral resistance* R_T through activation of smooth muscles in the peripheral arteries. The latter regulatory actions increase and thus *normalize the blood pressure* level. *Another example* of a feedback-based control loop is given by the regulation of the *core body temperature*, as discussed in Sect. 3.1.5 and shown in Fig. 3.23.

Mutual interrelations of physiologic functions and parameters mainly result from control loops, as already indicated by the above example of the blood pressure-controlled system (Fig. 3.26). That is, the control of the target parameter, e.g., blood pressure, *influences other parameters*, e.g., f_C, heart contractility, and R_T. By contrast, other parameters *influence* the *target* and *controlled parameters*, e.g., breathing influences the blood pressure, for details see Sect. 3.2.1.

In particular, interrelations of physiologic parameters in the time domain or the *phase coupling* between the parameters are needed for an efficient use of energy resources in humans, see Sect. 3.2.3. Furthermore, *frequency coupling* is often observed between the fundamental frequencies of the periodic time courses of the parameters; it has functional and diagnostic relevance concerning efficient use of the body's resources. In this regard, it is interesting to note that synchronization in the time and frequency domains shows the *lowest energy expenditure* while a *disturbed coupling* occurs in *pathological* cases. *Normalized coupling* can be re-established during active rehabilitation (Hildebrandt et al. 1998).

Figure 3.27 exemplifies mutual *interrelations* of the main vital physiological phenomena such as *respiration*, *heartbeat*, and *blood circulation* during *inspiration*; i.e., during a relatively *short time period* (Sect. 3.2.3). That is,

- In terms of *cardiorespiratory* interrelations, see Sect. 3.2.1, an increase in the inspired air volume temporally parallels both a decrease in the left ventricular

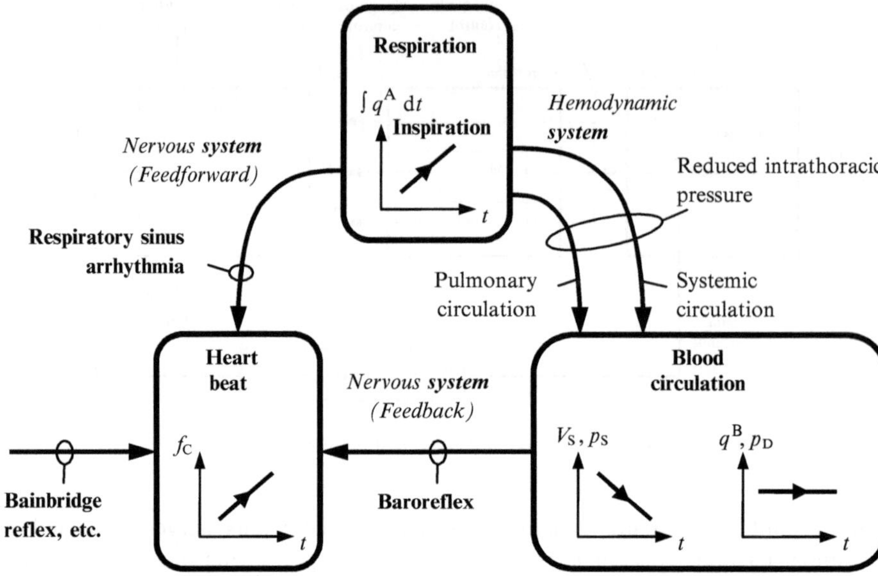

Fig. 3.27 Schematic example of short-term interrelations of main vital physiological phenomena during the inspiration phase of respiration. Mutual control of their descriptive parameters is depicted; compare Fig. 3.29. Timely behavior of the heartbeats is described by the heart rate f_C whose increase over time corresponds to an increase in inspired air volume $\int q^A dt$ (with q^A as air flow) and, on the other hand, to a corresponding decrease in the left ventricular stroke volume V_S and systolic blood pressure p_S. The blood flow q^B (or cardiac output) remains balanced during inspiration according to the product $f_C \cdot V_S$ (= q^B), see (2.30), while the diastolic blood pressure p_D remains balanced through reduced diastolic (filling) time, i.e., increased f_C

stroke volume V_S and an increase in f_C to level off the cardiac output q and thus the efficient blood supply.

- In addition, in terms of *cardiovascular* interrelations, see Sect. 3.2.2, a decrease in the systolic blood pressure p_S coincides with an increase in f_C to level off the blood pressure.

It will be shown that cardiorespiratory and cardiovascular interrelations are governed by a complex interplay between the *circulatory and pulmonary systems* (Fig. 2.39), with the *hemodynamic system* (e.g., pressure gradients along vessels) and the *nervous system* (e.g., neurogenic control of f_C) being tightly involved.

3.2.1 Cardiorespiratory Interrelations

Respiration is a prominent and vital physiological phenomenon, the periodic rhythm of which encompasses the entire body and *impacts* other body phenomena of diagnostic relevance.[223] Interestingly, even the (most vital) *heartbeat* is influenced in its main characteristics of strength and periodicity, the *blood circulation* is modulated, and—if the respiration is (periodically) ceased in the case of apneas (Sect. 3.1.2)—the *blood oxygenation* level also experiences temporal changes. As already noted, a high degree of synchronization between respiration and other physiological phenomena can be expected to be favorable for *ergonometric optimization* and physical recreation (during sleep).

3.2.1.1 Phenomenological Physiology

Normal Respiration

All anatomical structures in the thorax are exposed to pressure changes during respiration. During the *inspiration* phase, when inspiratory muscles contract and the *diaphragm lowers*, the thoracic volume increases. A partial vacuum is created in the thoracic cavity and the *intrathoracic pressure* decreases, allowing air to enter the lungs, compare Fig. 3.27 (Silbernagl and Despopoulos 2007).

The pressure in the great *intrathoracic veins* decreases (relative to atmospheric pressure) which yields an increase in the *pressure gradient* between the peripheral venous system (exposed to the atmospheric pressure) and the intrathoracic veins (below the atmospheric pressure). This causes blood to be drawn from the peripheral veins into the *intrathoracic venous vessels* which are highly compliant and expand to receive the blood[224]; the phenomenon is also referred to as the *respiratory pump*. In consequence, the volume of the *peripheral venous blood* decreases during inspiration and—as demonstrated in Fig. 3.28—the *peripheral vein circumference* strongly decreases due to large venous compliance. On the other hand, the right side of the heart experiences an enhanced filling from the intrathoracic veins (compare Fig. 2.39) which leads to an increased *right ventricular stroke volume* by about 15% and to a *more forceful contraction*[225] (Hayano et al. 1996; Karam et al. 1984).

[223] From a *historical perspective*, it is worth noting that systematic investigations into the effects of respiration on cardiovascular function date back even to famous experiments with horses by Stephen Hales in the first half of the eighteenth century, see Footnote 204. He may have been the first to record the inspiratory fall in arterial blood pressure (Olsen et al. 1985), see section "Normal Respiration."

[224] The increased return of venous blood during inspiration is actually facilitated by *venous valves*; thus, the blood return is increased more by inspiration than it is decreased by expiration (Footnote 117).

[225] This is the so-called *Frank–Starling law*, named after Otto Frank (1865–1944), a German physiologist, and Ernest Henry Starling (1866–1927), an English physiologist. The law states that the *contraction force* of the cardiac muscle is proportional to the *heart volume* (filling volume or end-diastolic volume), especially, to the length of the heart muscles before contraction (Footnote 104). Consequently, the end-diastolic volume of ventricle is directly proportional to the following *stroke volume* of this ventricle.

Fig. 3.28 Respiration induced changes in cross section of peripheral veins and arteries

In addition, the reduced pressure outside the right side of the heart favors *expansion of the right atrium and right ventricle*, which contributes to the increased right ventricular stroke volume.

The increased right ventricular stroke volume during *inspiration* evokes

- *Delayed increase* in V_S since the circulation system is a *closed loop system*[226] (Fig. 2.39). The delay results from resistance to flow and hydraulic capacitances between the right ventricular output and left ventricular filling (Clifford et al. 2006). Normally, the delay postpones the rise in V_S even until the *following expiration* phase so that V_S effectively and momentarily *decreases* during inspiration.
- Furthermore, *decrease* in V_S is also given during inspiration because strengthened filling of the right side of the heart physically *compresses the left ventricle* and displaces the *interventricular septum* toward the left (compare Fig. 2.32a). Consequently, the left ventricle becomes less compliant and the (end-diastolic) left ventricular geometry is mechanically reduced. That is, in terms of *ventricular interdependence*, an increased right ventricular volume (end-diastolic volume) can only occur at the expense of the space devoted to the left ventricle, i.e., the corresponding decrease in the left ventricular volume (end-diastolic volume).

In addition, a few other mechanisms contribute to the effective decrease in V_S during *inspiration*:

- *Increased* f_C—according to the so-called *respiratory sinus arrhythmia*, see below—reduces the diastolic filling time of the heart and thus the end-diastolic volume. Consequently, the increased f_C contributes to the decrease in V_S.

In other words, increasing systemic venous return, e.g., during *inspiration or changing from upright to supine* (gravitational force leading to an increase in the thoracic venous blood volume), increases the filling pressure of the right ventricle which leads to an increased right ventricular stroke volume and subsequently to a (delayed) increase in V_S and q [compare Footnote 226 and (2.30)]. Obviously, other compensatory mechanisms such as baroreflex (Sect. 3.2.2) counteract the potential imbalance of the latter physiological parameters.

[226]The systemic *venous return* must equal q when averaged over time because the circulation system is essentially a *closed loop system*, see Fig. 2.39 and Footnote 225.

- Moreover, *decreased intrathoracic pressure* temporarily decreases the intracavitary left ventricular ejection pressure and thus V_S (Olsen et al. 1985), in terms of *reverse thoracic pump*[227] (and in contrast to *thoracic pump*[228]).
- Lastly, *decreased intrathoracic pressure* increases pulmonary (compliant) venous capacitance, favors pooling of blood in expanded lungs, decreases blood flow from pulmonary veins to the left atrium, and diminishes left atrial (and ventricular) filling and thus V_S.

Therefore, the level of V_S decreases slightly by about 10% during *inspiration* (Elstad et al. 2001; Olsen et al. 1985; Guz et al. 1987; Silbernagl and Despopoulos 2007), compare Fig. 3.27. In consequence, the peripheral arterial volume (or peripheral blood perfusion), peripheral *pulsation* strength, and peripheral *arterial circumference* decrease, as schematically shown in Fig. 3.28. The arterial circumference decreases to a lesser extent compared to the veins because the arterial compliance is much lower than the venous compliance (Sect. 2.5.1). Actually, an *inspiratory* decline in V_S may even result in the so-called *pulsus paradoxus*.[229]

In consequence of respiratory modulation of V_S, respiration drives the blood pressure as well. The levels of p_S *and mean arterial pressure* are reduced by about 5% (Elstad et al. 2001) or by less than 10 mm Hg during inspiration because of the inspiratory decline of V_S; for the relevant interrelation between the two consult

---

[227]The *reverse thoracic pump* mechanism accounts for the influence of *intrapleural pressure* (Sect. 2.6.2) on the arterial blood flow (Olsen et al. 1985). Factors elevating intrapleural pressure, e.g., the expiratory phase of breathing, augment intracavitary *left ventricular pressure* referenced to atmospheric (peripheral) pressure, i.e., augment the effective left ventricular ejection pressure (=left ventricular pressure − atmospheric pressure). Conditions reducing intrapleural pressure, e.g., inspiratory phase, lower the left ventricular ejection pressure and thus the level of V_S. It should be noted that peripheral vasculature terminates the blood flow; consequently, the atmospheric (peripheral) pressure is the appropriate *reference pressure* for the effective ejection pressure.
Interestingly, the reverse thoracic pump or changing intrapleural pressure *does not affect* the *right ventricular volume*. During the right ventricular ejection, the right ventricle (input of the pulmonary circulation, Fig. 2.39) is in continuity with the left atrium (output), both of which are identically subjected to changes in the intrapleural pressure; i.e., to its expiratory rise and inspiratory fall.

[228]The *thoracic pump* mechanism clearly contrasts with the *reverse thoracic pump* (from Footnote 227). The thoracic pump refers to a blood pump to the extent that elevated intrathoracic pressure (e.g., coughing or chest compressions during cardiopulmonary resuscitation) tends to press the blood out of the pulmonary vessels into the heart and then onwards into the periphery. Therefore, the elevated intrathoracic pressure effectively increases V_S. In this regard, *inspiration* may be referred to as a *reverse thoracic pump* mechanism (Olsen et al. 1985), compare Footnote 227.

[229]The *pulsus paradoxus* describes the absence of peripheral pulse despite the *presence of cardiac contraction* (Barach 2000; Khasnis and Lokhandwala 2002). It is due to an increased (exaggerated) amplitude of the inspiratory fall in p_S by more than 10 mm Hg. Two conditions may evoke the pulsus paradoxus: large variations in *intrapleural pressure* (e.g., due to forced respiratory effort, severe asthma, or pulmonary embolism) or increased *coupling* between the right and left ventricles (e.g., acute right heart failure). Conversely, marked increases in the intrapleural pressure (e.g., coughing) may produce *reversed pulsus paradoxus*, i.e., the presence of a peripheral pulse in the *absence of cardiac contraction*.

section "Pulse Waveforms of Pressure and Flow." By contrast, the variation of the *diastolic blood pressure* p_D over the respiratory cycle is relatively small, since f_C increases during inspiration (see below) and thus compensates for the effect of decreased V_S; the duration of diastole is reduced and thus the time for the pressure to drop to p_D is also reduced (section "Pulse Waveforms of Pressure and Flow").

If f_C is assumed to be constant, then reduced V_S *during inspiration* would cause a reduced *left ventricular output* [$= q$, see (2.30)] with a reduced p_S and *mean arterial pressure*, compare Fig. 3.27. In fact, it would represent a disadvantageous situation for the body with vital physiological *parameters* being *imbalanced.* Actually, in the time frame of respiratory variations, only the varying q may trigger changes in the blood pressure, since possible changes in vascular R_T and aortic *stiffness* are too *slow*[230] (section "Pulse Waveforms of Pressure and Flow").

However, blood pressure receptors located in arteries and veins—the so-called *baroreceptors* (Sect. 3.2.2)—*alarm* the central nervous system about the imbalance of the blood pressure, i.e., the reduced blood pressure during *inspiration.*[231] In response, the level of f_C is increased via the (parasympathetic) vagus nerve which controls the pacemaker of the heart. This *compensatory mechanism* (partly) balances q, given as a product of the (inspiratory) decreased V_S and increased f_C (2.30), compare Fig. 3.27. In fact, the increased f_C then further decreases V_S by *reducing cardiac filling time.* During expiration, the reverse is true. Likewise, the right ventricular ejection is prolonged while the left ventricular ejection is shortened during inspiration.

The aforementioned phenomenon of *increased* f_C by about 10% during *inspiration* is called *respiratory sinus arrhythmia* (Silbernagl and Despopoulos 2007; Elstad et al. 2001), compare Fig. 3.27. The arrhythmia generally occurs from the influence of *breathing* on the sympathetic and parasympathetic efferent branches of the *autonomic nervous system* (Sect. 3.1.1) directed toward the sinoatrial node to initiate (normal) heartbeats (Sect. 2.4.2). The *parasympathetic* influence decreases during inspiration, whereas it increases during expiration (Footnotes 188 and 189). In particular, the respiratory sinus arrhythmia arises in the course of *multiple phenomena*:

- Direct command from the *central nervous system*, i.e., *feedforward* control from Fig. 3.27
- *Feedback*-based control of blood pressure within the scope of *baroreflex* (Sect. 3.2.2), i.e., compensatory mechanism from above

---

[230]The sympathetic *control of* R_T (e.g., increase in R_T) to balance the blood pressure is not likely to be involved in synchrony with the *respiration* cycle (e.g., inspiration), because the contraction of smooth muscles in the arterial wall needs a relatively long time to develop; compare Footnote 231.

[231]Here, it should be noted that multiple mechanisms of *sympathetic and parasympathetic* origins are involved in blood pressure control in terms of the so-called *baroreflex*, as will be discussed in Sect. 3.2.2.

- Mechanical impact of increased return of venous blood during inspiration in terms of the so-called *Bainbridge reflex*[232] (Zemaityte 1997)
- *Feedback*-based control of carbon dioxide in terms of *chemoreflex*, e.g., increasing hypercapnia (or increasing hypoxia) diminishes the respiratory sinus arrhythmia (Yasuma and Hayano 2004)
- Respiratory stimulation of stretch-sensitive *mechanoreceptors* in the lung and thorax (Zemaityte 1997)

However, the *primary control* of the heart rhythm is carried out by the parasympathetic efferent *vagus nerve* acting on the sinoatrial node. That is, the activity of the vagus nerve (i.e., impulse traffic or firing rate) is *impeded* during inspiration, which leads to *increase in* f_C; during expiration this pattern is reversed.

Respiratory-related *changes of* f_C are larger for *supine* than *standing* because of increased parasympathetic effect (Sect. 3.1.1). Furthermore, these changes are fortified with the duration of the respiratory cycle suggesting that *slow breathing* intensifies the parasympathetic effect; possibly through resonance with the low-frequency components of the HRV, i.e., with the Mayer waves from Sects. 3.1.1 and 3.2.2.

It is interesting to note that *respiratory sinus arrhythmia* mainly *buffers* respiration-synchronous fluctuations in the *mean arterial pressure*[233] (Elstad et al. 2001). In fact, balancing the mean arterial pressure seems to be vital for the body because the mean arterial pressure determines q (2.20) and thus the *effective blood supply*. On the other hand, respiratory arrhythmia seems to contribute to the *temporal match* of *blood perfusion* and *air ventilation* within each respiratory cycle (Yasuma and Hayano 2004). That is, *heartbeats clustering* together during inspiration with an increased right ventricular stroke volume matches increased pulmonary perfusion during inspiratory *increased lung volume*. Conversely, the arrhythmia suppresses unnecessary heartbeats during expiration with reduced lung volume to *save cardiac and respiratory energy*.

Since f_C is an inverse measure for the duration of the cardiac period, the *respiratory modulation of the cardiac period* should be quickly discussed. In particular, respiration-synchronous duration changes of specific phases of the cardiac cycle are of interest; for phase definitions see Sect. 2.4.2. The *ventricular filling phase* during *diastole* is the phase which is most effected by varying interbeat interval, i.e., this phase prolongs with increasing cardiac interval or decreasing f_C. Interestingly, if the

---

[232]The *Bainbridge reflex*, named after Francis Arthur Bainbridge (1874–1921), an English physiologist, is related to a decrease in the efferent (parasympathetic) *vagus nerve* activity in response to an increased *right atrial volume* (or pressure), namely, in response to elongation of stretch-sensitive mechanoreceptors in the right atria. Consequently, a rise in the central venous pressure (during inspiration) or in the total blood volume (buffered mainly in the venous system, compare Sect. 2.5.1) tends to provoke an increase in f_C to draw more blood out of the right atrium or to prevent the pooling of blood in the venous system, respectively.

[233]Interestingly, *respiratory sinus arrhythmia* seems to buffer respiration-synchronous fluctuations of p_S in the *tilted* position only while these fluctuations are even reinforced by the arrhythmia in the *supine* position (Elstad et al. 2001).

durations of different phases of the cardiac cycle are corrected for interbeat interval duration (e.g., through regression analysis), multiple adaptive and compensatory effects for decreased V_S during inspiration become prominent.[234]

After the qualitative behavior of *numerous cardiorespiratory interrelations* has been discussed, it is worth noting that respiration-synchronous fluctuation of a *physiological parameter* (e.g., f_C or V_S) is characterized by all three:

- Deflection *amplitude*
- Fluctuation *frequency*
- *Phase*[235] with respect to the respiration cycle

The cardiorespiratory behavior of the deflection *amplitude* has been exemplified in many cases, e.g., V_S decreases during inspiration. Moreover, various cardiorespiratory interrelations exist with respect to the *fluctuation frequency and phase* if *heartbeat and respiration* are considered in parallel.

That is, the *quotient* of f_C to the *respiratory rate* f_R deserves some notice. During sleep or in relaxed states the quotient approaches the value of 4; i.e., within a single respiratory cycle about four cardiac cycles take place (Hildebrandt et al. 1998; Moser et al. 2008). The quotient tends to increase during physical activity or stress while its typical values vary between 2 and 10. Actually, a *quotient of 4* is regarded as ideal for regeneration and recreation of the human body within the scope of biological rhythms, compare Sect. 3.2.3 (and Fig. 3.45).

Interestingly, *phase coupling* exists between *cardiac and respiratory cycles* to increase functional and energetic efficiency. For instance, the onset of the inspiration phase during *sleep* is observed to be coupled to or occur within the middle of the cardiac cycle (Hildebrandt et al. 1998). By contrast, loss of this phase synchronization has been reported during episodes of sleep *apnea* (Penzel et al. 2007).

With advancing *age*, inspiratory intrathoracic pressure becomes less negative due to the weakening of breathing muscles with age. The corresponding inspiratory decrease in p_S becomes less pronounced so that the pulsus paradoxus (Footnote 229) may even disappear (Shiomi et al. 1993). Aging also impairs autonomic regulatory

---

[234]When different *phases of cardiac cycle* are *corrected* for varying interbeat interval, the systolic (contraction and ejection) time intervals show a stronger dependency on *respiration phase* than the diastolic (relaxation and filling) time intervals (Leeuwen and Kuemmell 1987). Regarding the *systolic period*, the left ventricular *ejection time* decreases during inspiration while *isovolumetric contraction* time (or *pre-ejection period*) simultaneously increases (Sect. 2.4.2). This is because V_S decreases during inspiration, which shortens the ejection time and yields an earlier closure of the aortic valve but increases the pressure gradient between aortic pressure and (reduced) left ventricular pressure (*increased afterload during inspiration*). To overcome this relative increase in the aortic pressure before blood ejection, a longer period of pre-ejection and contraction is required. Regarding the *diastolic period*, a lengthening of ventricular *filling phase* was observed at the beginning of inspiration, which can be viewed as a compensatory effect that partly offsets the loss of V_S or the loss of efficiency of the left ventricular function during inspiration (Leeuwen and Kuemmell 1987).

[235]Formal *definitions* of amplitude, frequency, and phase are given in Footnote 145.

functions; the efficiency of the *respiratory sinus arrhythmia* is reduced along with diminished *baroreflex* (Sect. 3.2.2.1). In consequence, the HRV due to breathing decreases with age, as already noted in Sect. 3.1.1.

Ceased Respiration

In contrast to *normal respiration* from section "Normal Respiration" under Sect. 3.2.1.1, a complete (or partial) temporal cessation of respiration results from an abnormal situation (e.g., obstructive *apnea* from Sect. 3.1.2) or a deliberate action (e.g., *breath holding* from Fig. 3.20 in Sect. 3.1.4). The cardiorespiratory interrelations that arise during apneas are complex but highly informative; they convey valuable information on *interrelated physiological mechanisms* and provide a profound basis for the interpretation of *biosignals* formed.

Generally, *bodily defense mechanisms*—in the event of an abnormal situation—will attempt to maintain an appropriate level of q, whereas V_S and f_C are two fundamental factors in the modulation of q (2.30); compare with the compensatory mechanisms from section "Normal Respiration" under Sect. 3.2.1.1. Different types of regulatory mechanisms, e.g., mechanical and neuronal, are involved with their impact being altered by multiple factors, such as sleep state, hypoxemia, or (reduced) baroreflex responsiveness. Therefore, the defensive response of the body is quite complex and sometimes not even unidirectional (Footnotes 236 and 240).

During obstructive apnea, as shown in Fig. 3.29, strong (mechanical) inspiratory efforts occur due to collapsed upper airways. The *intrathoracic pressure* experiences dramatic swings (or periodical decreases) and the *intrapleural pressure* decreases significantly. In analogy to the inspiration phase during normal respiration (section "Normal Respiration" under Sect. 3.2.1.1), the *venous backflow* and the filling of the right ventricle increase, which lead to an increased *right ventricular stroke volume* (Schäfer 1996; Konietzko et al. 1998), compare Footnote 227.

The filling of the left side of the heart is impaired because of ventricular interdependence (section "Normal Respiration" under Sect. 3.2.1.1). The level of V_S temporally *decreases* during apnea (Trzebski and Smietanowski 2001; Garpestad et al. 1992; Konietzko et al. 1998), see Fig. 3.29. The level of f_C decreases below its normal level (bradycardia) because of an increased (parasympathetic) activity of the *vagus nerve* but not because of ongoing hypoxemia (Schäfer 1996). The bradycardia aims to *reduce oxygen consumption* in the heart muscle and to *prolong diastolic filling time* during metabolic (apneic) stress period (Zemaityte 1997); compare Footnote 239. In consequence, the *level of q*—given as a product of V_S and f_C (2.30)—decreases (Fig. 3.29). However, the apneic changes of V_S and f_C seem to be different for different sleep phases.[236]

---

[236]According to Stoohs and Guilleminault (1992), during *rapid eye movement* (REM) *phase* of sleep (Sect. 3.2.4) there is a decrease in f_C during *apneas* while V_S does not change significantly. It yields a *reduced q* (2.30) and a clear *dissociation* of f_C and V_S. On the other hand, *nonrapid*

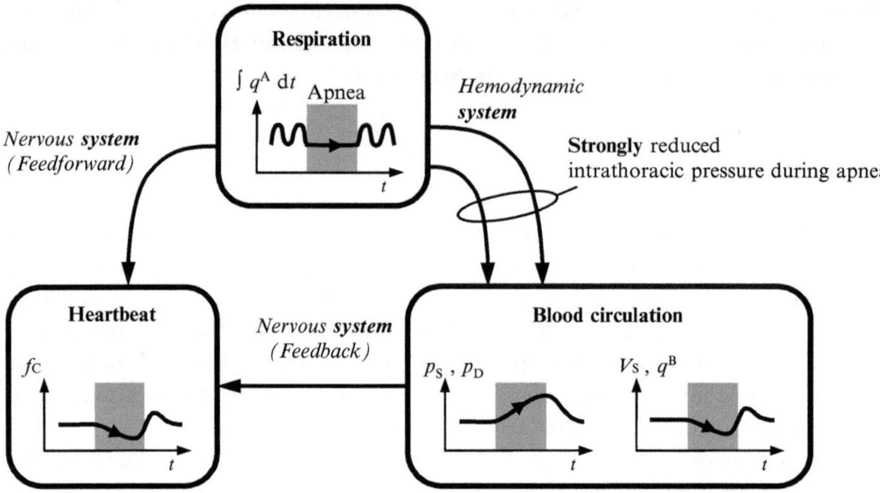

Fig. 3.29 Schematic example of short-term interrelations of the main vital physiological phenomena during obstructive apnea. Mutual control of their descriptive parameters is shown; compare Fig. 3.27. Heartbeats are described by the heart rate f_C, a decrease of which corresponds to ceased respiration with constant air volume $\int q^A dt$ (with q^A as air flow) and, on the other hand, to an increase in the systolic blood pressure p_S and diastolic blood pressure p_D, a decrease in the left ventricular stroke volume V_S and blood flow q^B (or cardiac output)

Because of the cessation of breathing, the level of *hemoglobin oxygen saturation* in blood decreases with a certain *time delay*[237] (Sect. 3.1.4), whereas the level of carbon dioxide increases. The appropriate chemical receptors, i.e., *chemoreceptors*[238], respond to this life-threatening imbalance (hypoxemia and hypercapnia) that *progressively* stimulates *sympathetic activity* to reestablish respiration (Somers et al. 1995; Bradley et al. 2003). The hypoxic stimulation

eye movement (NREM) *phase* shows significant V_S drops during apneas with a corresponding increase in f_C which actually compensates and *balances* q. The differing behavior of f_C and V_S during REM and NREM phases also demonstrates different control mechanisms of the autonomic nervous system in the course of apneas. It should be noted that REM phase normally comprises longer obstructive apneas, higher degree of desaturation, and less pronounced decreases in the intrathoracic pressure (compare reduced muscle tension in REM, Sect. 3.2.4) in comparison with NREM phase (Findley et al. 1985).

[237]The *delay* of the hemoglobin *oxygen desaturation* during apnea is determined by both intrinsic *physiological phenomena* and *methodological limitations* while measuring desaturation (Sect. 3.1.4). In short, the slow process of *handling oxygen* (storing in hemoglobin and delivery to tissues) over an intermediate buffer (hemoglobin) facilitates this time delay. In addition, peripheral locations of sensors (e.g., on a finger, as usually chosen to measure the saturation) contribute to this time delay because of the *inert blood flow* from the lungs to the finger and (potentially) *poor blood perfusion* in the periphery.

[238]The *chemoreceptors* involved, located in the carotid sinus (Footnote 248) and aortic arch, detect levels of oxygen and carbon dioxide.

seems to attenuate the arterial baroreflex (Trzebski and Smietanowski 2001), see Sect. 3.2.2.1. The sympathetic nerves reinforce *peripheral vasoconstriction* which progressively increases R_T and reduces peripheral blood volume. In particular, the so-called *selected vasoconstriction*[239] arises which *restricts* perfusion of nonvital organs (e.g., skeletal muscles) and *maintains* perfusion of vital organs (e.g., heart, brain). Sympathetic activation tends to increase f_C, whereas the actual course of f_C during apnea is *less regular* and depends on the changing balance between parasympathetic and sympathetic activities.

The levels of p_S and p_D continuously increase[240] during (voluntary) apneas (Trzebski and Smietanowski 2001; Pitson and Stradling 1998; Konietzko et al. 1998; Stoohs and Guilleminault 1992), as indicated in Fig. 3.29. Multiple factors contribute to the latter blood pressure changes, either as a compensatory action or as a consequence of an imbalance. Basically, the blood pressure increase, associated with reduced both V_S and q, reflects an increased R_T along with increased sympathetic activation. In addition, an increased opposition to systolic ejection (increased *afterload*) and an increased stretching of the ventricular muscles prior to contraction (increased *preload*) with considerably decreased intrathoracic pressure during apneas are also assumed to contribute to the blood pressure increase during apneas.

At the end of apnea, *hypoxemia* manifests. The apnea is terminated by a *respiratory arousal* if inspiratory efforts during apnea were not sufficient. This contributes to already high sympathetic activity which increases f_C above its normal level (*tachycardia*), leads to a peak in q and blood pressure [up to 25% above normal levels (Schäfer 1996) reflecting severe vasoconstriction at the end of apnea], and initiates hyperventilation to favor oxygen intake and abolish hypoxemia. After the *resumption of breathing*, a temporal drop in p_S and p_D can be usually observed, as also reported in Trzebski and Smietanowski (2001). It may be due to relieved vasoconstriction or even temporal vasodilation of peripheral vessels to accelerate ongoing compensation for the prior interruption of oxygenated blood supply. The oxygen level actually recovers (relatively) slowly after apnea termination (Lee-Chiong 2006); compare Footnote 237.

According to the impact of *aging* from Sect. "Normal Respiration" under Sect. 3.2.1.1, the intermittent decreases in the *intrathoracic pressure* during apneas can be expected to weaken with age along with the weakening of the apneas severity. For instance, the minimum level of *oxygen saturation* during apneas increases (improves) with age (Bixler et al. 1998), which indicates the decreasing severity of apneas. Furthermore, overall defense mechanisms which aim to maintain an appropriate (constant) level of q during apneas may be blunted with aging (Stoohs and Guilleminault 1992).

---

[239] All three, *breath holding*, *bradycardia*, and *selected vasoconstriction*, are usually referred to as *diving reflex*. Interestingly, face immersion into cold water fortifies the diving reflex through the activation of temperature receptors in the upper airways.

[240] During relatively *long obstructive apneas* with concurrent but weak intrathoracic pressure swings in terms of inspiratory efforts, the arterial pressure may also drop during apnea (Konietzko et al. 1998). It may be due to peripheral vasodilation provoked by advanced hypoxemia.

3.2.1.2 Biosignals and Parameters

As respiration-related changes of vital physiological parameters have been discussed—including normal and ceased respiration (Sect. 3.2.1.1)—the corresponding biosignals and derived parameters will be presented below. It should illustrate the *targeted use of biosignals* in assessing informative cardiorespiratory interrelations for diagnostic and therapeutic aims.

Normal Respiration

Figure 3.30 demonstrates basic *cardiorespiratory interrelations* on the basis of three *biosignals* and two derived *parameters*. The first biosignal is the barocardiogram (Fig. 3.30a), a *mechanic biosignal* based on the volume clamp method (Sect. 3.1.3.1), and conveys the *blood pressure* waveform in the radial artery. The successive cardiac pulses can be observed, with their upper envelope being periodically modulated; compare similar modulation in Fig. 3.11a. Actually, the upper and lower envelopes represent the sequences of p_S and p_D values (Fig. 3.30c), respectively. This modulation rhythm of p_S corresponds to the rhythm of another biosignal, the mechanorespirogram (Fig. 3.30e), a *mechanic biosignal* which results from *respiration-synchronous* circumference changes of the abdomen with the rate f_R (Sect. 3.1.2). *Residual cardiac components* can be recognized in the recorded circumference changes, compare Fig. 3.6a.

The time course of the resulting p_S (Fig. 3.30c) confirms its temporal decrease during the inspiration phase (section "Normal Respiration" under Sect. 3.2.1.1). By contrast, the level of p_D does not rhythmically change over the respiration cycle (Fig. 3.30c).

The surface *electrocardiogram* from Fig. 3.30b, an *electric biosignal*, reflects the electrical excitation of the heart (Sect. 4). Its waveform is closely related to that of the blood pressure (Fig. 3.30a), as described in Sect. 3.1.3.1. The envelope of the electrocardiogram is modulated with respiration; that is, the *amplitude of R peaks* decreases with inspiration and subsequently increases with expiration in the given case (compare also Fig. 3.32). *Two effects* mainly contribute to the *respiratory modulation* of the surface electrocardiogram:

- Mechanic *displacement* and *reorientation*[241] of the *heart* relative to the (electrocardiogram) electrodes on the skin because of respiratory movements in the inner thorax. In electrical terms, the total heart dipole (or mean electrical axis of the heart) experiences a spatial and cyclic displacement (and reorientation) over

[241] As shown in Bachta et al. (2009), the *motion of the myocardium* of the heart yields a cardiac component with f_C and a superimposed respiratory component with f_R. The ratio of the amplitude of the respiratory component to that of the cardiac component strongly depends on the motion direction, the ratio being the highest for motion along the interior–superior direction.

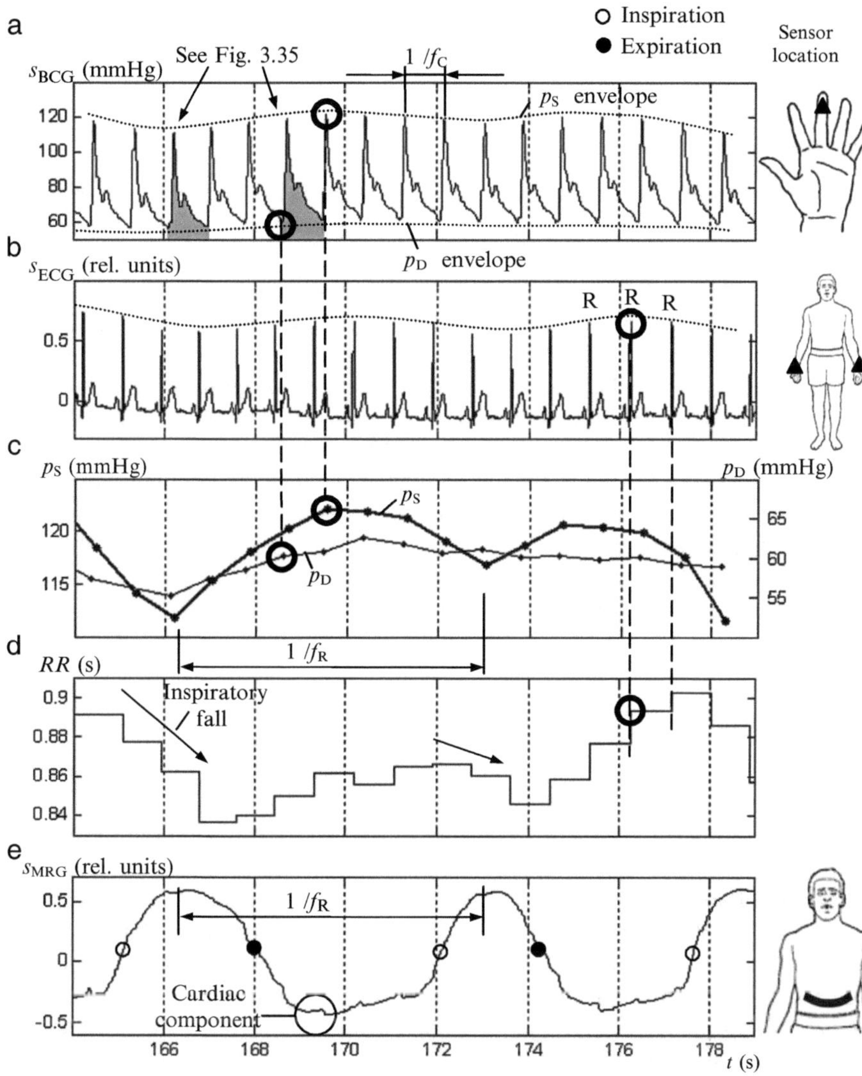

Fig. 3.30 Influence of respiration on cardiac and respiratory signals exemplified by synchronous (**a**) mechanic biosignal barocardiogram s_{BCG} (from a finger on the left hand), (**b**) electric biosignal electrocardiogram s_{ECG} (lead I Einthoven) with indicated R peaks, (**c**) systolic blood pressure p_S and diastolic pressure p_D derived from s_{BCG} (relevant pressure points indicated by *circles in bold*), (**d**) interbeat intervals RR $(= 1/f_C$ with f_C as instantaneous heart rate) derived from s_{ECG} (relevant time instants indicated by *circles in bold*), and (**e**) mechanic biosignal mechanorespirogram s_{MRG} (from abdominal circumference changes). The respiratory modulations of biosignals with respiratory rate f_R are shown. The pulses from s_{BCG} with a *grey background* are analyzed in Fig. 3.35

the respiratory cycle. It leads to periodic changes in the dipole projection on the electrode axis and thus to periodic changes in the electrocardiogram level.

- In addition, *electrical impedance* between the heart and electrodes changes over the respiratory cycle, contributing to the respiratory modulation of the voltage drop across the electrodes.

Figure 3.30d depicts the duration of successive *interbeat intervals* (inverse to the instantaneous f_C) as derived from the R peaks of the electrocardiogram (Fig. 3.30b). As expected from *respiratory sinus arrhythmia* (section "Normal Respiration" under Sect. 3.2.1.1), the interbeat intervals decrease during inspiration and f_C increases accordingly. Additional examples of interbeat intervals modulated by respiration can be seen in Figs. 3.3a and 3.33d.

Thus, *respiration* may be *derived out of the electrocardiogram*[242] in many ways, by tracking not only amplitude changes of R peaks (QRS complex, Fig. 3.30b) or temporal changes of *interbeat intervals* (Fig. 3.30d), but also amplitude changes of T waves or even area changes of *QRS complex* (Maier et al. 2009; Langley et al. 2010). Here, it should be noted that *thorax anisotropy* and *intersubject variability* of the *mean electrical axis* account for *different effects of respiration* on different electrocardiogram leads (Clifford et al. 2006). That is, the specific change (e.g., during inspiration) of the latter electrocardiogram characteristics (i.e., their increase or decrease) depends on the electrocardiogram *lead* used *and the subject* under investigation.

An interesting *age-related* issue should be mentioned within the context of the electrocardiogram derived respiration. The amplitude changes of *R peaks* do not depend on age, while the modulation strength of the *interbeat intervals* weakens with increasing age (section "Normal Respiration" under Sect. 3.2.1.1). Thus, the derived respiration (or even apneas detected from the derived respiration) is more robust among different ages if the implemented algorithms evaluate the amplitude of R peaks but not interbeat intervals (or instantaneous f_C modulation[243]).

While Fig. 3.30 contrasts *electric and mechanic* biosignals in terms of cardiorespiratory interrelations, Fig. 3.31 demonstrates interrelations between *electric, acoustic, and optic* biosignals. Besides the electrocardiogram in Fig. 3.31a, it depicts a *phonocardiogram* (Fig. 3.31b), an *acoustic biosignal*, which reveals heart sounds due to consecutive closures of heart valves (Sect. 5). In parallel, two *optoplethysmograms, optic biosignals*, from fingers of both hands are depicted in Fig. 3.31c, d reflecting pulsatile absorption of incident light by peripheral blood. It should be noted that intrinsic *interrelations* in between the above electric, acoustic, and optic

[242]Usually, the respiration activity derived from the electrocardiogram is referred to as EDR, i.e., electrocardiogram-derived respiration.

[243]From an engineering point of view, the *frequency-related* parameters are more robust than *amplitude related*. A possible impact of body movements or external noise on the electrocardiogram derived respiration is expected to be lower if *frequency modulation* is used instead of *amplitude modulation*.

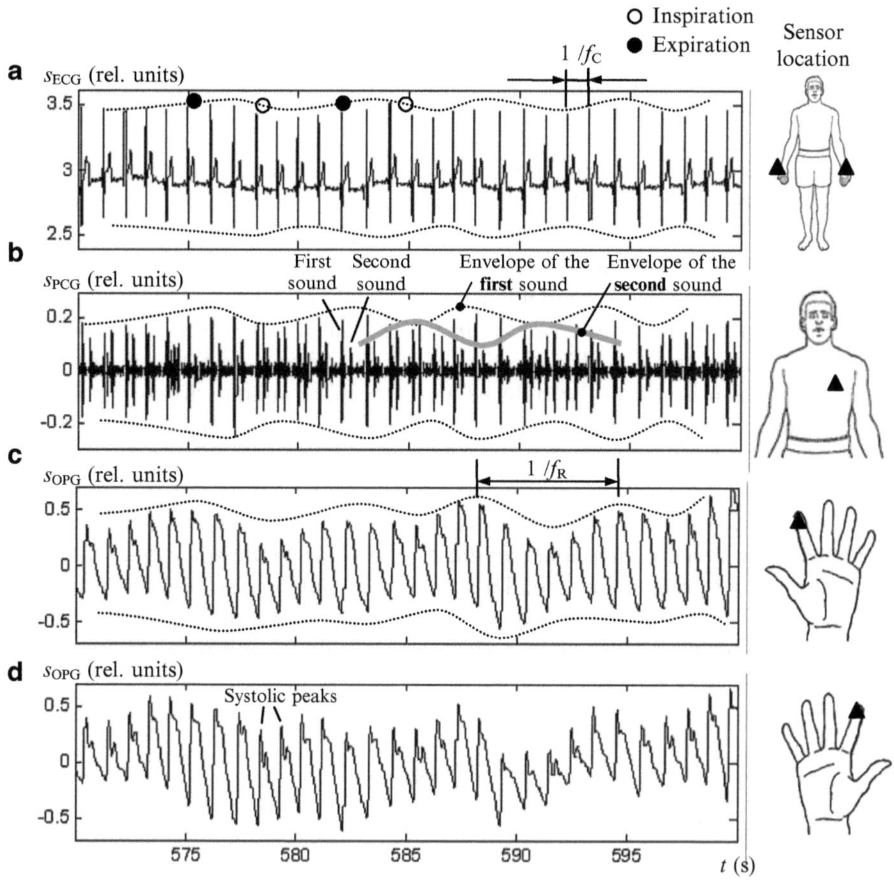

Fig. 3.31 Influence of respiration on cardiac signals exemplified by synchronous (**a**) electric biosignal electrocardiogram s_{ECG} (lead I Einthoven), (**b**) acoustic biosignal phonocardiogram s_{PCG} (from the heart region on the chest), (**c**) optic biosignal optoplethysmogram s_{OPG} (from a finger on the left hand), and (**d**) another optoplethysmogram s_{OPG} (from a finger on the right hand) with indicated heart rate f_{C} and amplitude modulation of cardiac deflections with respiratory rate f_{R}

biosignals of mainly *cardiac origin* have been already described in Sect. 3.1.1 from phenomenological and physiological points of view.

According to Fig. 3.31b, the amplitude of *heart sounds* is clearly modulated by respiration, with modulation disappearing when holding breath (compare Fig. 3.5b). During *inspiration* the modulation effects can be summarized as follows:

- *Intensification* of sounds from the right side of the heart, i.e., *right-sided heart sounds*, generated by closure of the right-sided tricuspid and pulmonary valve
- *Attenuation* of *left-sided heart sounds*, generated by closure of the left-sided mitral and aortic valve
- (Intensified) *Splitting* of the first and second heart sound

Generally, the changing volume of the lung influences the pressure conditions within the heart and those close to the heart, which in turn mechanically influences *intensity and timing* of the *valve's closure*. The dominance of the *right-sided heart sounds* can be explained by the increased right ventricular stroke volume during *inspiration* (section "Normal Respiration" under Sect. 3.2.1.1). The volume of the decelerated blood (ceased blood flow during the valve's closure) increases in the right side of the heart, which tends to fortify the intensity of the right-sided sounds, see generation mechanisms of heart sounds in Sect. 5. On the other hand, the amount of blood entering the left-sided chambers of the heart is decreased, which causes the *left-sided heart sounds* to decrease in intensity.

The reason for the *split heart sounds*, i.e., an *audible separation* between consecutive components within the first or second heart sound, can be attributed to the temporal increase in the right ventricular stroke volume during inspiration, which causes the pulmonary and tricuspid valves to stay open longer during ventricular systole (Sect. 2.4). Consequently, the delayed closure of the pulmonary valve gives rise to a *late sound* in the split second heart sound. In addition, a slightly earlier closure of the aortic valve due to the simultaneously decreased V_S yields an *early sound* component and thus a *gap between the early and late sound contributions*. As a result, the second heart sound is split more strongly at inspiration in comparison with expiration. Figure 3.32 illustrates the splitting of the second heart sound which is clearly widened during inspiration compared to expiration ($\Delta t_1 > \Delta t_2$ in Fig. 3.32b).

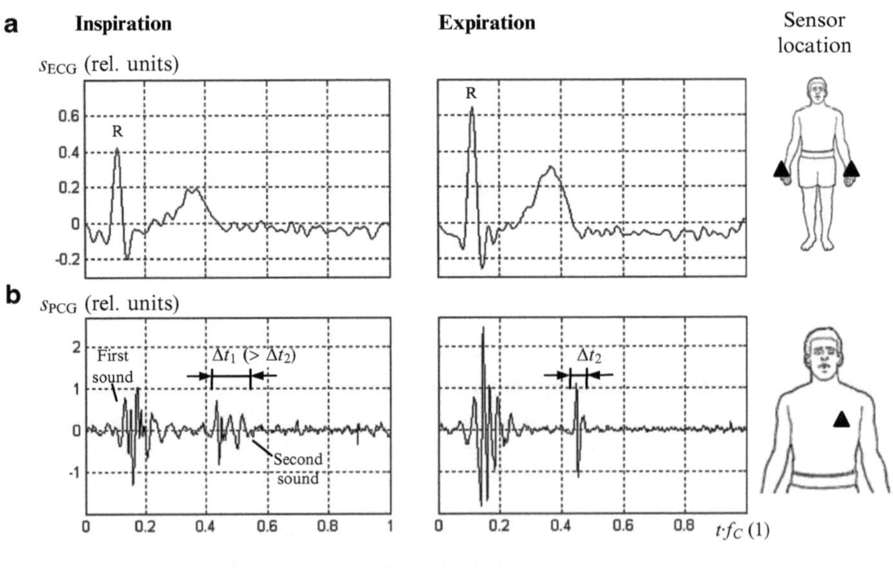

Fig. 3.32 Influence of respiration on splitting of the second heart sound. (**a**) Electric biosignal electrocardiogram s_{ECG} (lead I Einthoven) during inspiration and expiration. (**b**) Acoustic biosignal phonocardiogram s_{PCG} (from the heart region on the chest)

A closer look at the heart sounds in Fig. 3.31b reveals that the amplitude of the *first heart sound* (compare Fig. 3.2c) decreases during *inspiration* while that of the *second heart sound* increases, see the corresponding envelopes. That is, the amplification of the right-sided heart sounds is stronger in the second heart sound than the concurrent attenuation of the left-sided heart sounds. The reverse is true for the first heart sound. In particular, the decreased V_S and decreased left ventricular *contraction force* (compare Footnote 225) contribute to the attenuation of the first heart sound during inspiration, whereas an increased *pressure difference* between aortic pressure and left ventricular pressure (increased *afterload*) causes the second heart sound to be accentuated (Amit et al. 2009). In addition, the first and second heart sound was observed to be slightly delayed and advanced, respectively, during inspiration (Amit et al. 2009), which corresponds to the discussed changes in the stroke volumes.

However, the observed inspiratory behavior of both heart sounds may not generally be valid, as also reported in Amit et al. (2009). For instance, Fig. 3.5b demonstrates increasing first and second heart sounds during inspiration. Furthermore, decreased intensities of both heart sounds during inspiration were reported in Ishikawa and Tamura (1979) and can also be observed in Fig. 3.32b.

The deflection amplitude of the *optoplethysmogram* (Fig. 3.31c, d) shows a clear respiration-synchronous modulation. The deflection decreases with inspiration because of *inspiratory fall in* V_S and thus reduced pulsation of peripheral blood vessels. Similar amplitude modulation of the *optoplethysmogram* can be observed in Fig. 3.11d.

Shallow and *deep breathing* are compared in Fig. 3.33. Besides an electrocardiogram and mechanorespirogram, Fig. 3.33 includes a *mechanospirogram*, a *mechanic biosignal*, reflecting the effective oral *airflow*. The airflow (Fig. 3.33b) is positive during inspiration and negative during expiration, which corresponds to an increase and decrease in the abdominal circumference (Fig. 3.33c), respectively; compare Fig. 3.5a, c. In the case of *deep breathing*, the air flow and abdominal circumference changes are roughly doubled when compared to shallow breathing. In parallel, respiratory modulation of the electrocardiogram gets more pronounced and the amount of variation of the derived interbeat intervals roughly doubles, which exemplifies a *heightened respiratory sinus arrhythmia* during deep breathing. Not only deep breathing but also *slow breathing* potentiates the respiratory sinus arrhythmia; for instance, the oscillation amplitude of interbeat intervals increased from 40 to more than 80 ms (i.e., doubled) for $1/f_R = 4$ s and $1/f_R = 8$ s (doubled), according to Migeotte et al. (2009).

Figure 3.34 depicts the *transition from shallow to deep breathing* on the basis of abdominal circumference changes and *optoplethysmogram*. As soon as the breathing becomes deeper, the deflection amplitude of the *optoplethysmogram* is reduced due to a reduced pulsation of peripheral blood volume, compare with Fig. 3.31c, d. The reduced pulsation corresponds to an exaggerated decline of V_S during strengthened inspiration. Thus, deep breathing may diminish the peripheral pulse in terms of the *pulsus paradoxus* (Footnote 229).

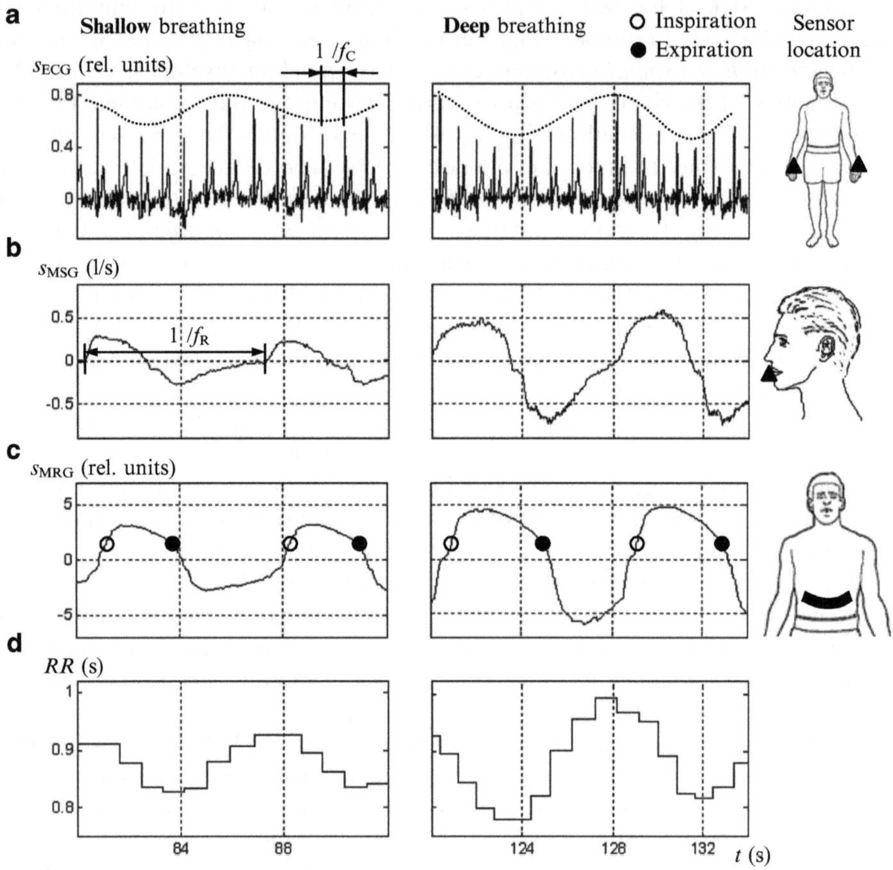

Fig. 3.33 Influence of shallow and deep breathing on diverse biosignals and their derivations. (**a**) Electric biosignal electrocardiogram s_{ECG} (lead I Einthoven). (**b**) Mechanic biosignal mechanospirogram s_{MSG} (from mouth airflow) with respiratory rate f_R indicated. (**c**) Mechanic biosignal mechanorespirogram s_{MRG} (from abdominal circumference changes). (**d**) Interbeat intervals RR ($= 1/f_C$ with f_C as instantaneous heart rate) derived from s_{ECG}

After *global behavior* of the different biosignals and their interrelations have been considered in the course of respiration, more *local changes* in the biosignal's waveform should be shortly discussed. Figure 3.35 demonstrates a conclusive example of a *blood pressure waveform*; in particular, respiratory changes of the waveform's *contour* and *duration* are shown from the end of expiration to the end of inspiration.

According to section "Pulse Waveforms of Pressure and Flow," the *waveform* of a pressure pulse is strongly affected by *reflection* phenomena which in turn depend on *arterial stiffness* and consequently on the actual level of *blood pressure* (compare Fig. 2.42b). Since the mean arterial blood pressure and p_S decrease

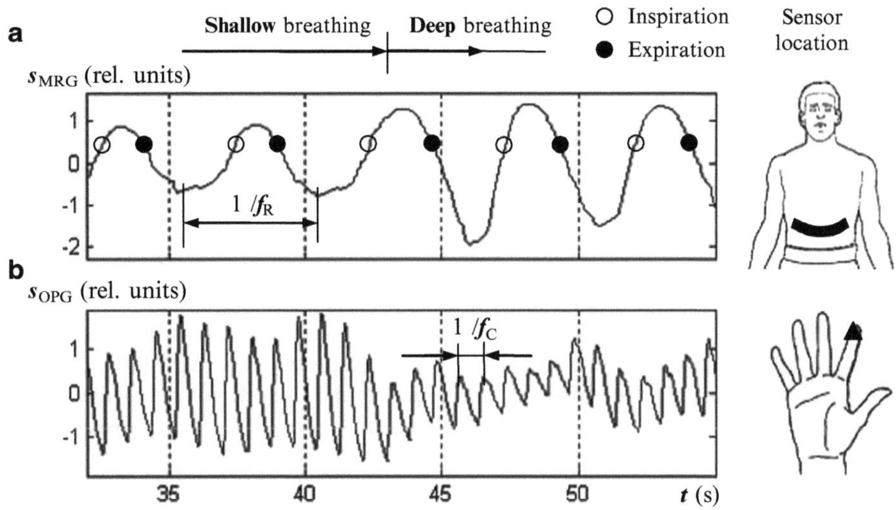

Fig. 3.34 Transition from shallow to deep breathing. (**a**) Mechanic biosignal mechanorespirogram s_{MRG} (from abdominal circumference changes) with indicated respiratory rate f_R. (**b**) Optic biosignal optoplethysmogram s_{OPG} (from a finger on the right hand) with indicated heart rate f_C

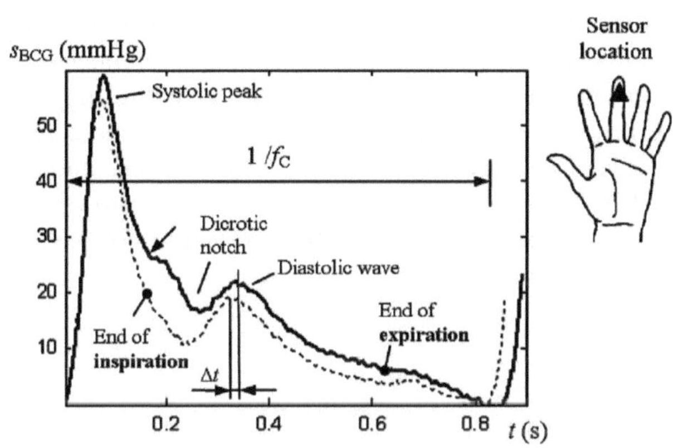

Fig. 3.35 Morphological changes in the pulse waveform of mechanic biosignal barocardiogram s_{BCG} (from a finger on the left hand) during the respiration cycle in a healthy young person with f_C as instantaneous heart rate. Offsets were subtracted from s_{BCG} to facilitate their comparison. Here, two cardiac pulses from Fig. 3.30a (indicated by *grey background*) are given, which respectively arise at the end of the inspiration phase and at the end of the expiration phase. An *arrow* marks an inflection point as an approximate start of the reflected wave

during inspiration (section "Normal Respiration" under Sect. 3.2.1.1), there must be corresponding changes in the waveform.

At the *end of inspiration*, see Fig. 3.35, there is no clear *inflection point* which would indicate the start of the reflected wave. The reflected wave propagates relatively slowly because of reduced blood pressure (section "Pulse Propagation") and seemingly merges with the pure (reflectionless) diastolic wave after the dicrotic notch. The presence of the *reflected wave* during *diastole* is supported by two additional observations. First, the amplitude of the total diastolic wave is higher at the end of inspiration than at the end of expiration; it indicates the presence of another superimposed wave besides the (reflectionless) diastolic wave. Second, the maximum of the total diastolic wave occurs earlier in time at the end of inspiration, see $\Delta t \neq 0$ in Fig. 3.35. The diastolic wave is advanced in time due to overlapping with the earlier reflected wave. By contrast, at the *end of expiration* a clear inflection point can be recognized, see arrow in Fig. 3.35. The inflection point indicates the onset of a fast-moving reflected wave, situated after the systolic peak but before the dicrotic notch. Furthermore, the *waveform duration* is shorter at the end of inspiration because of the respiratory sinus arrhythmia, as already shown in Figs. 3.30 and 3.33.

Besides respiration-synchronous changes of the blood pressure waveform, the *pulse waveform* of the *optoplethysmogram* is also subjected to respiratory impact, as can be observed in Figs. 3.31c, d and 3.36. The oscillations of blood pressure determine pulsatile extensions of peripheral blood vessels, though in a highly nonlinear way (Fig. 2.42b), and thus impact the absorption of incident light crossing pulsating blood vessels. Actually, numerous biosignals change their contour throughout the respiration cycle as soon as they are interrelated with the blood pressure, compare Fig. 3.14.

According to Fig. 3.31c, d, the *second wave* within a single cardiac cycle (the wave following the systolic peak, Fig. 3.31d) becomes prominent during inspiration phase, which confirms the discussion from above related to Fig. 3.35. In addition, Fig. 3.36 depicts different phases of the respiration cycle in more detail. The second wave is progressively delayed in time with *ongoing inspiration*, i.e., moves to the right in Fig. 3.36, which proves that the *reflected wave* slows down. Obviously, the reverse is true for expiration. As already observed in Fig. 3.35, the duration of the cardiac cycle is lowest at the end of inspiration.

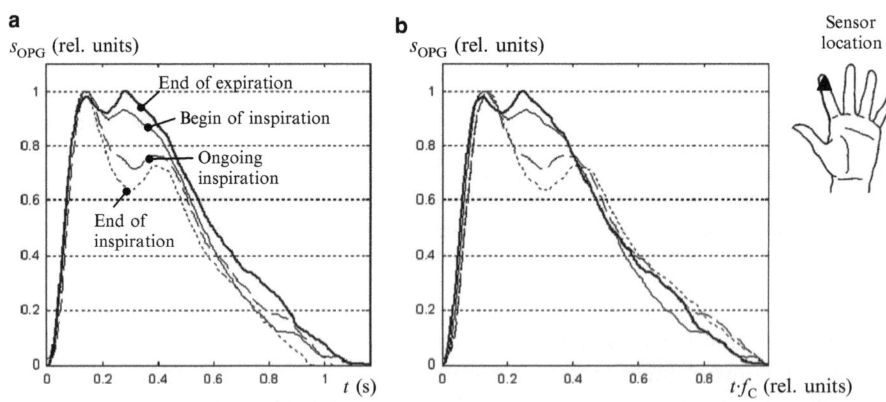

Fig. 3.36 Morphological changes in the pulse waveform of optic biosignal optoplethysmogram s_{OPG} (from a finger on the left hand) during a respiratory cycle in a healthy young person on (**a**) absolute time t-axis and (**b**) relative t-axis with f_C as instantaneous heart rate

Ceased Respiration

In this chapter, *cardiorespiratory interrelations* from section "Ceased Respiration" under Sect. 3.2.1.1 will be exemplified during voluntary *apnea*. Figure 3.37 demonstrates the course of abdominal circumference changes in synchrony with p_S and p_D, interbeat intervals, and *optoplethysmogram* recording. During apnea, the *respiratory component* of the circumference changes disappears, while the *residual cardiac component* remains, as indicated in Fig. 3.37a (compare Figs. 3.6a and 3.30e). The levels of p_S and p_D temporally increase during apnea and temporally drop below their pre-apneic levels, shortly after the resumption of breathing (Fig. 3.37b).

Respiratory sinus arrhythmia can be easily discerned in Fig. 3.37c when respiration—as a force modulating *interbeat intervals*—is effectively present, i.e., before and after the depicted apnea. During apnea, the heart rhythm speeds up a bit and then slows down in the second half of the apneic period, as also observed in Trzebski and Smietanowski (2001). Interestingly, a peak of f_C occurs after the resumption of breathing, probably as a compensatory reaction to counteract synchronous drops in p_S and p_D; compare Fig. 3.37b, c.

The envelope width of the *optoplethysmogram* (Fig. 3.37d) is markedly reduced in the second part of the apneic period as well as during breathing events terminating the apnea. The reduced variation of the pulsatile light absorption indicates an increased *peripheral vasoconstriction* and reduced pulsatile blood volume in the periphery. That is, *sympathetic activation* is suggested by decreased pulsatile deflection of the *optoplethysmogram* during apnea (Gil et al. 2009). A few breaths after the apnea terminates, a *pronounced vasodilation* of peripheral vessels can be recognized when compared with the pre-apneic period. The vasodilation may compensate for the temporally interrupted oxygen supply during the preceding apnea and *selected vasoconstriction* of peripheral vessels.

As illustrated in Fig. 3.20, the *oxygenation* level temporarily decreases during voluntary apnea. The hemoglobin oxygen saturation shows a temporal drop with a significant delay of about 20 s with respect to the interval of breath holding. In the case of *repetitive apneas*, as shown in Fig. 3.9, there is even a fluctuation of the oxygenation level from one apneic event to another; in addition, the level of f_C oscillates from bradycardia to tachycardia (section "Ceased Respiration" under Sect. 3.2.1.1).

There are numerous *other biosignals and parameters* which specifically vary in the course of apneic events and thus are relevant for diagnosis of the apnea. For instance, the *pulse running time*, a parameter which can be relatively easily estimated in an unobtrusive way (section "Estimation from Pulse Running Time"), decreases at the end of the apneic event due to enhanced sympathetic activation (Gil et al. 2009). This is because arterial stiffness increases which raises the velocity of propagating blood pressure waves and thus reduces the pulse (or wave) running time. To give another example, a changing *area of the QRS complex* from the electrocardiogram may facilitate apnea detection since it reflects the respiration cycle according to section "Normal Respiration" under Sect. 3.2.1.2. In addition, a strong *phase shift* occurs between the sequences of QRS areas from *different*

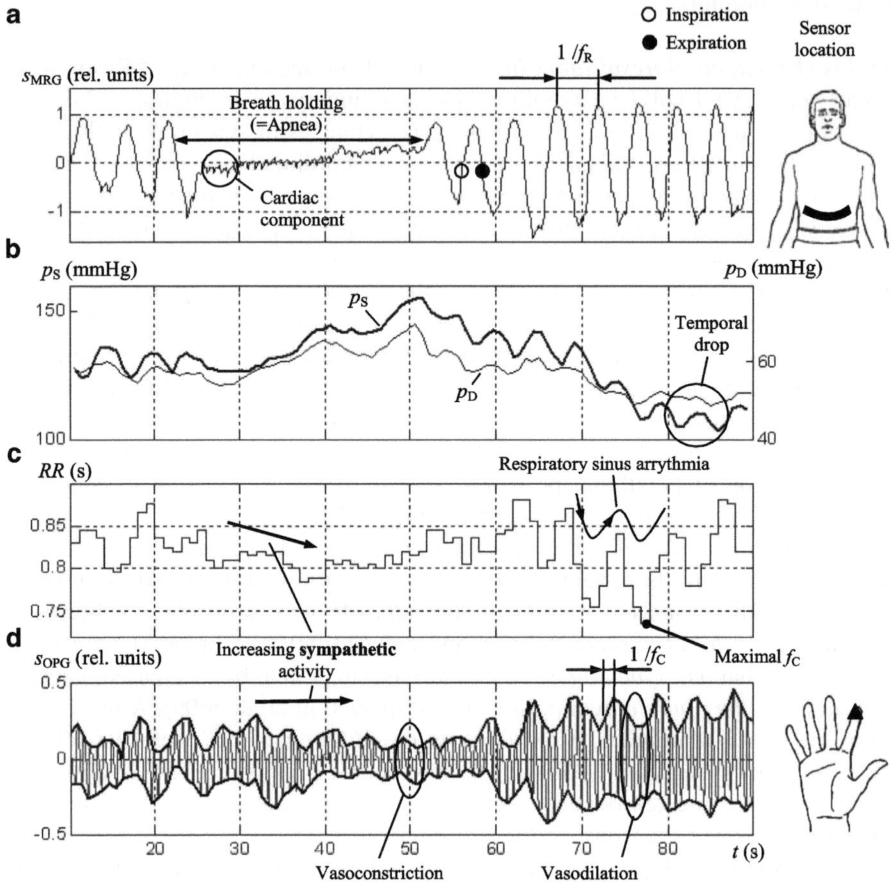

Fig. 3.37 Influence of voluntary breath holding on diverse biosignals and their derivations. (**a**) Mechanic biosignal mechanorespirogram s_{MRG} (from abdominal circumference changes) with indicated respiratory rate f_{R}. (**b**) Systolic blood pressure p_{S} and diastolic blood pressure p_{D} derived from mechanic biosignal barocardiogram (from a finger on the left hand); compare with Fig. 3.30a, c. (**c**) Interbeat intervals RR ($= 1/f_{\text{C}}$ with f_{C} as instantaneous heart rate) derived from electric biosignal electrocardiogram (lead I Einthoven); compare with Fig. 3.30b, d. (**d**) Optic biosignal optoplethysmogram s_{OPG} (from a finger on the right hand) enclosed by envelopes

electrocardiogram leads after the transition from (obstructive) apnea to normal breathing (Maier et al. 2009). The phase shift mainly results from a changing respiratory modulation of the electrocardiogram at the latter transition, whereas the modulation impacts change depending on the different body parts the leads are applied to. Namely, during *normal breathing* the chest and abdomen move nearly in phase while during *apnea* they move out of phase (*paradoxical respiration*, Fig. 3.9), compare Sect. 3.1.2 and section "Normal Respiration" under Sect. 3.2.1.2.

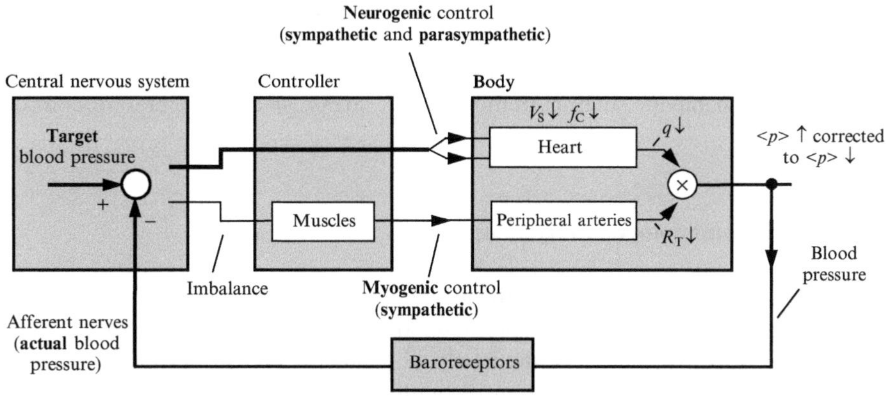

Fig. 3.38 Feedback-based control of blood pressure p (i.e., systolic, diastolic, and mean arterial pressure), compare Fig. 3.26. The responses of the left ventricular stroke volume V_S, heart rate f_C, cardiac output $q(= V_S \cdot f_C)$, see (2.30), and total peripheral resistance R_T are indicated for increased (imbalanced) mean arterial pressure $\langle p \rangle (= q \cdot R_T)$, see (2.20)

3.2.2 Cardiovascular Interrelations

While *cardiorespiratory interrelations* (Sect. 3.2.1) could be seen in the light of the body's attempts to maintain an appropriate *level of q* during the respiration cycle, *cardiovascular interrelations* mainly target maintaining an appropriate *blood pressure* with *neurocardiovascular reflexes*. However, the cardiorespiratory and cardiovascular systems cannot be strictly separated for they depend highly on each other. For instance, blood pressure is strongly *interrelated* with *respiration*, as abundantly noted in Sect. 3.2.1.

The cardiovascular system can be basically viewed as a *feedback-based and pressure-controlled system*, as illustrated in Fig. 3.38. The *target level* of blood pressure is determined by the central nervous system, whereas the target is related to all important characteristics of the blood pressure, i.e., p_S, p_D, and mean arterial pressure (section "Pulse Waveforms of Pressure and Flow"). A potential imbalance between the target and actual levels serves as the input to a controller. The *controller* mainly comprises *neurogenic control* and *myogenic control*, i.e., pressure control throughout the autonomic nervous system and smooth muscles, respectively. In the former case the *target organ* of the controller is the *heart*, while in the latter case it is the walls of *peripheral arteries*. The levels of q and R_T are adjusted, respectively, whose product mainly determines the actual (mean arterial) blood pressure (2.20). *Stretch-sensitive receptors* (baroreceptors) monitor the actual blood pressure and relay this information as *feedback* to the central nervous system. If there is a misalignment between the target and actual blood pressure, a readjustment of neurogenic or myogenic actions is performed.

Moreover, *chemoreceptors* monitoring the *oxygen* and *carbon dioxide* levels in blood (Footnote 238) may also play a significant role in the control of the *cardiovascular* system. For instance, a reduced oxygen level usually evokes a rise in q, compare peaks of q at the respiratory arousal terminating apnea (section "Ceased Respiration" under Sect. 3.2.1.1). In addition, many *more phenomena* influence the cardiovascular behavior such as humoral state, stress, sleep, temperature control, and others.

3.2.2.1 Phenomenological Physiology

Arterial blood pressure underlies the biological control system that provides a *short-term*[244] *regulation* of blood pressure. If summarized from section "Normal Respiration" under Sect. 3.2.1.1, the levels of p_S and *mean arterial blood pressure* fluctuate throughout the *respiratory cycle* with a time shift of a few seconds, falling with inspiration and rising with expiration. This behavior can be explained by intrathoracic pressure changes during breathing and accompanying neuronal control from the central nervous system. By contrast, the level of p_D does not markedly change with respiration because duration changes of the diastolic phase tend to compensate for respiratory changes of V_S and p_S (section "Pulse Waveforms of Pressure and Flow").

Generally, *passive* and *active* mechanisms contribute to the *blood pressure control*:

- *Cushioning reservoir* of large arteries (or Windkessel, Sect. 2.5.2.1)—a *passive mechanism*—determines that the accumulated (exponential) decrease in blood pressure during diastolic phase is proportional to the duration of the actual interbeat interval because there is more time for blood to flow out of the arteries (section "Pulse Waveforms of Pressure and Flow"). In consequence, the level of p_D progressively decreases with increasing *pulse interval* ($= 1/f_C$).
- *Contractility of heart muscles*—an *active mechanism*—evokes an increase in the strength of the following ventricular contraction after a relatively long pulse interval ($= 1/f_C$), according to Frank–Starling law (Footnote 225). The *end-diastolic filling*[245] of the ventricles is augmented after the long pulse interval, leading to a more forceful ventricular contraction and thus to an increase in the *systolic–diastolic pressure* (or pulse pressure $p_S - p_D$). In addition, the ventricles are more relaxed after a longer filling phase, which tends to fortify the following contraction even more. In consequence, the systolic–diastolic pressure increases with increasing *pulse interval*.
- *Baroreflex*, i.e., an *active mechanism* of blood pressure control, is based on a negative feedback loop (Fig. 3.38) with all characteristics of blood pressure being controlled: p_S, p_D, and *mean arterial pressure*.

---

[244] The *long-term regulation* of arterial blood pressure relies on specific *slow* hormonal and renal mechanisms which primarily affect *blood volume* (Silbernagl and Despopoulos 2007).

[245] It should be noted that *end-diastolic volume* (proportional to V_S) is a major *limiting factor* of increased q when f_C is elevated (2.30).

Baroreflex yields efficient blood pressure regulation[246] within a *few cardiac cycles* with a *resolution* of about <5 mm Hg (Zemaityte 1997). The main target of the regulation is *buffering* of temporal blood pressure changes by an appropriate governing

- *Sympathetic* nervous activity (Footnote 188)
- *Parasympathetic* nervous activity (Footnote 189)

compare Fig. 3.38 and Sect. 3.1.1. If an *increased blood pressure* is assumed, mechanical receptors for monitoring the blood pressure level become progressively activated; compare section "Normal Respiration" under Sect. 3.2.1.1. The latter *baroreceptors* are mainly located in arteries and veins, as detailed below. The frequency of action potentials of baroreceptors increases, with this information being forwarded to the central nervous system, namely, the cardiovascular control centre in the brain. In order to lower (i.e., normalize) the blood pressure, the levels of $q(= V_S \cdot f_C)$, see (2.30), and/or R_T have to be reduced by efferent sympathetic and parasympathetic signals from the central nervous system. The corresponding efferent signals are directed toward *active tissues* of the cardiovascular system, namely, the *heart* and *peripheral vessels*. Here, it should be noted that q and R_T comprise the only control parameters for *active short-term control* of the blood pressure, see section "Pulse Waveforms of Pressure and Flow."

In response to increased blood pressure, the *level of q* is reduced by appropriate neuronal *sympathetic* and *parasympathetic* signals to the *heart muscle* and *sinoatrial node* that reduce *contractility of the heart* (reduce V_S) and decrease f_C, respectively. The latter *neurogenic control* is shown in Fig. 3.38. On the other hand, the level of R_T is reduced by appropriate neuronal *sympathetic* signals to *smooth muscles* that withdraw vasoconstriction of blood vessels. The constriction of arterial and venous diameters is weakened, allowing for more blood to flow out of the (large) arteries (Sect. 2.5.2.2); this effect is summarized under *myogenic control* in Fig. 3.38.

In terms of physiology, the *sympathetic activity* has to be inhibited and/or *parasympathetic activity* accelerated to compensate for the increased blood pressure. Both branches of the nervous system actually have opposing but *complementary effects* on blood pressure within the scope of baroreflex (compare Footnote 190); consequently, both branches should always be *considered together* in their complementary effects.

Generally, *accelerated sympathetic activity increases blood pressure* through the following mechanisms:

- Elevation of R_T via increased *vasoconstriction* of peripheral vessels
- Elevation of V_S [and q, see (2.30)] via strengthened *contractility* of the heart muscles

[246] It is interesting to note that while *baroreflex* controls the systemic *blood pressure*, i.e., increasing pressure leading to decreased f_C (see text), the *Bainbridge reflex* (Footnote 232) controls the *blood volume*, i.e., increasing volume leading to increased f_C.

- Elevation of f_C (and q) via (1) activation of beta receptors in the *sinoatrial node* of the heart, quickening the pace of the heart, (2) increased propagation speed of the excitation through the *atrioventricular node* in the heart (see Sect. 2.4.2), and (3) increased rate of relaxation of the *heart muscles*.

By contrast, an *accelerated parasympathetic activity decreases blood pressure* through the following mechanisms:

- *Lowering of* f_C (and q) via (1) increased activation of the efferent *vagus nerve*, i.e., increased impulse traffic (or firing rate) of the nerve toward the *sinoatrial node* and (2) slowing of the propagation speed of the excitation through the *atrioventricular node*
- *Lowering* of *contractility* of the right *atrium*

It should be stressed that when the parasympathetic activity is rising, the sympathetic activity simultaneously decreases; the *antagonistic impact* of the dominant parasympathetic activity diminishes the prevailing sympathetic level, see Footnote 190. In a first approximation, *sympathetic activation* dominates in the *speripheral vessels* and *heart muscles* mainly determining R_T and V_S, whereas *parasympathetic activation* dominates in the modulation of the *sinoatrial node* mainly determining f_C.

It is worth noting that *sympathetic* acceleration is followed by a relatively *slow increase* in f_C while *parasympathetic* acceleration causes a relatively *rapid decrease* in f_C; thus, *reaction speeds* differ for both branches of the nervous system (compare Footnote 190). In other words, an increase in blood pressure initiates a more rapid slowing of the heartbeat than its quickening with a pressure decrease. This yields an *asymmetric behavior* between changes in blood pressure and f_C.

In terms of the *baroreflex* control of the *blood pressure characteristics* (section "Pulse Waveforms of Pressure and Flow"),

- The *level of* p_S is mainly adjusted via V_S, whereas both *sympathetic and parasympathetic* nervous systems are involved in an antagonistic way (Footnote 190). Actually, there is a strong relationship between p_S and f_C because of the *antagonistic control*. For instance, the balancing of a temporally increased blood pressure (e.g., due to increased sympathetic and impeded parasympathetic activities) parallels a diminished f_C because of decelerated sympathetic and *reciprocally* accelerated parasympathetic activities, as a *baroreflex response*.
- The *level of* p_D is mainly controlled via R_T throughout the *sympathetic* system. Obviously, changes of f_C also impact p_D, as already mentioned in the case of the cushioning reservoir.
- Correspondingly, the *mean arterial pressure* is controlled by V_S, f_C [or q, (2.30)] and R_T (2.20), since it represents a surrogate measure for p_S, p_D, and pressure pulse waveform (compare Footnote 140).

Baroreceptors respond to stretching of vessels walls so that if pressure rises, the walls expand and the *stretch-sensitive mechanoreceptors* (primary receptors, Sect. 2.2.1) within increase their firing rate. Not only is the mean arterial pressure registered but also the rate of pressure change, i.e., the *receptor's response* is modulated[247] by *tonic* and *reflexive* (or phasic) components of blood pressure. The receptors are located in the *low-pressure* and *high-pressure* parts of the circulatory system, involving *venous* and *arterial* vessels, compare Fig. 2.39. Low-pressure baroreceptors are found in large systemic *veins* and in the *right atrium* of the heart; these baroreceptors are mainly involved in the regulation of *blood volume* in the (highly elastic) venous side where most of the blood is stored (Sect. 2.5.1). High-pressure baroreceptors are largely embedded within the *carotid sinus*[248] and *aortic arch* to monitor the pressure of the blood flow delivered to the brain and the rest of the body, respectively. In fact, the carotid sinus is quantitatively the most important location for regulating arterial pressure around its normal mean level of about 100 mm Hg (Zemaityte 1997).

It has been shown that blood pressure control has a significant impact on f_C. However, the *target blood pressure* cannot immediately become the *actual blood pressure*, as is sensed by the baroreceptors, compare Fig. 3.38. There is an *intrinsic delay* in *hemodynamic response*, afferent–efferent *neural pathways*, central processing, and effector response. In particular, the delay dominates in the *sympathetic part* of the control loop. In addition, though sympathetic nerves rapidly conduct targeted changes of smooth muscle contraction toward peripheral arteries, the targeted mechanical effects of vasoconstriction or vasodilation need a certain amount of time to be developed. Thus, the *compensatory effect* (or regulating effect) of the control loop is *not immediate* but comes up with a *latency* of a few beats or after 2–5 s (Seydnejad and Kitney 2001). The levels of q and R_T are balanced with some delay to normalize blood pressure (Fig. 3.38). The delay inevitably provokes *oscillations of blood pressure* and f_C around their target values - oscillations of blood pressure always leading - because the target blood pressure always lags behind the actual blood pressure. Consequently, *resonant behaviors* of blood pressure and f_C become prominent.

The levels of blood pressure and f_C are resonantly modulated at about 0.1 Hz, i.e., the modulation rate is usually lower than the normal respiratory rate; compare Fig. 3.4a from Sect. 3.1.1. In particular, it is called a *Mayer wave arrhythmia* (Seydnejad and Kitney 2001) and is generally brought about by

- Oscillations in control systems with baroreceptors and chemoreceptors in *feedback loops*
- *Centrogenic rhythm* in the brain stem with interconnection to respiratory oscillators

[247]The *baroreceptors* comprise two types of nerve fibers, myelinated fibers for a rapid impulse conduction, the so-called *A-delta* fibres, and unmyelinated *C fibres* for a relatively slow impulse conduction, compare Sect. 2.1.4.1. In fact, the A-delta fibres mainly transmit *rapid changes* of blood pressure while the C fibres the (relatively) *slowly changing mean* arterial pressure. Interestingly, the efferent vagus nerve is more strongly controlled by the C fibres in comparison with the A-delta fibres (Zemaityte 1997) which seems to confirm the relevance of the *mean arterial pressure control*, compare section "Normal Respiration" under Sect. 3.2.1.1.

[248]The *carotid sinus* refers to the dilation of the internal *carotid artery* close to the bifurcation of external and internal carotids, compare Fig. 2.40.

- *Autorhythmicity* of vascular smooth muscles

Typically, an *increased baroreflex sensitivity*—for its experimental assessment see Sect. 3.2.2.2—accounts for a slow *heartbeat* (low f_C), increased HRV, and consequently accounts for decreased *blood pressure variability* and even decreased *mean arterial blood pressure*. Moreover, the increased baroreflex sensitivity may facilitate protection of the heart from temporal *pathologies* such as arrhythmic events (Zemaityte 1997).

The aforementioned regulating *cardiovascular mechanisms* are active in the *normal state* with physiological parameters kept in an appropriate *balance*, e.g., in terms of continuous q or balanced blood supply. In addition, numerous cardiovascular interrelations are only triggered by *abnormal events* (e.g., ectopic beats) aiming to compensate for the event's abnormal influence. For instance, such interrelations include the so-called *heart rate turbulence* which refers to a short-term fluctuation in f_C triggered by a single ventricular *ectopic beat* (Smith et al. 2010). That is, the ectopic beat induces a drop in p (and q) which is compensated by increased instantaneous f_C within a few normal beats following the ectopic beat [compare (2.30)]; then the heartbeat decelerates to its original level. Thus, the heart rate turbulence is a blood pressure *regulating mechanism*; likewise, the absence of turbulence after an ectopic beat indicates autonomic dysfunction.

Healthy aging is marked by increased *variability of blood pressure* because of reduced sensitivity of chemoreceptors and baroreceptors. Baroreflexes show a decreased ability to buffer changes in systemic blood pressure, whereas the ability to rapidly accelerate the heartbeat in response to acute alterations of blood pressure decreases as well. The arterial *baroreflex weakens* in the elderly partly because of vascular stiffening which decreases arterial compliance in regions containing baroreceptors. In addition, progressing neural deficits, reduced responsiveness of the sinoatrial nodc, and oxidative stress contribute to progressive impairment of the baroreflex (Monahan 2007). For instance, there is a greater incidence of hypotension with advanced age during the application of physiological stressors.

The *cardiovagal component* of the baroreflex sensitivity, i.e., vagally mediated reflex response (see Sect. 3.2.2.2), becomes impaired with *advancing age*. That is, cardiovagal sensitivity and age are inversely related; for instance, the sensitivity in elderly persons (60–80 years old) is about 7 ms/mm Hg and amounts to less than half the value among young persons (20–40 years old) of about 16 ms/mm Hg (Monahan 2007). Moreover, the sympathetic branch influencing f_C also seems to be impaired with age, because of the aforementioned antagonistic control. By contrast, the *sympathetic component* reflecting sympathetic outflow toward peripheral smooth muscles seems not to be impaired with age (Monahan 2007) and can be well maintained in healthy individuals even into the seventh decade (Ebert et al. 1992).

3.2.2.2 Biosignals and Parameters

The quantitative assessment of *baroreflex sensitivity* (efficiency) is highly relevant, for it reflects the status of autonomic regulatory functions. From an experimental point of view, two complementary baroreflex sensitivities can be differentiated and estimated:

- *Cardiovagal sensitivity*
- *Sympathetic sensitivity*

Cardiovagal sensitivity describes the *vagally mediated* reflex response of the cardiac period ($= 1/f_{\mathrm{C}}$). That is, induced changes of interbeat intervals are evaluated to be a *reflex response* while changes of p_{S} or mean arterial pressure can be seen as a *reflex stimulus*. It should be stressed that the relationship between the response and stimulus reflects the balance of *parasympathetic and sympathetic* activities, i.e., reflects their antagonistic behavior. Generally, as described in Sect. 3.2.2.1, the level of f_{C} *decreases to counteract increased blood pressure*.

A similar but more specific approach in the assessment of cardiovagal sensitivity comprises separate experimental evaluation of

- *The mechanical component* of cardiovagal sensitivity, revealing the mechanical relation between p_{S} (as the *stimulus*) and the systolic diameter of the carotid artery
- *The neural component* of cardiovagal sensitivity, reflecting the neuronal pathway from the systolic diameter of the carotid artery (with embedded *baroreceptors*) to the interbeat interval (as the *response*)

The product of the mechanical component and the neural component yields the integrated *cardiovagal sensitivity*, quantified as the ratio of the interbeat interval to p_{S} (Hunt et al. 2001). The cardiovagal sensitivity appears to exhibit *hysteresis*, i.e., the sensitivity shows higher values when blood pressure is rising than falling (Rudas et al. 1999). The hysteresis may arise due to the *different reaction speeds* of sympathetic and parasympathetic systems (Sect. 3.2.2.1).

It should be noted that the *cardiovagal sensitivity* varies over the *respiratory cycle*, since f_{C} (as reflex response) is also strongly influenced by respiration phase (section "Normal Respiration" under Sect. 3.2.1.1). The sensitivity is higher for expiration while decelerating the heartbeat (Zemaityte 1997), which may be also ascribed to the *different reaction speeds* of both nervous branches.

The *sympathetic sensitivity* accounts for sympathetic outflow to peripheral smooth muscles, which yields *vasoactive effects*. That is, sympathetically induced muscle activity is evaluated as a *reflex response* in relation to p_{D} as a *reflex stimulus*. The muscle activity can be experimentally derived from an electrode inserted into the peroneal nerve. Instead of the sympathetic outflow, peripheral q and R_{T} can be used as *surrogate measures*; in fact, the levels of q and R_{T} are inversely and directly related to the sympathetic activity, respectively. As described in Sect. 3.2.2.1, the level of vasoconstriction is lowered to counteract an increased blood pressure. In other words, there is an *inverse relationship between the sympathetic outflow and* p_{D}, in which a decrease in the sympathetic outflow dilates arterial vessels and compensates for increased p_{D}.

From an *experimental* point of view, assessment of baroreflex sensitivity requires

- *Natural* (spontaneous) blood pressure changes or
- *Induced* blood pressure changes

Natural or *spontaneous changes* arise during *respiration* and allow for the estimation of *cardiovagal sensitivity*. As described in section "Normal Respiration" under Sect. 3.2.1.1 and shown in Fig. 3.30c, d, changes in heartbeat intervals are synchronized with spontaneous beat-to-beat changes in blood pressure during the respiratory cycle. Namely, shortened heartbeat intervals are associated with decreased p_S during inspiration because of the *baroreceptor reflex*. The corresponding *quantitative assessment* of baroreflex sensitivity can be performed in

- Time domain
- Frequency domain

In the *time domain*, spontaneous and monotonic *up sequence (or down sequence) of interbeat intervals* throughout three or more consecutive heartbeats is considered for numeric analysis, which timely corresponds to up sequence (or down sequence) of p_S. That is, continuous changes of interbeat intervals are depicted over the corresponding continuous evolution of p_S. The resulting *gain* (or slope) quantifies the *cardiovagal sensitivity* with its typical values of about 10 ms/mm Hg. In addition to this static property (gain), also a dynamic property like the *time delay of the reflex loop* can be experimentally determined; for instance, as the delay between corresponding up sequences of interbeat intervals and p_S values.

The fluctuations in interbeat intervals and p_S are transformed from the time domain into the *frequency domain*, either by the *Fourier transformation* (Footnote 150) or by the calculation of the power spectral density (Footnote 193). The spectral transfer function is estimated between interbeat intervals and p_S. The corresponding *gains* of the *transfer function* at *specific frequencies*—usually around 0.1 Hz corresponding to Mayer waves (Sect. 3.2.2.1) and around the respiratory rate—facilitate a quantitative estimation of the *cardiovagal sensitivity*.

The *induced changes* of blood pressure are usually based on pharmacological interventions which target *vasoactive manipulation* of blood pressure. In addition, different *breathing maneuvers* can be performed to vary blood pressure in a limited range, e.g., the Valsalva maneuver (Kamiya et al. 2000), see Footnote 173. During the induced changes, both *cardiovagal and sympathetic sensitivities* can be assessed by measuring *reflex changes* in *pulse interval* (i.e., ratio of interbeat interval to p_S) and *smooth muscle* activity (i.e., ratio of muscle activity to p_D), respectively. In particular, sympathetic sensitivity is quantitatively assessed as the *gain* (or slope) between continuous *down sequences (or up sequences) of the sympathetic outflow* and the corresponding continuous *up sequences (or down sequences) of* p_D.

Figure 3.39 illustrates experimental data on the cardiovagal and sympathetic sensitivities of the baroreflex. The *cardiovagal sensitivity* is demonstrated during *normal respiration* (Fig. 3.39a). Three spontaneous up and down sequences of interbeat intervals and the corresponding p_S are shown throughout four up to five heartbeats. It can be observed that the mean gain of the depicted sequences amounts to about 5 ms/mm Hg.

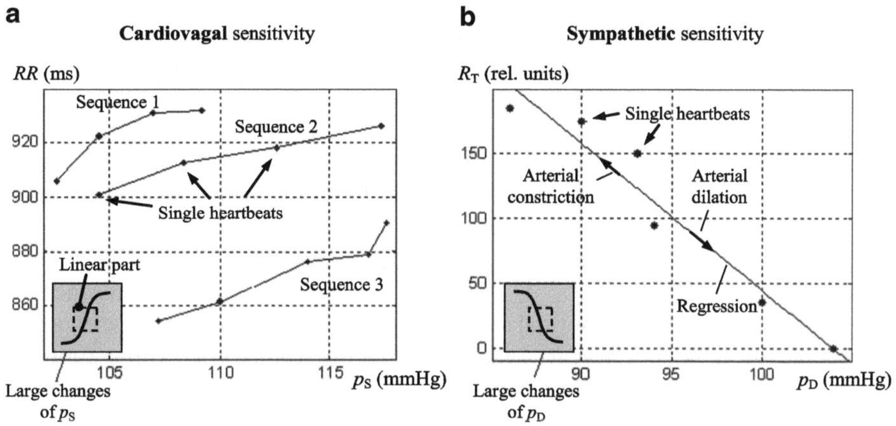

Fig. 3.39 Baroreflex sensitivity. (**a**) Cardiovagal sensitivity during normal breathing. The interbeat intervals RR (as baroreflex response) derived from electrocardiogram (lead I Einthoven) are depicted over systolic blood pressure p_S (as baroreflex stimulus) derived from barocardiogram (from a finger on the left hand); compare Fig. 3.30a, b. Three spontaneous up and down sequences of RR and p_S are shown during four to five consecutive heartbeats. (**b**) Sympathetic sensitivity during the Valsalva maneuver (Footnote 173) which provokes a sympathetically induced variation of diastolic blood pressure p_D. The peripheral resistance R_T (as response) estimated from optoplethysmogram (from a finger on the right hand) is depicted over p_D (as stimulus) derived from the barocardiogram. In both cases (**a**) and (**b**), sigmoid patterns are indicated, which would arise for large changes of the stimuli p_S and p_D

Sympathetic sensitivity was experimentally assessed during the *Valsalva maneuver* with R_T being estimated as the baroreflex response (Fig. 3.39b). In particular, the values of R_T and p_D were taken during the early period of blood pressure decrease during the Valsalva maneuver (Kamiya et al. 2000). The level of R_T was estimated from the *optoplethysmogram*, reflecting absorption of incident light by blood volume in the finger, compare Figs. 3.31c, d or 3.36. The estimation was based on the fact that the *baseline* of the *optoplethysmogram* is related to local nonpulsatile blood volume in the finger (Nitzan et al. 1998). This volume decreases with constriction and increases with relaxation of blood vessels innervated by sympathetic nerves. As expected, the level of R_T in Fig. 3.39b decreases with increasing p_D. The superimposed arrows over the regression line indicate regulatory *constriction* and *dilation* in response to decreased p_D and increased p_D, respectively.

Actually Fig. 3.39 depicts only the *linear parts* of the baroreflex-mediated responses pertaining to interbeat interval ($= 1/f_C$) and R_T. The applied *stimulus* here—given by changes of *blood pressure*—is relatively *weak*; the stimulus is only limited to *spontaneous* physiological perturbation which is induced by *normal respiration*. For a large blood pressure stimulus, the *cardiovagal sensitivity* demonstrates a *sigmoid pattern* while the *sympathetic sensitivity* an *inverted sigmoid pattern*; see sigmoid patterns in Fig. 3.39 (Monahan 2007; Rea and Eckberg 1987; Moser et al. 2008).

Figure 3.40 illustrates the *cardiovagal sensitivity* for *large changes of blood pressure*. In comparison with Fig. 3.39a, the sigmoid pattern is reversed because interbeat intervals (Fig. 3.39a) are substituted by their inverse characteristic f_C (Fig. 3.40). As discussed in Sect. 3.2.2.1, the delayed compensatory effect of the baroreflex yields an *oscillatory response* in f_C (Mayer waves) if an imbalanced p_S is given as (*oscillatory*) *stimulus*. The resultant oscillations in f_C comprise an inherent part of the *heart rate variability* (Sect. 3.1.1). Actually, the steepest point of this regulatory curve or the *highest gain* of the cardiovagal sensitivity is located near the *normal* p_S (Monahan 2007; Rea and Eckberg 1987), see Fig. 3.40. Thus, a preferably *high variability of* f_C indicates the relevance (and the presence) of *reflexive baroreceptors*. The normal regulation of the blood pressure can be expected to become unstable if afferent signaling of baroreceptors concerning reflexive changes of blood pressure disappeared. In other words, the highest gain assures *stability* of the blood pressure regulation and the *lowest variability of the blood pressure*. The *highest gain* implies that relatively large f_C changes compensate for relatively small p_S changes.

Changes with *aging* of the *cardiovagal regulatory curve* are indicated in Fig. 3.40, which account for the impact of aging on cardiovascular structure and function (Sect. 3.1.3). The curve is shifted to larger p_S values and becomes less steep, indicating a *reduced cardiovagal sensitivity* (Sect. 3.2.2.1). In consequence, the mean arterial blood pressure and variability of blood pressure are larger for the aged population if an identical HRV is assumed for both regulatory curves; compare larger Δp_S^2 for the aged population in relation to Δp_S^1 in Fig. 3.40.

Baroreflex sensitivity is a highly *sensitive diagnostic measure* reflecting actual physiologic state and general health. To exemplify this, it is worth recalling that *respiration* with a cyclic period of less than 10 s *modulates the baroreflex sensitivity* even within this short time frame.

Generally, the *baroreflex sensitivity* decreases with *standing*, physical, and mental *exercise* (Zemaityte 1997). In the latter cases, metabolic needs mainly determine

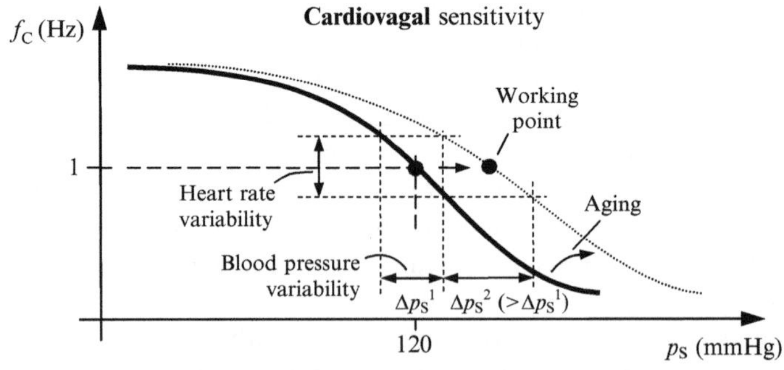

Fig. 3.40 Schematic cardiovagal sensitivity illustrated by the relation between heart rate f_C and systolic blood pressure p_S; compare Fig. 3.39a. The impact of aging is indicated for the above relation and changing blood pressure variability Δp_S

blood pressure, f_C, and other parameters. For increased stress, the working point from Fig. 3.40 is dislocated to another less steep regulatory curve, whereas f_C loses its variability and its relevance as the control parameter for the short-term regulation of the blood pressure. *Sickness* usually decreases the sensitivity as well; for instance, *hypertension* (Footnote 166) moves the curves from Fig. 3.39a to the right while the gain is reduced, similar to the aging impact from Fig. 3.40. Interestingly, the *baroreflex sensitivity* increases during *sleep* and is depressed in OSA. In particular, the *sympathetic sensitivity* of the baroreflex tends to lower during sleep, especially in the NREM phase (Sect. 3.2.4). Conversely, the *cardiovagal sensitivity* has a tendency to increase during the latter sleep phase (Nakazatoa et al. 1998).

3.2.3 Biological Rhythms

Cyclic or periodic phenomena of living organisms or organs are known as biological rhythms. These rhythms are ways to *integrate* and *coordinate* body functions and to anticipate *environmental rhythms* around the body. Tuning and synchronization of rhythms *reduces energy* needs, especially during rest or sleep (see also Sect. 3.2). The necessity of these rhythms can be explained by the fact that the organism needs to give a special *performance* and, on the other hand, operating efficiency should be assured by *regeneration*, given the *limited space* in a cell, organs, and body. Both aims cannot be attained at the same time and space; thus, a *temporal compartmentalisation* is needed to allow different environments to occur in the *same space but at different times* (Moser et al. 2008). To give some examples, inspiration and expiration, systole and diastole, wakefulness and sleep cannot arise efficiently at the same time. A *rhythmic stratification* is needed which encompasses different time scales, as shown in the following explanation.

Generally, there are two types of biological rhythms, exogenous and endogenous. The *exogenous rhythms* are controlled directly by the environment around the body, e.g., the presence of light, while the *endogenous rhythms* are driven by internal biological clocks. In practice, mixed types of rhythms occur with the biological clocks *entrained* by different timers, as described below. Figure 3.41 demonstrates different rhythms over different time scales by a *multiclock of live*:

- *Less than 24-h rhythm*: the shortest cycles originate from *ultradian rhythms* (latin *ultra* more than and *dies* day), including
  - *Millisecond* rhythms which become apparent in the modulation of neuronal and muscular actions as derived from an electroencephalogram or modulation of heart sounds (compare phonocardiogram in Fig. 1.15c).
  - *Second* rhythms yielding, for instance, modulation of blood pressure in vessels, known as propagating pulse waves along the vessels (*optoplethysmogram* in Fig. 1.15c).
  - *Minute* rhythms which manifest as recurrent modulation of biochemical, physiological, psychological parameters, such as hormonal activity, blood

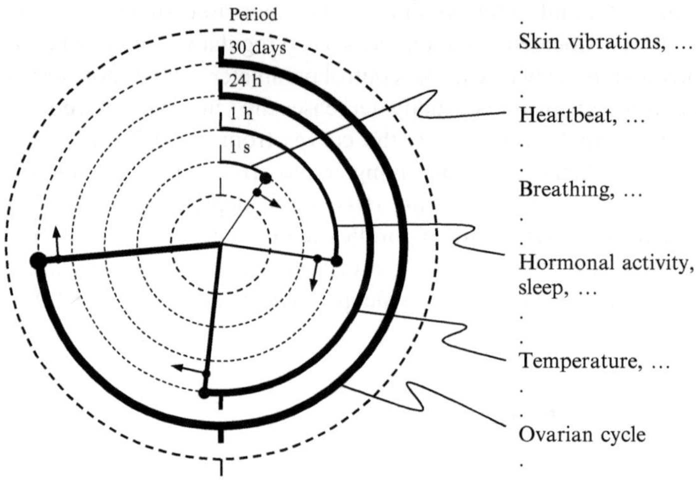

Fig. 3.41 Biological rhythms over different time scales—a multiclock of life

pressure, blood perfusion, heart rate, activation of smooth muscles of arterial vessels, and breathing (mechanorespirogram in Fig. 1.15c).
  - *Hour* rhythms, e.g., cycling of the sleep stages during the night (compare Fig. 3.48) and change of the pain threshold during the day with a period of about 2 h.

- *24-h rhythm*: the *circadian rhythms* (latin *circa* around and *dies* day) comprising biochemical, physiological, and psychological parameters, such as alertness, body temperature, growth hormone, and cortisol (compare core body temperature in Fig. 1.15b)
- *More than 24-h rhythm*: the *infradian rhythms* (latin *infra* below and *dies* day) showing periods greater than the circadian rhythm, including
  - *Week* rhythms lasting for about 7 days, e.g., periodic swelling of wounds, fever increase, immune reactions, and diverse physiological parameters (e.g., blood pressure) in premature infants and newborns.
  - *Month* rhythms lasting for about a month, e.g., menstruation cycle in the human adult female or lunar-related changes of color sensitivity in the human eye.
  - *Year* rhythms with cycle durations of about a year, e.g., human adaptation to environmental changes.

In particular, the *ultradian rhythms* allow the body to harmonize quickly with *superior (body) rhythms*, e.g., adaptation of heart rate, stroke volume, or blood pressure to the superior breathing cycle, see Sect. 3.2.1. In many cases such *harmonization* is vitally important for the human body in order to continuously *adapt* to environmental changes (within the body) by minimizing signs of wear and *energy expenditure*. For instance, an increase in the heart rate during inspiration counterbalances a concurrent decrease in the left ventricular stroke volume in order to keep the blood flow constant (Sect. 3.2.1). The latter example also demonstrates that fast rhythms, e.g., heartbeat, may be seen as *sampler* of slow rhythms, e.g., respiration, with information on the slower rhythms embedded into that of the faster rhythms.

In contrast to the ultradian rhythms, the *infradian rhythms* encompass harmonization of the body with relatively long *environmental rhythms*. For instance, the environmental rhythms are due to the monthly rotation of the moon around the earth or yearly rotation of the earth around the sun. In the former case, the monthly menstruation cycle is a prominent example while in the latter case, an adaptation of the body to seasonal changes of temperature and light occurs, which is obviously more prevalent in animals living in the wild than in modern humans.

In between the ultradian and infradian rhythms, there are the *circadian rhythms*. The circadian rhythms reflect *adaptation of the body to our daily life* by alternating between *performance* and *resting phases*, mainly determined by the rotation of the earth in relation to the sun. Figure 3.42 demonstrates the behavior of a few important physiological parameters over 24 h. *Cortisol*, a stress hormone, is produced in the second half of the night to prepare the body for waking up each and every morning (Fig. 3.42a). Another hormone, the so-called "darkness" hormone melatonin, causes

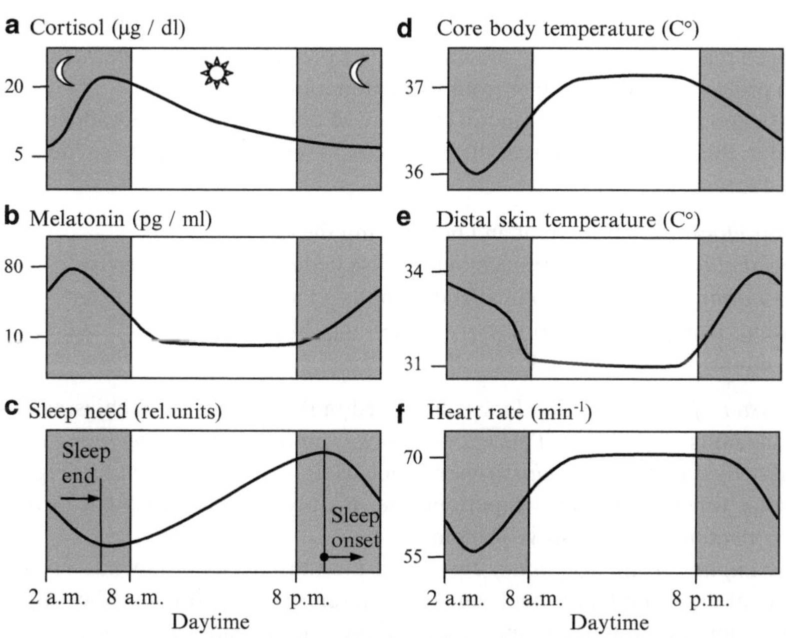

Fig. 3.42 Sketched circadian rhythms of different body parameters—a qualitative and quantitative representation. The relative amount of hormones (**a**) cortisol and (**b**) melatonin in comparison with (**c**) sleep need. The behavior of vital biosignals such as (**d**) core body temperature, (**e**) distal skin temperature, and (**f**) heart rate

drowsiness, lowers body temperature, and strengthens the immune system by suppressing free radicals (Hildebrandt et al. 1998) and is activated by darkness and inhibited by light. Thus, its onset parallels nightfall, as shown in Fig. 3.42b, with peaks in the middle of the night and with a gradual drop off during the second half of the night. Both hormones help in regulating the circadian cycle. The subjectively reported *sleep pressure* (the ease with which to fall asleep) increases throughout the waking hours and is replenished during sleep (Fig. 3.42c). An inverse behavior of the core body temperature and distal skin temperature can be observed in Fig. 3.42d, e. That is, a decrease in the *core body temperature* and a corresponding increase in the distal *skin temperature* not only promote sleep in the evening, but also reflect adaptation of the core and skin temperatures to environmental temperature (Figs. 3.21 and 3.25), compare Sects. 3.1.5 and 3.2.4. Last but not least, the heart rate is reduced during the night, since the parasympathetic activity is increased leading to body relaxation and recovery (compare Sect. 3.1.1).

Obviously, the aforementioned *physiologic parameters* in Fig. 3.42 under circadian control are not complete. Other parameters to be mentioned within this scope are *systolic blood pressure*, with a minimum in early morning hours at about 3 a.m. amounting to about −3% of the mean level, *breathing depth*, slightly increased during the night, *reaction time*, prolonged by about 10% in the early morning hours, and *physical work capacity*, with a maximum of about 10% above average in the early morning hours [percentage values are approximated from Hildebrandt et al. (1998)]. Interestingly, this *paradoxical behavior* of increased physical capacity in parallel to increased reaction time during the *night* shows prevailing *parasympathic activation* which protects this potent state from its exploitation.

To be more precise, the temporal structure of our *daily life* (=circadian rhythms) is under the control of three different *clocks* (Kräuchi 2007), as depicted in Fig. 3.43a:

- *Solar clock*, providing light and heat during the day
- *Social clock*, which determines our work schedule and daily activity
- *Circadian clock* (=biological/endogenous clock), which is essential for the timing of physiological processes across the 24 h, such as sleep and release of hormones

The *master of the circadian clock* is localized in the nucleus suprachiasmatic in the hypothalamus of the brain. This *master clock is entrained* (i.e., *synchronized*) to the 24-h day by different *secondary timers*, such as light, heat, and feeding (Fig. 3.43b). The solar timer is the most important one, mainly governed by light through the eyes, activating photosensitive proteins.

The *impact of the secondary timers* on the master clock depends on both the strength of the secondary timers, e.g., the differences in amplitude of day–night light intensity, and on how much and in what direction the endogenous period deviates from the 24-h solar cycle; in other words, how much the daily light signal has to advance or delay the circadian rhythm (Kräuchi 2007).

For instance, a transient misalignment between the circadian system and the solar cycle is observed during *jet lag*, whereby an abrupt shift in the environmental time

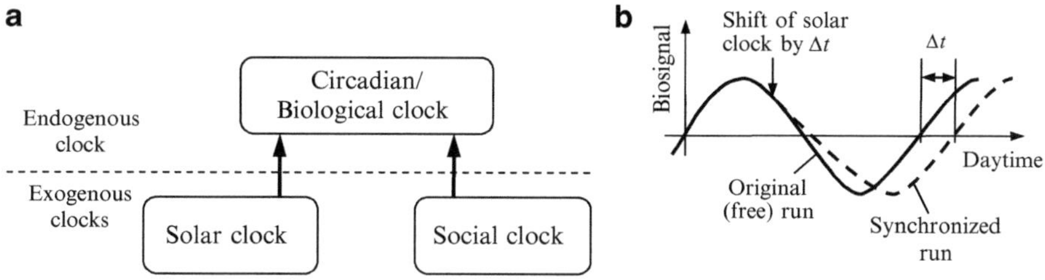

Fig. 3.43 Structure of (**a**) circadian regulation and (**b**) clock entrainment for a shift of a secondary timer by time Δt

occurs and the circadian clock takes several days to be re-entrained to the new light–dark cycle (compare Fig. 3.43b). By contrast, in night *shift work* the wake/sleep cycle is shifted, making a re-entrainment of the circadian clock necessary. All these cases yield (temporal) loss of synchronization which could subsequently lead to reduced sleep efficiency, increased sleep onset latency as well as disturbed endocrine functions.

Actually, without any synchronization the circadian clock would *run free* with a *period* of between 24 and 25 h according to the inner rhythm of cells in the nucleus suprachiasmatic. Individuals may have different free run periods, for instance, because of genetic differences. Subjects who prefer to go to sleep and get up early, the so-called *larks*, tend to have a shorter free run period than those who prefer to sleep later, the so-called night *owls*. The free run period can be observed, for instance, in temporally isolated volunteers or fully blind persons (Wever 1979; Peter et al. 2007).

In general, the *complexity* and *extent* of *rhythmic change* increases with increasing period duration (Hildebrandt et al. 1998), compare Fig. 3.44:

- The *shortest ultradian* rhythms comprise small structures, such as single *cells* (e.g., action impulses of nerves, Sect. 2.1.3.2) or *tissues* (electroencephalogram of brain, Sect. 3.2.4) with substances of low molecular weight involved (Na^+, K^+ ions). These rhythms serve as the basis for *body information systems.*
- The *longer ultradian* rhythms encompass larger structures such as *organs* (cardiac contractions, Sect. 3.1.1) and the entire cardiovascular system (blood pressure variation, Sect. 3.2.2), which serve as *transport and distribution systems* in the body.
- The *circadian* rhythms already affect the *entire body* (wake/sleep cycle, Sect. 3.2.4), which are mainly based on *metabolic processes* of high molecular substances.
- The longest *infradian* rhythms are related to human *population behavior.*

From a formal point of view, *short rhythms* exhibit periods of short impulses (=relaxation oscillation) and are more variable in their *frequency* than in their *amplitude*; see Footnote 145 for basic characteristics of time functions. An extreme case is given by the action impulse of a receptor cell, as shown in Fig. 3.44a, with

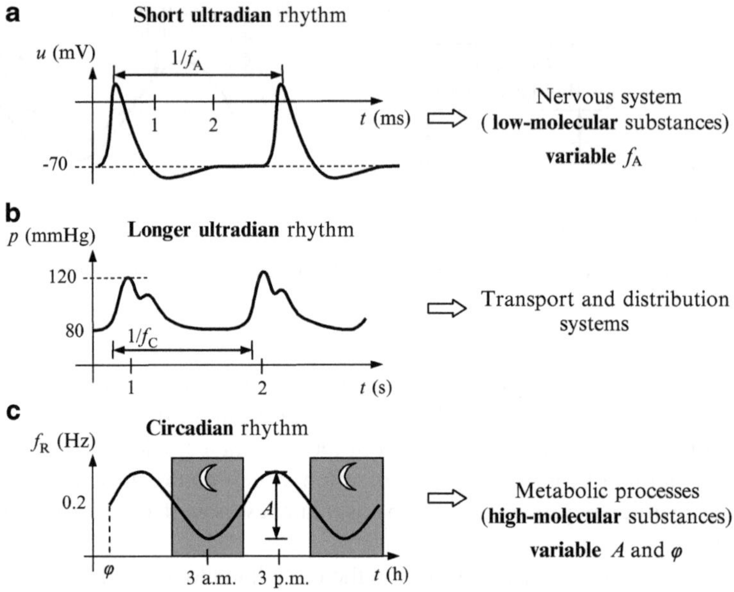

Fig. 3.44 Comparison of ultradian and circadian rhythms with respect to their waveforms and origin. (**a**) Sketched spiky action impulses of membrane voltage u with indicated activation rate f_A. (**b**) Periodic variation of blood pressure p over a cardiac cycle with indicated heart rate f_C. (**c**) Sinusoidal changes of respiratory rate f_R over 24 h with indicated amplitude A and phase φ. Data given in accordance with Hildebrandt et al. (1998)

a *constant impulse amplitude* according to the all-or-none law and only the *impulse frequency varying* according to the receptor stimulus (Sect. 2.2.2). By contrast, *longer rhythms* show a smoother form, as demonstrated by blood pressure changes in Fig. 3.44b over the cardiac cycle. Lastly, a sinusoidal waveform (like swinging pendulum) results for the *longest rhythms*, which are more stable in *frequency* than in *amplitude and phase*, as demonstrated in Fig. 3.44c.

It is important to note that *biological rhythms* are not independent but usually *synchronized* in *phase* and/or *frequency*, which has a strategic functional relevance. In general, the fast rhythms may generate a phase change of the slow rhythms (e.g., jet lag from above, Fig. 3.43b), whereas the slow rhythms may affect the fast rhythms by synchronizing them in terms of frequency modulation (e.g., heart rate modulation by breathing, Sect. 3.2.1). Both mechanisms serve the economic needs of nature, while the *synchronized state* shows the *lowest energy expenditure* and the most efficient use of body resources.

Generally, synchronization in phase or *phase coupling* occurs between heterogeneous physiological phenomena, such as heartbeat, respiration, blood circulation, and peripheral blood perfusion (Fig. 3.27). More precisely, as described in Sect. 3.2, changes of the heart rate, mean arterial blood pressure, and systolic blood pressure are locked to the phase of the respiration cycle, while the diastolic blood pressure is related to the total peripheral resistance. The synchronization in frequency or

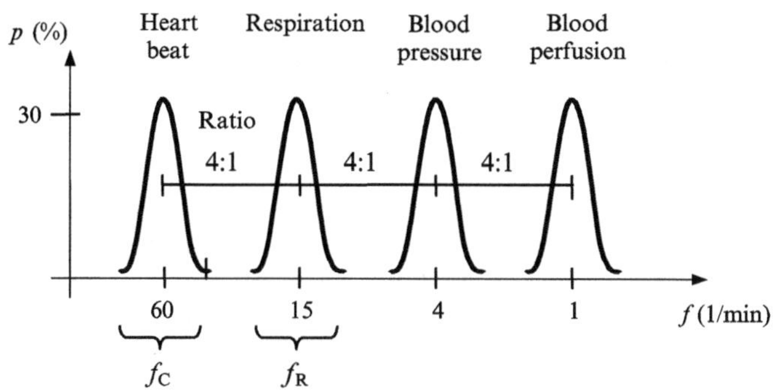

Fig. 3.45 Schematic representation of the probability p distributions of fundamental frequency f of the cardiac activity with f_C as heart rate, respiration activity with f_R as respiratory rate, blood pressure changes, and perfusion changes. Approximate ratios of the respective mean values of f are indicated. Data taken partially from Hildebrandt et al. (1998)

frequency coupling is demonstrated in Fig. 3.45. The mean value of the heart rate is ideally matched to the mean respiratory rate with the ratio of about 4:1. Surprisingly similar ratios exist between the mean respiratory rate, mean rate of (mean arterial) blood pressure changes, and mean rate of the blood perfusion changes (or that of the peripheral resistance), see also Sects. 2.3.2, 3.2.1, and 3.2.2.

Actually, establishment of the phase coupling and the frequency coupling is intensified in resting states, such as sleep (Sect. 3.2.4), whereas during stress or exercise the relationships between multiple physiological frequencies seem to disappear (Moser et al. 2008; Hildebrandt et al. 1998). Thus, a *continuous interplay* occurs between vital physiological rhythms (Fig. 3.27) in terms of their *phase and frequency*.

In terms of (energetically) optimal synchronization of physiological body parameters, synchronization of the rhythms may entangle not only the fundamental harmonic of physiological parameters, as demonstrated in Figs. 3.42 and 3.44, but also *synchronization* of their *higher harmonics*. A convincing example is given by the electro-mechanical coupling between ventricular action of the heart and the vascular network during exercise in comparison to rest, when aortic pressure p and aortic flow q are considered (Nichols and O'Rourke 2005), compare Figs. 2.40 and 2.48. *At rest*, the value of f_C, i.e., the frequency of fundamental and (energetically) dominant harmonic of p and q, corresponds to the modulus of the input impedance $\underline{Z}_I$ *at a relatively high value* (f_C<4 Hz, see Footnote 153 in Sect. 2.5.2). It indicates a relatively high ventricular load to generate *high p amplitude* for a given q amplitude within the dominant fundamental harmonic. Tachycardia resulting from *exercise* shifts f_C to higher frequencies at which $|\underline{Z}_I|$ *is relatively low*, facilitating lower p amplitude for the given q amplitude. Thus, the arterial system appears to deliver optimal performance during exercise and suboptimal performance during rest due to synchronization along different frequencies.

There is increasing evidence that *disappearance of the rhythms* and their synchronization results in the *loss of health*, as demonstrated by the aforementioned night shift work or jet lag (Moser et al. 2008; Hildebrandt et al. 1998). In addition, modern life impedes rhythmicity because of artificial light, controlled ambient temperature, and reduced outdoor activities. The future medical profession will increasingly utilize human rhythms to make therapies more efficient.

The circadian rhythm undergoes significant changes *with age*. The amplitude and stability of the circadian rhythm are reduced in the elderly, even up to a relative absence of the rhythm (Sack et al. 2007a, b). In addition, the rhythm advances along the time scale (Van Someren et al. 2002; Lee-Chiong 2006), for instance, in the case of the body core temperature (Fig. 3.42d). The advancement results in a pattern in which older persons usually get sleepy early in the evening and then wake up early in the morning; in terms of Fig. 3.43b it would correspond to a shift of the oscillation to the left. Also a re-entrainment efficiency of the circadian clock is reduced in response to night shift work or jet lag. The arising circadian rhythms are associated with more awakenings during the night (compare later Fig. 3.48c), i.e., associated with a manifold disruption of the circadian cycle, resulting in a need for more naps during the day and increased daytime drowsiness. These age-related changes in circadian rhythms occur for several physiological reasons, including changes in hormonal balance. For instance, spending less time outdoors or eye diseases in the elderly may block the light needed to govern the solar timer, reducing the necessary entrainment of the master clock (Fig. 3.43a) and thus destabilizing the circadian rhythm.

3.2.4 Sleep

Sleep[249] is a state of rest which normalizes physiological parameters destabilized during the day and is essential for survival.[250] It is not a passive state but rather a *night factory* with intrinsic rhythmic behavior, as will be shown later. Sleep involves not only the brain but also the entire body. In particular, *sleep* comprises the so-called

- *Rapid eye movement* (REM) phase
- *Nonrapid eye movement* (NREM) phases.
- Together with *awake* phase

---

[249] Sleep has always been a mystery and a source of inspiration for all artists and researchers throughout the centuries (Fig. 3.46). According to the words of Derek Walcott "All-humbling sleep, whose peace is sweet as death, whose silence has all the sea's weight and volubility, who swings this globe by a hair's trembling breath (Hamner 1997)."

[250] As reviewed in Shepard et al. (2005), epidemiologic data on more than one million subjects have shown that mortality rates were significantly increased for subjects reporting less than 4 h or more than 10 h of sleep per night. The consistency of manifold reports suggests that deviations in sleep duration from the norm, i.e., insufficient or excessive sleep, may adversely influence human longevity.

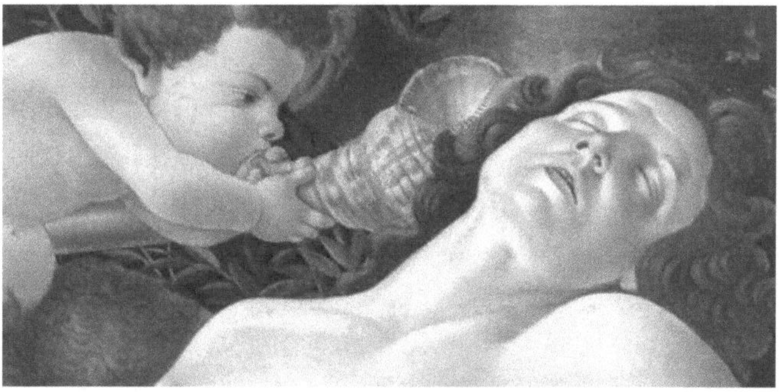

Fig. 3.46 Sleeping, dreaming, and awakening. A detail from Botticelli (1483)

thus, the *three basic states* of the central nervous system and body functions are constituted. Sleep is tightly coupled with the *circadian rhythm* (Sect. 3.2.3) and is composed of successive cycles of REM and NREM phases, compare Figs. 3.47b and 3.48. The NREM sleep is further differentiated into *shallow sleep*, including NREM stages 1 and 2, and *deep sleep*, including NREM stages 3 and 4.

The particular classification of the sleep phases and stages is based on different biosignals, whereas the *electroencephalogram,*[251] *electromyogram* for the assessment of muscle tension, and *electrooculogram* for the recording of eye

---

[251]The *electroencephalogram* refers to the recording of the brain's spontaneous electrical activity by the use of multiple scalp electrodes. The potential differences are registered which arise because of equalizing currents in the extracellular space during neuronal activity (compare Sect. 2.1.4 and Sect. 4). The more neurons that are timely synchronized in their firing, the higher is the amplitude of the electroencephalogram; the contribution of a single neuron activity to the electroencephalogram is very small because of a strong signal attenuation in the tissue, scalp, and skin. Usually the course of the electroencephalogram exhibits rhythmic behavior, which justifies an establishment of predefined *frequency bands* being particularly relevant for *sleep staging*. That is, the so-called

- *Beta waves* correspond to rhythmic electroencephalogram behavior in the range of 13–30 Hz reflecting an active or busy human state.
- *Alpha waves* are in the range of 8–13 Hz showing relaxation or closed eyes.
- *Theta waves* fill the range of 4–8 Hz and indicate drowsiness.
- *Delta waves* comprise the range up to 4 Hz usually denoting sleep.

The number of *electrodes* ranges from two, e.g., only for discrimination in between wake and sleep, up to a few hundred for research purposes. Usually the application of the scalp electrodes is based on the international *10–20 system* using *letter–number* designation (Malmivuo and Plonsey 1995). The 10 and 20 refer to percentage distances from reference points or neighbouring electrodes. The reference points are the nasion, the point between the forehead and the nose, and the inion, the lowest point of the skull on the midline. From these points, the skull perimeters are measured in the transverse and median planes with the electrodes being placed at the distances 10%-20%-20%-20%-20%-10% along both perimeters. The letter of the electrode position designates the anatomical area, whereas the odd numbered electrodes are placed on the left and the even numbered electrodes on the right. In total 21 electrodes are applied, including two on both ear lobes (A1 and A2). For instance, a derivation C4-A1 denotes the potential difference between the central region to

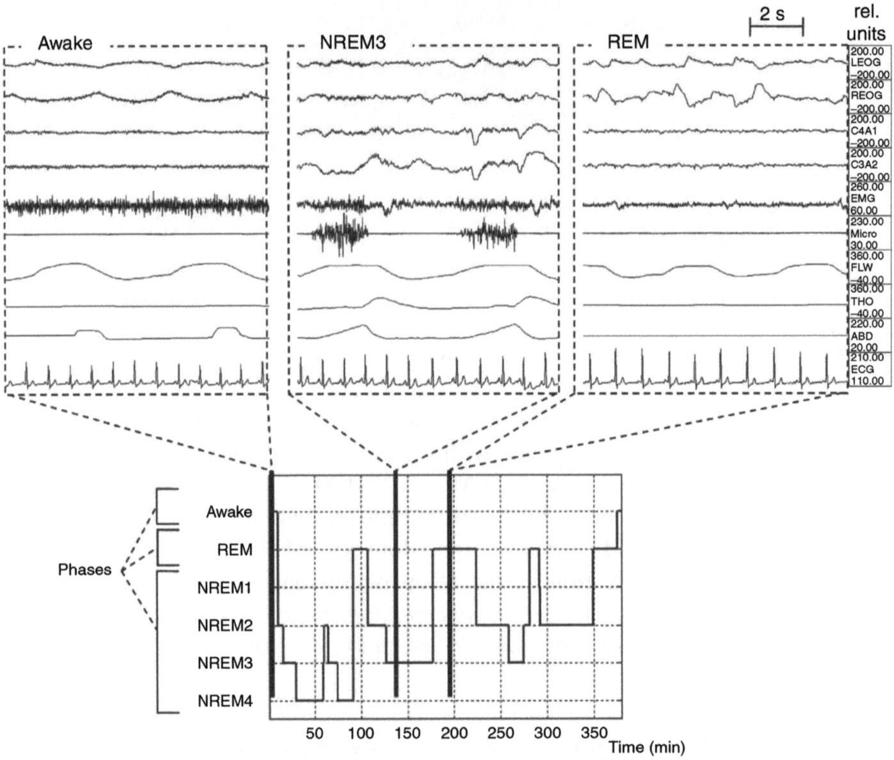

Fig. 3.47 Typical examples of multichannel monitoring within the scope of polysomnography (PSG). (**a**) The PSG signals during, from left to right, awake phase, deep sleep, i.e., nonrapid eye movement (NREM) phase stage 3, and paradoxical sleep, i.e., rapid eye movement (REM) phase with an identical resolution of respective signals. The depicted signals include, from top to bottom, electrooculogram from the left eye, electrooculogram from the right eye, electroencephalogram derivation C4-A1 (see Footnote 251), electroencephalogram derivation C3-A2, electromyogram, breathing sounds amplitude, airflow, chest extension, abdominal extension, and electrocardiogram. (**b**) The corresponding somnogram of the subject showing successive awake/sleep phases and stages

movements occupy a central role within, as originally introduced in Rechtschaffen and Kales (1968). To be more precise, the *phases and stages are defined* as follows (Saletu and Saletu-Zyhlarz 2001; Lee-Chiong 2006), compare Fig. 3.47:

- The *awake phase* is dominated by (relatively fast) alpha and beta waves in an electroencephalogram (see Footnote 251), accompanied by normal eye movements and high muscle tension.

---

the right in the frontal plane and the left ear lobe, whereas C3-A2 denotes the potential difference between the central region to the left in the frontal plane and the right ear lobe (compare Fig. 3.47).

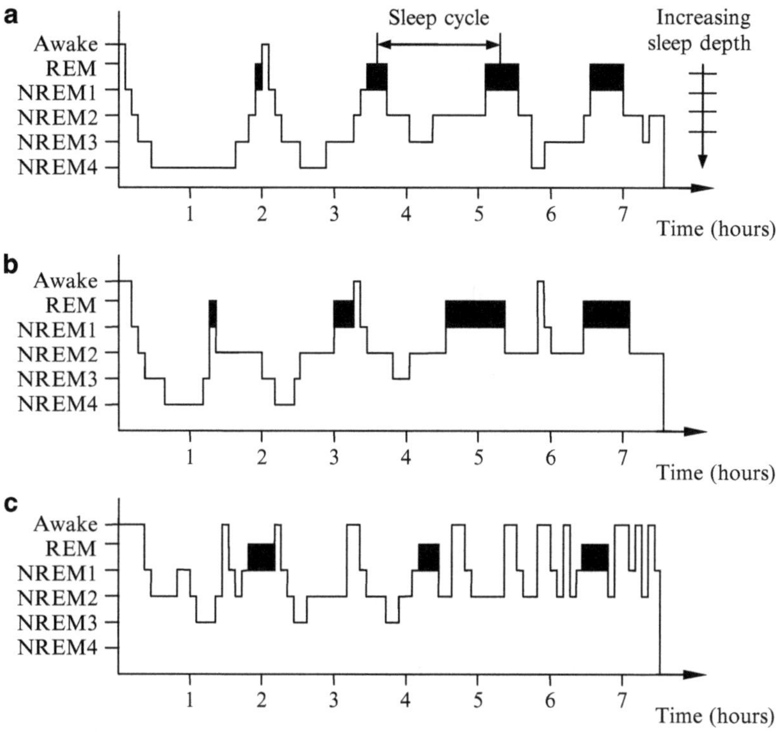

Fig. 3.48 The distribution of sleep stages (somnograms) for different ages. A schematic somnogram of (**a**) children, (**b**) young people, and (**c**) old people with indicated rapid eye movement (REM) phases and four non-REM (NREM) phases. Data taken partially from Klösch et al. (2007)

- The *REM phase*, the so-called paradoxical (fast) sleep, is denoted by REM s in parallel with a nearly absent muscle tension. The heartbeat and respiration are accelerated while the blood pressure is increased. In general, a heightened mental activity can be detected counteracting daily life's wearing effects, including realistic rehearsals of threatening events, accompanied by memorable (lucid) *dreaming*. It performs the processing/saving of implicit memory tasks, e.g., motoric skills or cycling a bicycle, and normally prevails in the second half of the night (compare Fig. 3.48a, b).
- The *NREM phase* corresponds to revitalization of the body (with about 40% drop in glucose consumption), plays an important role in processing of explicit memory tasks, e.g., daytime learning of vocabulary is consolidated by nighttime sleeping, and normally prevails in the first half of the night.
  - The *NREM stage 1*, the *onset of sleep*, is marked by reduction in alpha waves and an increase in (slower) theta waves. Slow eye movements prevail with slowly vanishing muscle tension. A partial consciousness may exist.

- The *NREM stage 2*, the *shallow sleep*, is marked by theta waves and upcoming (even slower) delta waves (up to 20% of the total wave patterns). Here, eye movements are missing and the muscle tension continues to decrease. The person is unconscious though awakened easily.
- The *NREM stage 3*, the *deep sleep*, is given with an even stronger domination of delta waves (up to 50%) of strongly increased amplitude. Eye movements are absent and the muscle tension continues to decrease.
- The *NREM stage 4*, the *deepest sleep*, begins when domination of the delta waves exceeds 50%. Eye movements are missing and the muscle tension attains a low but still tangible value.

As a rule of thumb, the *REM and NREM phases* show rapid and slow/absent eye movements, respectively. On the one hand, the *sleep depth* in the NREM phase (Fig. 3.48a) increases with the electroencephalogram becoming slower in time and stronger in amplitude, while, on the other hand, the sleep depth increases with decreasing muscle tension.

Besides the aforementioned changes in the electroencephalogram, electromyogram, and electrooculogram over different sleep phases and stages, numerous other *biosignals* vary in a specific way *during the sleep phases and stages* as well as during the *wake/sleep cycle*, compare Sect. 3.2.3. For instance, the *heart rate* (Sects. 3.1.1 and 3.2.3), systolic *blood pressure* (section "Pulse Waveforms of Pressure and Flow"), and *sympathetic activity* decrease with increasing sleep depth; for instance, the systolic blood pressure may drop by about 10–20% during deep sleep (Schäfer 1996). The *stroke volume* and *cardiac output* drop with sleep onset (Varoneckas et al. 1999). With increasing sleep depth, the *stroke volume* increases mainly due to reduced heart rate (see section "Normal Respiration" under Sect. 3.2.1.1); the *cardiac output*, i.e., the effective blood supply (2.20), remains nearly constant due to opposite (adapting) changes of the stroke volume and heart rate, see (2.30).

HRV is increased during sleep as compared with awake but decreases with increasing sleep depth (Sect. 3.1.1). The *respiratory rate* and its *variability* are reduced in deep NREM sleep while the respiratory volume slightly increases, compare Fig. 3.44c. For circadian changes of various biosignals see also Sect. 3.2.3.

A diagnostic assessment of sleep, including its phases and stages, has been based on the so-called *polysomnography* (PSG) forming the framework upon which the field of sleep disorders has been built over the last 40 years (Ferber et al. 1994). The PSG is an attended monitoring of sleep requiring an overnight hospitalization in a sleep laboratory (Fig. 3.49). It includes recording a large number of *brain*, *cardiac*, and *respiratory parameters* by the use of the following physiological signals (ASDA 1994):

- Electroencephalogram
- Electrooculogram
- Chin electromyogram
- Electrocardiogram
- Airflow

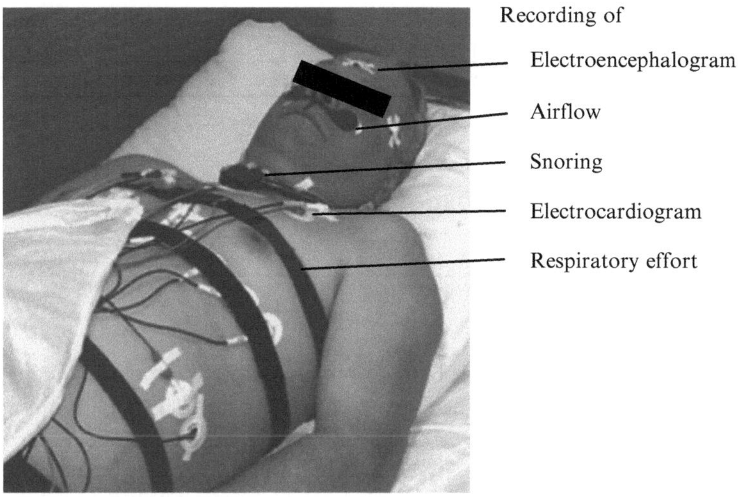

Fig. 3.49 Multichannel monitoring of a patient under polysomnography in sleep lab

- Respiratory effort
- Oxygen saturation
- Body position

Figure 3.47a demonstrates the most important *PSG signals* for sleep staging during the awake phase, deep sleep (NREM stage 3), and paradoxical sleep (REM phase). As already described, specific waveforms of electrooculogram, electroencephalogram, and electromyogram allow for a determination of an actual sleep phase or stage.

The corresponding Fig. 3.47b shows the *cyclic changes of the REM and NREM* phases over the entire night, with a resulting duration of the *sleep cycle* of *about 2 h*. In fact, this plot of the PSG scored phases and stages is called a *somnogram* (latin *somnus* sleep and greek *gramma* letter, character). It can be observed that the sleep depth (Fig. 3.48a) decreases over the night with the deep sleep (and the deepest sleep) dominating in the first half of the night. By contrast, the paradoxical sleep (REM phase) and shallow sleep (NREM stage 2) dominate in the second half. According to Table 3.1, the awake phase is the shortest one, while the REM phase and NREM stage 2 account for the most of the sleep time in normal adult persons.

A body that is asleep is in the most relaxed state of normal daily life, and this relaxed state in turn influences the *thermoregulatory system*. Usually in the evening *before sleep*, the heat (=blood) is redistributed from the core (heat source) to the shell (heat sink or radiators) which *downregulates core body temperature* (CBT), as shown in Figs. 3.21 and 3.42d, e. That is, the CBT decreases during *sleep onset* while the corresponding *distal skin temperature increases*. In other words, the body looses in the thickness of its heat shell (see Fig. 3.21b). In the *morning hours*, the reverse is true (Fig. 3.42d, e), the *CBT increases* while the *distal*

Table 3.1 Relative approximate durations of the sleep phases and stages for healthy young persons (about 25 years) and elderly persons (80 years)

Sleep phases/stages	Young adult persons (%)	Elderly persons (%)
Awake	< 5	14
REM	25	18
NREM stage 1	4	8
NREM stage 2	50	54
NREM stage 3	6	4
NREM stage 4	13	2

Data compiled from Schäfer (1996), Konietzko et al. (1998), Lee-Chiong (2006)

temperature decreases. The latter behavior appears to have functional significance for physiology at the level of molecules to cognition, e.g., supports immunological defense mechanisms due to increased blood volume in the periphery during sleep (Van Someren et al. 2002), and obviously leads to a larger difference between the diurnal maximum and nocturnal minimum values of the CBT than *without sleep* (Kräuchi 2007).

It is interesting to note that the CBT and especially its change are crucial for *sleep induction*. Sleep under entrained conditions is *initiated* on the declining portion of the CBT curve when both its *rate of change* and *body heat loss* are maximal. Typically, the maximum value of CBT occurs in the early evening and the minimum in the second half of the nocturnal sleep (Fig. 3.42d).

From a *practical point of view*, falling asleep can be promoted by lying down relaxed in a comfortable thermal environment, warm drinks as well as hypnotic suggestions of warmth, autogenic training, switching lights off to permit nocturnal melatonin to rise (compare Fig. 3.42b), or the intake of melatonin or classical sleeping pills (Kräuchi 2007). Most of the latter measures evoke an increase in distal skin temperatures and a decrease in the CBT, promoting sleep. In other words, the body heat loss (before lights are turned off) via *nonpharmacological relaxation-induced vasodilation* of distal skin regions promotes drowsiness and a rapid onset of sleep.

The above *circadian pattern of the CBT* results from the balance between heat production and heat loss (Sect. 3.1.5). Actually, the *distal skin* regions show the lowest *mean temperature* if compared to the *proximal sites*, e.g., 31°C for toe and 35°C for forehead (Kräuchi 2007), compare Figs. 3.21 and 3.25. By contrast, the *oscillation amplitudes* of temperature over 24 h are higher in *distal* skin regions (about 3°C for finger measured in the lab) than in *proximal* sites (about 1°C for abdomen in the lab). Interestingly, these oscillation amplitudes in ambulatory conditions or home environments are higher by a factor of 2–3 than in laboratory settings, because external timers—compare Sect. 3.2.3—are missing in the laboratory.

Thus, the distal skin temperature of hands and feet are crucial parameters. For instance, women, who suffer from cold hands and feet (known as *vasospastic syndrome*), as well as persons under *chronic stress*, who suffer from physiologically manifested vasospastic syndrome, will exhibit not only a lower *capacity to lose heat* during the daytime but also—as a disadvantage—a prolonged *sleep onset latency* (=time duration from full wakefulness to sleep) and a disturbed *phase of entrainment* (=phase in between the circadian clock and the wake/sleep cycle) (Kräuchi 2007).

Somnograms for *different ages* are depicted in Fig. 3.48. The subfigures compare the cyclic changes of the sleep phases and stages over varying age, i.e., from children, to young people, up to the elderly. In children and young people, 3–5 cycles of about 2 h in duration each can be observed during the night. In particular, *children* (Fig. 3.48a) exhibit a lot of *deep sleep* and a lot of *REM phases*. With *increasing age* (Fig. 3.48b), deep sleep begins to dominate in the first half of the night, while the REM phase dominates in the second half. In contrast to children and young people, sleep is strongly fragmented in the *elderly* (Fig. 3.48c) with prevailing shallow sleep. Here, nocturnal awakenings occur more frequently, especially in the morning. Table 3.1 emphasizes the aging effect on the relative changes in the durations of the sleep phases and stages. In comparison with young persons, the deepest sleep is dramatically reduced in the elderly, while the awake phase and sleep onset are strongly prolonged, which deteriorates the quality of sleep in the elderly.

References

G. Amit, K. Shukha, N. Gavriely, N. Intrator: Respiratory modulation of heart sound morphology. American Journal of Physiology. Heart and Circulatory Physiology 296(3), 796–805 (2009).

H.H. Asada, P. Shaltis, A. Reisner, R. Sokwoo, R.C. Hutchinson: Mobile monitoring with wearable photoplethysmographic biosensors. IEEE Engineering in Medicine and Biology Magazine, 22(3), 28–40 (2003).

J. Aschoff: Temperature regulation (in German: Temperaturregulation). Eds. O.H. Gauer, K. Kramer, R. Jung: Energy budget and temperature regulation. Physiology of human (in German: Energiehaushalt und Temperaturregulation. Physiologie des Menschen). Publisher Urban & Schwarzenberg, 43–112 (1971).

ASDA (Standards of practice committee of the American Sleep Disorders Association): Practice parameters for the use of portable recording in the assessment of obstructive sleep apnea. Sleep 17(4), 372–377 (1994).

W. Bachta, P. Renaud, L. Cuvillon, E. Laroche, A. Forgione, J. Gangloff: Motion prediction for computer-assisted beating heart surgery. IEEE Transactions on Biomedical Engineering 56(11), 2551–2563 (2009).

A. Baharav, S. Kotagal, V. Gibbons, B.K. Rubin, G. Pratt, J. Karin, S. Akselrod: Fluctuations in autonomic nervous activity during sleep displayed by power spectrum analysis of heart rate variability. Neurology 45(6), 1183–1187 (1995).

P. Barach: Pulsus Paradoxus. Hospital Physician, 49–50 (2000).

D. Barschdorff, M. Erig: Continuous blood pressure monitoring during stress ECG. Biomedizinische Technik 43(3), 34–39 (1998).

D.E. Bebout, P.D. Mannheimer, C.C. Wun: Site-dependent differences in the time to detect changes in saturation during low perfusion. Critical Care Medicine 29(12), 115 (2001).

J.T. Bigger: Overview of RR variability. www.nerveexpress.com (2006).

E.O. Bixler, A.N. Vgontzas, T. Ten Have, K. Tyson, A. Kales: Effects of age on sleep apnea in men. Americal Journal of Respiratory and Critical Care Medicine 157(1), 144–148 (1998).

S. Botticelli: Mars and Venus. Oil painting from National Gallery in London (1483).

T.D. Bradley, R. Tkacova, M.J. Hall, S. Ando, J.S. Floras: Augmented sympathetic neural response to simulated obstructive apnoea in human heart failure. Clinical Science 104, 231–238 (2003).

P. Busek, J. Vankova, J. Opavsky, J. Salinger, S. Nevsimalova: Spectral analysis of the heart rate variability in sleep. Physiological Research 54(4), 369–376 (2005).

Y. Chen, L. Li, C. Hershler, R.P. Dill: Continuous non-invasive blood pressure monitoring method and apparatus. Patent Nr. US 2003 0199770 (2003).

Z. Chen, E.N. Brown, R. Barbieri: Characterizing nonlinear heartbeat dynamics within a point process framework. IEEE Transactions on Biomedical Engineering 57(6), 1335–1347 (2010).

G.D. Clifford, F. Azuaje, P.E. McSharry: Advanced methods and tools for ECG data analysis. Artech House (2006).

W.A. De Backer: Central sleep apnoea, pathogenesis and treatment: An overview and perspective. European Respiratory Journal 8(8), 1372–1383 (1995).

T.J. Ebert, B.J. Morgan, J.A. Barney, T. Denahan, J.J. Smith: Effects of aging on baroreflex regulation of sympathetic activity in humans. American Journal of Physiology, Heart and Circulatory Physiology 263, 798–803 (1992).

M. Elstad, K. Toska, K.H. Chon, E.A. Raeder, R.J. Cohen: Respiratory sinus arrhythmia: Opposite effects on systolic and mean arterial pressure in supine humans. Journal of Physiology 536.1, 251–259 (2001).

R. Ferber, R. Millman, M. Coppola, J. Fleetham, C.F. Murray, C. Iber, V. McCall, G. Nino-Murcia, M. Pressman, M. Sanders, K. Strohl, B. Votteri, A. Williams: Portable recording in the assessment of obstructive sleep apnea. American Sleep Disorders Association (ASDA) standards of practice. Sleep 17(4), 378–392 (1994).

L.J. Findley, S.C. Wilhoit, P.M. Suratt: Apnea duration and hypoxemia during REM sleep in patients with obstructive sleep apnea. Chest 87(4), 432–436 (1985).

D. Franchi, R. Bedini, F. Manfredini, S. Berti, G. Palagi, S. Ghione, A. Ripoli: Blood pressure evaluation based on arterial pulse wave velocity. Computers in Cardiology, 397–400 (1996).

M. Garbey, N. Sun, A. Merla, I. Pavlidis: Contact-free measurement of cardiac pulse based on the analysis of thermal imagery. IEEE Transactions on Biomedical Engineering 54(8), 1418–1426 (2007).

E. Garpestad, H. Katayama, J.A. Parker, J. Ringler, J. Lilly, T. Yasuda, R.H. Moore, H.W. Strauss, J.W. Weiss: Stroke volume and cardiac output decrease at termination of obstructive apneas. Journal of Applied Physiology 73(5), 1743–1748 (1992).

L.A. Geddes, M.H. Voelz, S. James, D. Reiner: Pulse arrival time as a method of obtaining systolic and diastolic blood pressure indirectly. Medical and Biological Engineering and Computing 19(5), 671–672 (1981a).

L.A. Geddes, M.H. Voelz, C.F. Babbs, J.D. Bourland, W.A. Tacker: Pulse transit time as an indicator of arterial blood pressure. Psychophysiology 18(1), 71–74 (1981b).

L.A. Geddes, R.A. Roeder: Where do ideas come from? IEEE Engineering in Medicine and Biology Magazine 28(5), 60–61 (2009).

E. Gil, M. Mendez, J.M. Vergara, S. Cerutti, A.M. Bianchi, P. Laguna: Discrimination of sleep-apnea-related decreases in the amplitude fluctuations of PPG signal in children by HRV analysis. IEEE Transactions on Biomedical Engineering 56(4), 1005–1014 (2009).

H. Golub: Non-invasive cuff less determination of blood pressure. Patent Nr. CA 2268073 (1998).

E.F. Greneker: Radar sensing of heartbeat and respiration at a distance with application of the technology. Radar 97(449), 150–154 (1997).

W. Greubel, A.A.C. Müller, H. Stein, R. Wieczorek: Method of continuous measurement of blood pressure in humans. Patent Nr. US 5237997 (1993).

B. Gribbin, A. Steptoe, P. Sleight: Pulse wave velocity as a measure of blood pressure change. Psychophysiology 13(1), 86–90 (1976).

A. Guz, J.A. Innes, K. Murphy: Respiratory modulation of left ventricular stroke volume in man measured using pulsed Doppler ultrasound. Journal of Physiology 393, 499–512 (1987).

R.D. Hamner: Critical perspectives on Derek Walcott. Lynne Rienner Publishers (1997).

H.H. Hardy, R.E. Collins: On the pressure-volume relationship in circulatory elements. Medical and Biological Engineering and Computing 20(5), 565–570 (1982).

J. Hayano, F. Yasuma, A. Okada, S. Mukai, T. Fujinami: Respiratory sinus arrhythmia. A phenomenon improving pulmonary gas exchange and circulatory efficiency. Circulation 94(4), 842–847 (1996).

G. Hildebrandt, M. Moser, M. Lehofer: Chronobiology and chronomedicine. Biological rhythms and medical consequences (in German: Chronobiologie und Chronomedizin. Biologische Rhythmen und medizinische Konsequenzen). Hippokrates Publisher (1998).

E. Hill, M.D. Stoneham: Practical applications of pulse oximetry. World Anaesthesia Online 11(4), 1–2 (2000).

HRV task force: Task Force of the European Society of Cardiology and the North American Society of Pacing and Electrophysiology: Heart rate variability: standards of measurement, physiological interpretation, and clinical use. Circulation 93(5), 1043–1065 (1996).

B.E. Hunt, L. Fahy, W.B. Farquhar, J.A. Taylor: Quantification of mechanical and neural components of vagal baroreflex in humans. Hypertension 37, 1362–1368 (2001).

K. Ishikawa, T. Tamura: Study of respiratory influence on the intensity of heart sound in normal subjects. Angiology 30(11), 750–755 (1979).

V. Kamat: Pulse oximetry. Indian Journal of Anaesthesia 46(4), 261–268 (2002).

A. Kamiya, S. Iwase, H. Kitazawa, T. Mano, O.L. Vinogradova, I.B. Kharchenko: Baroreflex control of muscle sympathetic nerve activity after 120 days of 6° head-down bed rest. American Journal of Physiology. Regulatory, Integrative and Comparative Physiology 278(2), 445–452 (2000).

E. Kaniusas, H. Pfützner, L. Mehnen, J. Kosel, J.C. Téllez-Blanco, G. Varoneckas, A. Alonderis, T. Meydan, M. Vázquez, M. Rohn, A.M. Merlo, B. Marquard: Method for continuous non-disturbing monitoring of blood pressure by magnetoelastic skin curvature sensor and ECG. IEEE Sensors Journal 6(3), 819–828 (2006a).

E. Kaniusas: Multiparametric physiological sensors. Habilitation theses at the Vienna University of Technology (2006b).

M. Karam, R.A. Wise, T.K. Natarajan, S. Permutt, H.N. Wagner: Mechanism of decreased left ventricular stroke volume during inspiration in man. Circulation 69, 866–873 (1984).

L.R. Keytel, J.H. Goedecke, T.D. Noakes, H. Hiiloskorpi, R. Laukkanen, L. Merwe, E.V. Lambert: Prediction of energy expenditure from heart rate monitoring during submaximal exercise. Journal of Sports Sciences 23(3), 289–297 (2005).

A. Khasnis, Y. Lokhandwala: Clinical signs in medicine: Pulsus paradoxus. Journal of Postgraduate Medicine 48(1), 46–49 (2002).

G. Klösch, P. Anderer, J. Zeitlhofer: Changes of sleep over aging (in German: Veränderungen des Schlafes im Alter). Presentation from annual conference of Austrian Sleep Research Association (2007).

R. Kolluri, R. Medero, B. Friedman, L. Hersch: Continuous, non-invasive technique for measuring pressure using impedance plethysmography. Patent Nr. US6648828 (2003).

N. Konietzko, H. Teschler, L. Freitag: Sleep apnea (in German: Schlafapnoe). Springer Publisher (1998).

K. Kräuchi: Physiological review: The thermophysiological cascade leading to sleep initiation in relation to phase of entrainment. Sleep Medicine Reviews 11, 439–451 (2007).

P. Langley, E.J. Bowers, A. Murray: Principal component analysis as a tool for analyzing beat-to-beat changes in ECG features: Application to ECG-derived respiration. IEEE Transactions on Biomedical Engineering 57(4), 821–829 (2010).

T.L. Lee-Chiong (ed.): Sleep. A comprehensive handbook. Wiley Publisher (2006).

P.V. Leeuwen, H.C. Kuemmell: Respiratory modulation of cardiac time intervals. British Heart Journal 58(2), 129–135 (1987).

M.N. Levy, H. Zieske: Autonomic control of cardiac pacemaker activity and atrioventricular transmission. Journal of Applied Physiology 27(4), 465–470 (1969).

Y.H. Lin, P.C.I. Ko, H.Y. Wang, T.C. Lu, Y.Y. Chen, I.C. Jan, G.J. Jan, N.K. Chou: Estimation of beat-to-beat systolic blood pressure using pulse arrive time and pulse width derived from

the photoplethysmogram. Proceedings of the 26th annual EMBS International Conference 2, 3456–3458 (2004).
F.A. Mahomed: The physiology and clinical use of the sphygmograph. Medical Times and Gazette 1, 62–64 (1872).
C. Maier, V. Rodler, H. Wenz, H. Dickhaus: ECG fingerprints of obstructed breathing in sleep apnea patients. IEEE Engineering in Medicine and Biology magazine 28(6), 41–48 (2009).
J. Malmivuo, R. Plonsey: Bioelectromagnetism, Principles and Applications of Bioelectric and Biomagnetic Fields. Oxford University Press (1995).
P.F. Migeotte, N. Pattyn, R. Vanspauwen, X. Neyt, M. Acheroy, P. Van de Heyning, F. Wuyts: Respiratory sinus arrhythmia on the ESA-short-arm human centrifuge. IEEE Engineering in Medicine and Biology magazine 28(6), 86–91 (2009).
W.R. Milnor: Hemodynamics. Williams & Wilkins Publisher (1989).
K.D. Monahan: Effect of aging on baroreflex function in humans. American Journal of Physiology. Regulatory, Integrative and Comparative Physiology 293(1), 3–12 (2007).
M. Moser, M. Frühwirth, T. Kenner: The symphony of life. Importance, interaction, and visualization of biological rhythms. IEEE Engineering in Medicine and Biology Magazine 27(1), 29–37 (2008).
T. Nakazatoa, T. Shikamaa, S. Tomab, Y. Nakajimab, Y. Masudaa: Nocturnal variation in human sympathetic baroreflex sensitivity. Journal of the Autonomic Nervous System 70(1–2), 32–37 (1998).
M.R. Neuman: Measurement of vital signs: Temperature. IEEE Pulse 1(2), 40–49 (2010).
M.R. Neuman: Measurement of blood pressure. IEEE Pulse 2(2), 39–44 (2011).
W.W. Nichols, M.F. O'Rourke: McDonald's blood flow in arteries: Theoretical, experimental and clinical principles. Hodder Arnold Publication Publisher (2005).
M. Nitzan, A. Babchenko, B. Khanokh, D. Landau: The variability of the photoplethysmographic signal—a potential method for the evaluation of the autonomic nervous system. Physiological Measurement 19(1), 93–102 (1998).
M. Okada: Possible determinants of pulse-wave velocity in vivo. IEEE Transactions on Biomedical Engineering 35(5), 357–361 (1988).
C.O. Olsen, G.S. Tyson, G.W. Maier, J.W. Davis, J.S. Rankin: Diminished stroke volume during inspiration: A reverse thoracic pump. Circulation 72, 668–679 (1985).
T.J. Pedley: The fluid mechanics of large blood vessels. Cambridge University Press (1980).
T. Penzel, A. Bunde, L. Grote, J.W. Kantelhardt, J.H. Peter, K. Voigt: Heart rate variability during sleep stages in normals and in patients with sleep apnea. Medical Infobahn for Europe 77, Studies in Health Technology and Informatics, IOS Press, 1256–1260 (2000).
T. Penzel, N. Wessel, M. Riedl, J.W. Kantelhardt, S. Rostig, M. Glos, A. Suhrbier, H. Malberg, I. Fietze: Cardiovascular and respiratory dynamics during normal and pathological sleep. Chaos 17(1), 015116–1 - 015116–10 (2007).
J.H. Peter, U. Koehler, L. Grote, T. Podszus: Manifestations and consequences of obstructive sleep apnea. The European Respiratory Journal 8(9), 1572–1583 (1995).
H. Peter, T. Penzel, J. H. Peter: Encyclopaedia of sleep medicine (in German: Enzyklopädie der Schlafmedizin). Springer Publisher (2007).
D.J. Pitson, J.R. Stradling: Value of beat-to-beat blood pressure changes, detected by pulse transit time, in the management of the obstructive sleep apnea/hypopnea syndrome. The European Respiratory Journal 12, 685–692 (1998).
R.F. Rea, D.L. Eckberg: Carotid baroreceptor-muscle sympathetic relation in humans. American Journal of Physiology. Regulatory, Integrative and Comparative Physiology 253(6), 929–934 (1987).
A. Rechtschaffen, A. Kales: A manual of standardized terminology, techniques and scoring system for sleep stages of human subjects. National Institutes of Health Publication, Bethesda, Maryland (1968).
A. Riftine: Theoretical review of the nerve-express system with sample cases. www.nerveexpress.com (2006).

L. Rudas, A.A. Crossman, C.A. Morillo, J.R. Halliwill, K.U.O. Tahvanainen, T.A. Kuusela, D.L. Eckberg: Human sympathetic and vagal baroreflex responses to sequential nitroprusside and phenylephrine. American Journal of Physiology, Heart and Circulatory Physiology 276(5), 1691–1698 (1999).

R.L. Sack, D. Auckley, R.R. Auger, M.A. Carskadon, K.P. Wright, M.V. Vitiello, I.V. Zhdanova: Circadian rhythm sleep disorders: Part I, Basic principles, shift work and jet lag disorders. American Academy of Sleep Medicine Review. Sleep 30(11), 1460–1483 (2007a).

R.L. Sack, D. Auckley, R.R. Auger, M.A. Carskadon, K.P. Wright, M.V. Vitiello, I.V. Zhdanova: Circadian rhythm sleep disorders: Part II, Advanced sleep phase disorder, delayed sleep phase disorder, free-running disorder, and irregular sleep-wake rhythm. American Academy of Sleep Medicine Review. Sleep 30(11), 1484–1501 (2007b).

B. Saletu, G.M. Saletu-Zyhlarz: What you always want to know about the sleep (in German: Was Sie schon immer über Schalf wissen wollten). Ueberreuter Publisher (2001).

J. Schäfer: Snoring, sleep apnea, and upper airways (in German: Schnarchen, Schlafapnoe und obere Luftwege). Georg Thieme Publisher (1996).

D.T. Schmitt, P.K. Stein, P.C. Ivanov: Stratification pattern of static and scale-invariant dynamic measures of heartbeat fluctuations across sleep stages in young and elderly. IEEE Transactions on Biomedical Engineering 56(5), 1564–1573 (2009).

S.R. Seydnejad, R.I. Kitney: Modeling of Mayer waves generation mechanisms. IEEE Engineering in Medicine and Biology Magazine 20(2), 92–100 (2001).

Y.W. Shau, C.L. Wang, J.Y. Shieh, T.C. Hsu: Noninvasive assessment of the viscoelasticity of peripheral arteries. Ultrasound in Medicine & Biology 25(9), 1377–1388 (1999).

J.W. Shepard, D.J. Buysse, A.L. Chesson, W.C. Dement, R. Goldberg, C. Guilleminault, C.D. Harris, C. Iber, E. Mignot, M.M. Mitler, K.E. Moore, B.A. Phillips, S.F. Quan, R.S. Rosenberg, T. Roth, H.S. Schmidt, M.H. Silber, J.K. Walsh, D.P. White: History of the development of sleep medicine in the United States. Journal of Clinical Sleep Medicine 1(1), 61–82 (2005).

T. Shiomi, R. Stoohs, C. Guilleminault: Aging, respiratory efforts during sleep, and pulsus paradoxus. Lung 171(4), 203–211 (1993).

S. Silbernagl, A. Despopoulos: Pocket-atlas of physiology (in German: Taschenatlas Physiologie). Georg Thieme Publisher (2007).

D.U. Silverthorn: Physiology (in German: Physiologie). Pearson Studium (2009).

D. Smith, K. Solem, P. Laguna, J.P. Martinez, L. Sörnmo: Model-based detection of heart rate turbulence using mean shape information. IEEE Transactions on Biomedical Engineering 57(2), 334–342 (2010).

E.J.W. Van Someren, R.J.E.M. Raymann, E.J.A. Scherder, H.A.M. Daanen, D.F. Swaab: Circadian and age-related modulation of thermoreception and temperature regulation: mechanisms and functional implications. Ageing Research Reviews 1, 721–778 (2002).

V.K. Somers, M.E. Dyken, M.P. Clary, F.M. Abboud: Sympathetic neural mechanisms in obstructive sleep apnea. Journal of Clinical Investigation 96(4), 1897–1904 (1995).

R. Stoohs, C. Guilleminault: Cardiovascular changes associated with obstructive sleep apnea syndrome. Journal of Applied Physiology 72(2), 583–589 (1992).

Y. Sugo, R. Tanaka, T. Soma, H. Kasuya, T. Sasaki, T. Sekiguchi, H. Hosaka, R. Ochiai: Comparison of the relationship between blood pressure and pulse wave transit times at different sites. Proceedings of the joint EMBS/BMES Conference 1, 222 (1999).

A. Trzebski, M. Smietanowski: Non-linear dynamics of cardiovascular system in humans exposed to repetitive apneas modeling obstructive sleep apnea: Aggregated time series data analysis. Autonomic Neuroscience: Basic and Clinical 90, 106–115 (2001).

M. Tsunoda, E. Takuro, S. Hashimoto, S. Honma, K.I. Honma: Effects of light and sleep stages on heart rate variability in humans. Psychiatry and Clinical Neurosciences 55(3), 285–286 (2008).

S.S Ul'Yanov, V.V. Tuchin: Pulse-wave monitoring by means of focused laser beams scattered by skin surface and membranes. Proceedings of International Society for Optics and Photonics (SPIE) 1884, 160–167 (1993).

K. Umetani, D.H. Singer, R. McCraty, M. Atkinson: Twenty-four hour time domain heart rate variability and heart rate: relations to age and gender over nine decades. Journal of the American College of Cardiology 31(3), 593–601 (1998).

G. Varoneckas, R. Dauksys, L. Zakarevicius, D. Zemaityte: Restoration of cardiovascular function during sleep in healthy subjects and ischemic heart disease patients. In D. Zemaityte (Ed.): Heart rate variability and cardiovascular patology. Kaunas University of Medicine Press, 192–200 (1999).

G. Varoneckas, D. Zemaityte: Quantitative evaluation of autonomic heart rate control during sleep stages using poincare plots. In D. Zemaityte (Ed.): Heart rate variability: poincare plots in evaluation of heart rate autonomic control. Kaunas University of Medicine Press, 75–93 (2002).

G. Varoneckas. Heart rate variability in the evaluation of functional status during training. Eds. G.R.J. Hockey et al., IOS Press, 90–106 (2003).

M.P. Villa, G. Calcagnini, J. Pagani, B. Paggi, F. Massa, R. Ronchetti: Effects of sleep stage and age on short-term heart rate variability during sleep in healthy infants and children. Chest 117(2), 460–466 (2000).

K.H. Wesseling, J.R. Jansen, J.J. Settels, J.J. Schreuder: Computation of aortic flow from pressure in humans using a nonlinear, three-element model. Journal of Applied Physiology 74(5), 2566–2573 (1993).

R.A. Wever: The circadian system of man: results of experiments under temporal isolation. Springer Publisher (1979).

M. Wibmer: One-dimensional simulation of arterial blood flow with applications. Dissertation at the Vienna University of Technology (2004).

M.W. Wukitsch, M.T. Petterson, D.R. Tobler, J.A. Pologe: Pulse oximetry: Analysis of theory, technology, and practice. Journal of Clinical Monitoring 4(4), 290–301 (1988).

F. Yasuma, J. Hayano: Respiratory Sinus Arrhythmia, Why does the heartbeat synchronize with respiratory rhythm? Chest 125(2), 683–690 (2004).

D. Zemaityte, G. Varoneckas, E. Sokolov: Heart rhythm control during sleep. Psychophysiology 21(3), 279–289 (1984).

D. Zemaityte: Autonomic regulation of heart rhythm: mechanisms, registration, clinical value (in Lithuanian: Sirdies ritmo autonominis reguliavimas: mechanismai, vertinimas, klinikine reiksme). Kaunas Medical Academy Publisher (1997).

Index

E. Kaniusas, *Biomedical Signals and Sensors I*, Biological and Medical Physics, Biomedical Engineering, DOI 10.1007/978-3-642-24843-6,

FSC
www.fsc.org
MIX
Papier aus verantwortungsvollen Quellen
Paper from responsible sources
FSC® C105338

If you have any concerns about our products,
you can contact us on
ProductSafety@springernature.com

In case Publisher is established outside the EU,
the EU authorized representative is:
Springer Nature Customer Service Center GmbH
Europaplatz 3, 69115 Heidelberg, Germany

Printed by Libri Plureos GmbH
in Hamburg, Germany